Lecture Notes in Computational Science and Engineering

28

Springer-Verlag Berlin Heidelberg GmbH

Carsten Carstensen
Stefan Funken
Wolfgang Hackbusch
Ronald H.W. Hoppe
Peter Monk
Editors

Computational Electromagnetics

Proceedings of the GAMM Workshop
on Computational Electromagnetics,
Kiel, Germany, January 26-28, 2001

With 48 Figures and 19 Color Plates

Springer

Editors

Carsten Carstensen

Institute for Applied Mathematics
Vienna University of Technology
Wiedner Hauptstraße 8-10
1040 Vienna, Austria
e-mail: carsten.carstensen@tuwien.ac.at

Stefan Funken

Institute of Mathematics
University of Erlangen-Nürnberg
Bismarckstraße 1 1/2
91054 Erlangen, Germany
e-mail: funken@mi.uni-erlangen.de

Wolfgang Hackbusch

Max-Planck-Institute for Mathematics
in the Sciences
Inselstraße 22
04103 Leipzig, Germany
e-mail: wh@mis.mpg.de

Ronald H.W. Hoppe

Department of Mathematics
University of Augsburg
Universitätsstraße 14
86159 Augsburg, Germany
e-mail: hoppe@math.uni-augsburg.de

Peter Monk

Department of Mathematical Sciences
University of Delaware
Newark, DE 19717, USA
e-mail: monk@math.udel.edu

Cataloging-in-Publication Data applied for

Bibliographic information published by Die Deutsche Bibliothek

Die Deutsche Bibliothek lists this publication in the Deutsche Nationalbibliografie;
detailed bibliographic data is available in the Internet at <http://dnb.ddb.de>.

Mathematics Subject Classification (2000):
65N30, 65N38, 65N55, 78A25, 78A45, 78A50, 78A55

ISSN 1439-7358

ISBN 978-3-540-44392-6 ISBN 978-3-642-55745-3 (eBook)
DOI 10.1007/978-3-642-55745-3

http://www.springer.de

Originally published by Springer-Verlag Berlin Heidelberg New York in 2003

Cover Design: Friedhelm Steinen-Broo, Estudio Calamar, Spain
Cover production: *design & production*
Typeset by the authors using a Springer TEX macro package

Printed on acid-free paper SPIN: 10885062 46/3142/LK - 5 4 3 2 1 0

Preface

> The dimmed outlines of phenomenal things all merge into one another unless we put on the focusing-glass of theory, and screw it up sometimes to one pitch of definition and sometimes to another, so as to see down into different depths through the great millstone of the world
>
> *James Clerk Maxwell (1831 – 1879)*

For a long time after the foundation of the modern theory of electromagnetism by James Clerk Maxwell in the 19th century, the mathematical approach to electromagnetic field problems was for a long time dominated by the analytical investigation of Maxwell's equations.

The rapid development of computing facilities during the last century has then necessitated appropriate numerical methods and algorithmic tools for the simulation of electromagnetic phenomena. During the last few decades, a new research area "Computational Electromagnetics" has emerged comprising the mathematical analysis, design, implementation, and application of numerical schemes to simulate all kinds of relevant electromagnetic processes. This area is still rapidly evolving with a wide spectrum of challenging issues featuring, among others, such problems as the proper choice of spatial discretizations (finite differences, finite elements, finite volumes, boundary elements), fast solvers for the discretized equations (multilevel techniques, domain decomposition methods, multipole, panel clustering), and multiscale aspects in microelectronics and micromagnetics.

Motivated by the need to reflect the most significant developments, the editors of this volume have organized a workshop on Computational Electromagnetics held at the Christian-Albrechts University at Kiel, January 26-28, 2001. The aim was to bring together leading experts to present and discuss the state-of-the-art as well as future directions of research.

The contributions in these proceedings cover a wide area of contemporary research activities in electromagnetic field computation. They clearly underline the important role of the interaction of modeling, analysis, and simulation as well as the significance of interdisciplinary cooperations to provide powerful tools for today's problems and the challenges of tomorrow.

The analysis of Maxwell's equations, which is a prerequisite for the appropriate development of numerical solution techniques, is addressed in the contribution by *Annalisa Buffa* about the proper specification and analysis of traces of vector fields from $H(\mathbf{curl}; \Omega)$ in case of polyhedral domains Ω.

For non convex computational domains with reentrant corners, it is known that nodal finite element approximations may fail. Remedies are suggested by *Oszkar Biro* and *Kurt Preis* who highlight the role of gauged vector fields for the 3D eddy current equations.

Daniele Boffi deals with new families of edge elements for the time-harmonic case with regard to optimal error estimates by the application of nonlinear approximation theory.

A convergence proof based on duality arguments for edge element discretized problems is presented by *Peter Monk.*

Within the framework of efficient numerical solvers, *Leszek Demkowicz* provides a progress report on adaptive hp-methods featuring the issues of error estimates and convergence analysis of multigrid solvers. The domain decomposition methodology is applied in the contribution by *Annalisa Buffa, Yvon Maday,* and *Francesca Rapetti* on mortar edge element approximations of electromagnetic devices with moving parts whereas *Ana Alonso, Paolo Fernandes,* and *Alberto Valli* introduce and analyse a two-domain formulation for the time-harmonic eddy current problem.

A very popular and widely used discretization scheme is the Finite Integration Technique. *Thomas Weiland's* overview includes its characteristic features, its range of applications, and some recent advances with regard to nonlinear coupled problems. The use of such methods in the development of a parallel algorithm for the coupled computation of electromagnetic eigenmodes and field distributions is addressed by *Hans-Walter Glock, Karsten Rothemund,* and *Ursula van Rienen.*

Considerable attention is also paid to boundary integral techniques and boundary element methods. *Penny Davies* and *Dugald Duncan* deal with the stability of numerical schemes for retarded potential integral equations, analyze instabilities and suggest remedies to get rid of them. *Ralf Hiptmair* presents a symmetry preserving FEM-BEM technique for the eddy current equations which yields quasi-optimal convergence.

Computational methods for micromagnetic problems relying on a hybrid FEM-BEM approach for the magnetostatic field and a BDF method for the time evolution of the magnetization are presented by *Thomas Schrefl, Dieter Suess, Werner Scholz, Hermann Forster, Vassilios Tsiantos,* and *Josef Fidler.*

The editors would like to express their sincere thanks to some individuals and institutions that have significantly contributed to the organization of the workshop and the preparation of these proceedings. Particular thanks go to the Christian-Albrechts-University at Kiel that provided the excellent conference facilities. Moreover, we are greatly indebted to the assistants of the Chair of Scientific Computing and especially to Gudrun Thiel for all their efforts before and during the conference. Likewise, we highly appreciate the assistance by Yuri Iliash in the compilation of these proceedings.

We are grateful to the "Deutsche Forschungsgemeinschaft (DFG)" and the "Gesellschaft für Angewandte Mathematik und Mechanik (GAMM)" for their financial support without which the workshop would not have taken place.

Last, but not least we are indebted to the editors of the "Lecture Notes in Computational Science and Engineering" for considering these proceedings as a volume within this series and to Martin Peters, Thanh-Ha Le Thi, and Leonie Kunz of Springer-Verlag for the continuous advice and support during the preparation and production of this volume.

Vienna, Erlangen, Leipzig,
Augsburg, Newark

August 2002

Carsten Carstensen
Stefan Funken
Wolfgang Hackbusch
Ronald Hoppe
Peter Monk

Table of Contents

Gauged Current Vector Potential and Reentrant Corners in the FEM Analysis of 3D Eddy Currents
Oszkár Bíró, Kurt Preis 1

Finite Elements for the Time Harmonic Maxwell's Equations
Daniele Boffi 11

Trace Theorems on Non-Smooth Boundaries for Functional Spaces Related to Maxwell Equations: an Overview
Annalisa Buffa 23

Applications of the Mortar Element Method to 3D Electromagnetic Moving Structures
Annalisa Buffa, Yvon Maday, Francesca Rapetti 35

Numerical Stability of Collocation Schemes for Time Domain Boundary Integral Equations
Penny Davies, Dugald Duncan 51

***hp*-Adaptive Finite Elements for Maxwell's Equations**
Leszek Demkowicz 69

Coupled Calculation of Eigenmodes
H.-W. Glock, K. Rothemund, U. van Rienen 85

Boundary Element Methods for Eddy Current Computation
Ralf Hiptmair 103

A Simple Proof of Convergence for an Edge Element Discretization of Maxwell's Equations
Peter Monk 127

The Time-Harmonic Eddy-Current Problem in General Domains: Solvability via Scalar Potentials
Ana Alonso Rodríguez, Paolo Fernandes, Alberto Valli 143

Finite Element Micromagnetics
Thomas Schrefl, Dieter Suess, Werner Scholz, Hermann Forster, Vassilios Tsiantos, Josef Fidler 165

Finite Integration Method and Discrete Electromagnetism
Thomas Weiland 183

Appendix. Color Plates 199

Gauged Current Vector Potential and Reentrant Corners in the FEM Analysis of 3D Eddy Currents

Oszkár Bíró and Kurt Preis

IGTE, Graz University of Technology, Kopernikusgasse 24, A-8010 Graz, Austria

Summary. The nodal finite element realization of the $\mathbf{T} - \Omega$ method involving a gauged current vector potential, $\mathbf{T}$, is shown to yield erroneous results if applied to 3D eddy current problems with reentrant corners in the conducting region. The reason for the problem is pinpointed to be the implicit gauging of the vector potential. A remedy of using no gauge in elements around the reentrant corners is suggested.

1 Introduction

The so called $\mathbf{T}-\Omega$ method [1] is a powerful tool for solving three-dimensional eddy current problems. It ensures the exact satisfaction of the normal component of the current density vanishing on conductor/nonconductor interfaces. In a nodal finite element (FEM) context, the gauging of the current vector potential, $\mathbf{T}$, is indispensable in order to achieve numerical stability [2]. If edge finite elements are used to represent $\mathbf{T}$, gauging is not necessary and a singular equations system is obtained. The convergence of the solver is acceptable then provided the right hand side vector is consistent [3]. Still, the nodal approach is superior as regards speed of solution.

It is demonstrated in this paper that the gauged nodal element version of the $\mathbf{T} - \Omega$ method leads to large errors in the vicinity of reentrant corners in the conducting region. These errors prove to be a consequence of the gauging term in the differential equation. Eliminating this term in the finite elements near the reentrant corner is shown to get rid of the problem but to result in a numerically less robust formulation. An algorithm is presented for identifying reentrant corners and investigations are carried out concerning the accuracy of the solution and the convergence of the solver in dependence on the number of elements left ungauged.

2 Failure of the Gauged $\mathbf{T} - \Omega$ Method with Reentrant Corners

Consider a copper cylinder closed at its base and top by circular discs made of copper and placed within a concentric cylindrical coil carrying a sinusoidal current of 50 Hz. The thickness of the conducting parts is 20 mm, the inner radius of the cylinder is 20 mm, its height is 100 mm. The radius of the discs is 60 mm. The inner and outer radii of the coil are 50 and 60 mm, respectively and its height is 60 mm. The amplitude of the uniform current density in the coil is 1 A/mm^2.

The problem being axisymmetric, it is easily solved by a 2D FEM eddy current code based on a single component magnetic vector potential. The solution is illustrated in Fig. 1 showing the distribution of the maximal value of the current density in the upper half of the closed conducting cylinder.

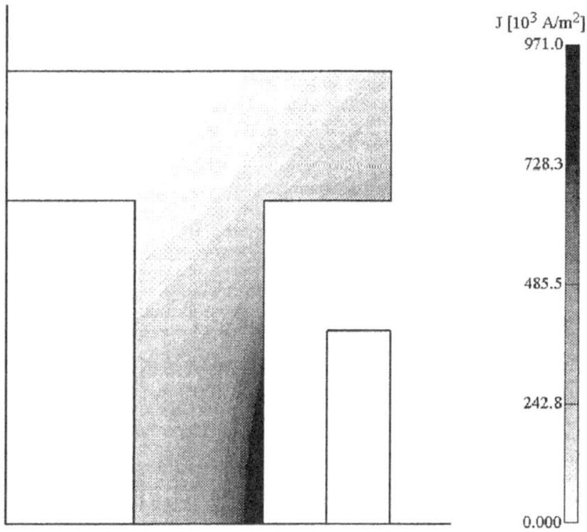

Fig. 1. Amplitude of the current density. 2D axisymmetric model

Applying the gauged $\mathbf{T} - \Omega$ formulation with nodal finite elements to a three-dimensional model of one eighth of the arrangement, the obviously wrong result in Fig. 2 is obtained.

If the circular cover is removed, leaving an open cylinder, the 2D and 3D results practically coincide (in the 2D case it is then necessary to treat the eddy current region as a conductor with zero net current). This indicates that the problem lies with the reentrant corner in the original problem.

Applying an edge element approach to the ungauged formulation [3], the right result is obtained as shown in Fig. 3. This would settle the case, the verdict being: use edge elements and not nodal ones, if it were not for the fact that a preconditioned conjugate gradient solver needs more than four times as many iterations in the edge element case as in the gauged nodal element approach (Table 1).

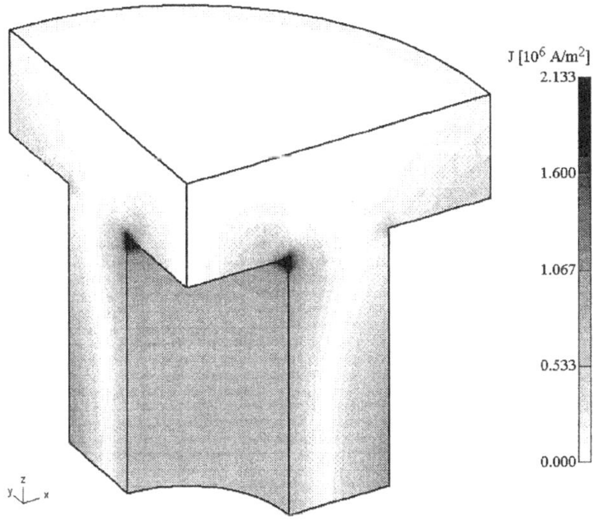

Fig. 2. Amplitude of the current density. 3D model, **T** represented by nodal elements and fully gauged

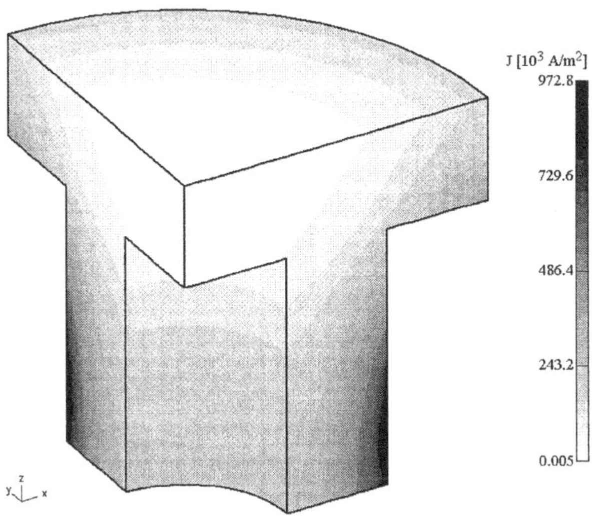

Fig. 3. Amplitude of the current density. 3D model, **T** represented by edge elements

Using nodal elements with the ungauged formulation needs more than eight hundred times as many iterations (Table 1) leading to the right answer shown in Fig. 4. This convergence rate is obviously not tolerable.

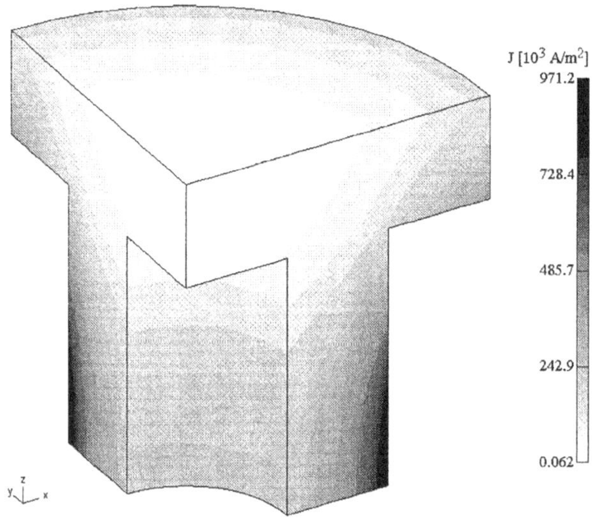

Fig. 4. Amplitude of the current density. 3D model, **T** represented by nodal elements and no gauge enforced

The above results indicate that the gauging of the vector potential leads to large errors if reentrant corners in the eddy current region are present.

3 The Reason for the Failure

The boundary value problem corresponding to the gauged $\mathbf{T}-\Omega$ formulation in a conducting region Ω bounded by the surface Γ is:

$$curl\,(\rho curl\mathbf{T}) - grad\,(\rho div\mathbf{T}) + j\omega\mu\,(\mathbf{T} - grad\Phi) = \mathbf{0}, \tag{1}$$

$$j\omega div\mu\,(\mathbf{T} - grad\Phi) = 0 \qquad \text{in } \Omega, \tag{2}$$

$$\mathbf{T}\times\mathbf{n} = \mathbf{0}, \tag{3}$$

$$\rho div\mathbf{T} = 0 \qquad \text{on } \Gamma. \tag{4}$$

where the sinusoidal time variation has been taken into account by using complex notation $\left(j = \sqrt{-1}\right)$, ρ is the resistivity, μ is the permeability, ω is the angular frequency and Φ is the magnetic scalar potential (sometimes denoted by Ω, hence the name $\mathbf{T} - \Omega$ formulation). Building the divergence of (1) and taking account of (2) yields the Laplace equation for $\rho div\mathbf{T}$ which, along with (4), implies the Coulomb gauge, i.e. $\rho div\mathbf{T} = 0$ in Ω.

However, both (1) and (4) are satisfied weakly only in a finite element Galerkin realization [2], leading to a not exactly zero divergence. This error is negligible, unless reentrant corners are present in Γ. At such corners, the divergence of $\mathbf{T}$ exhibits a singular behavior. This is a consequence of the boundary condition (3) in conjunction with the Helmholtz equation (1): Basically the same two equations are satisfied by the electric field intensity, $\mathbf{E}$, in a cavity surrounded by perfect conductors and $div\mathbf{E}$ is then indeed singular at reentrant corners, since it must equal the infinite charge density there.

This reasoning is supported by Fig. 5 showing the distribution of $div\mathbf{T}$ in a cross section of the 3D model. The singularity at the reentrant corners is clear. The maximal value of $div\mathbf{T}$ is in the same order of magnitude as the current density $curl\mathbf{T}$, which has a devastating effect on the result as shown in Fig. 2: the term $-grad(\rho div\mathbf{T})$ in (1) becomes appreciable, and thus (1) is not a good approximation of Faraday's law as it would be if the Coulomb gauge were satisfied.

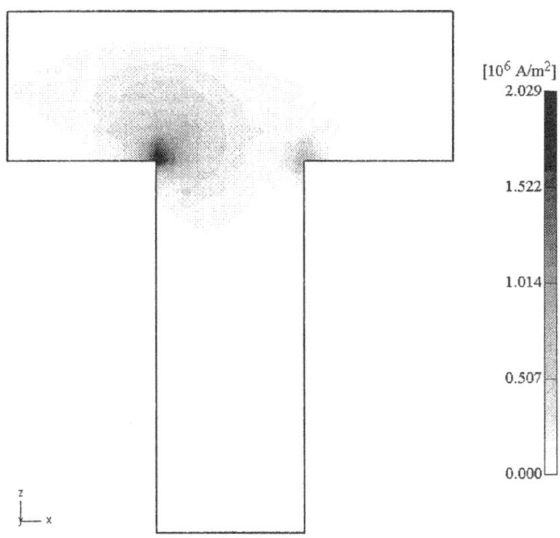

Fig. 5. Amplitude of $div\mathbf{T}$ in a cross section of the 3D model. $\mathbf{T}$ represented by nodal elements and fully gauged

4 A Remedy

Since the error in the solution obtained by the gauged $\mathbf{T} - \Omega$ formulation has been seen to be due to the term $-grad\,(\rho div\mathbf{T})$ in (1) becoming large in the vicinity of reentrant corners it seems to be natural to get rid of this term. This means abandoning the gauge and this is not feasible if carried out completely: the numerical stability becomes catastrophic as mentioned above and seen in Table 1.

It suffices, however, if the gauge is relinquished in the elements around the reentrant corner only. Using the ungauged differential equation obtained from (1) by eliminating the term $-grad\,(\rho div\mathbf{T})$ on the left-hand-side in the elements indicated in Fig. 6, results in the current density distribution shown in Fig. 7. This is seen to be practically identical with the ungauged solutions of Figs. 3 and 4, and also with the 2D solution of Fig. 1.

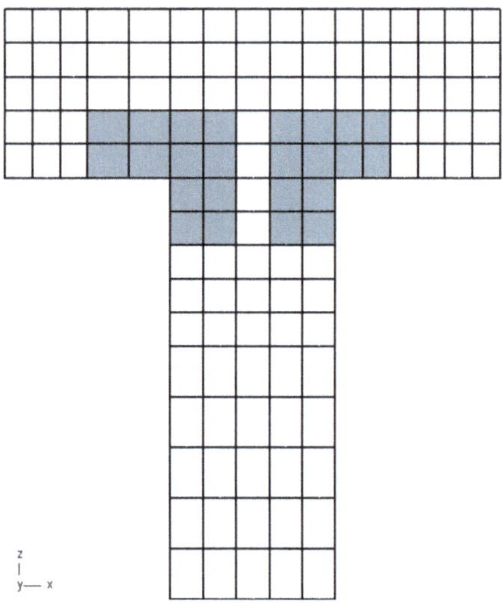

Fig. 6. Elements not gauged shown shaded in a cross section of the 3D model

5 Identification of Regions near Reentrant Corners

The question remains, how to identify reentrant corners in the conducting region and then the finite elements in their vicinity where the gauge should be relinquished.

Due to the boundary condition (3), the nodes on the interface Γ between the conducting and nonconducting regions have to be identified in order to

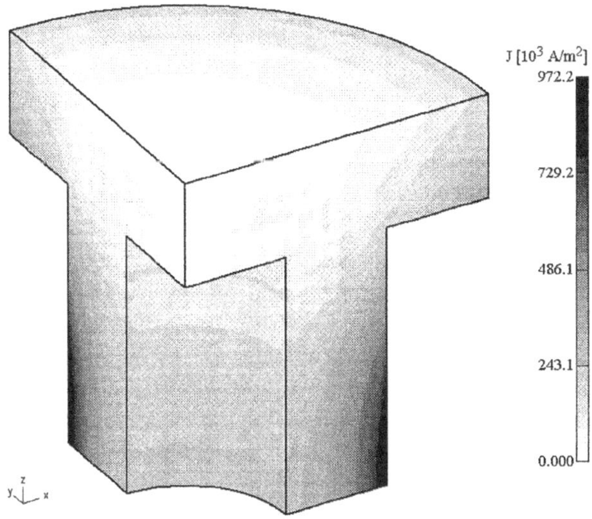

Fig. 7. Amplitude of the current density. 3D model, **T** represented by nodal elements and partially gauged. The ungauged elements are shown in Fig. 6

set the tangential components of **T** to zero in them. It is also necessary to compute the normal unit vector in these nodes, so that (3) can be enforced. This is achieved by scanning the element surfaces on Γ and computing the normal vectors to Γ in each node on them. If the normal vectors computed in the same node from different finite elements enclose an angle greater then a preset value α (e.g. α=10 degrees), the node lies on a corner. In this case, all three components of **T** have to be set to zero here. This means that all corner nodes are known in any implementation of the $\mathbf{T} - \Omega$ method.

To establish whether a corner node n is on a reentrant corner, all finite elements in the conducting region are identified which contain n. A plane Γ_0 is now determined which crosses through n and has a normal vector in the direction defined as the cross product of the two normal vectors to Γ having been found to enclose an angle greater than α as n has been identified to be a corner node. In effect, the normal vector of Γ_0 is the direction along the edge forming the corner. For all finite elements containing n and cutting Γ_0, the internal angles at n between the cuts are now computed and added. If the sum of these angles exceeds 180 degrees, n is on a reentrant corner, otherwise it is on a nonreentrant one.

Having identified all nodes on reentrant corners, let us define the level of the ungauged region as follows: A region of level 1 is formed by all finite elements sharing a reentrant-corner node. Level 2 is obtained by including all neighbors of these elements as well, and so on. For example, the region shown shaded in Fig. 6 is of level 2.

The effect of the level of the ungauged region on the convergence of the iterative solver and on the accuracy of the solution is investigated in the following section.

6 Numerical Investigations

A finite element model of the problem of the closed conducting cylinder enclosed by a cylindrical coil described in Sect. 2 consists of 5,490 second order hexahedral 20-node finite elements and 27,790 global nodes. This discretization referred to as coarse grid in the following can be seen in Fig. 6 in a cross section of the model. The results shown in Figs. 2, 3, 4, 5 and 7 have been obtained using this mesh.

A finer discretization has also been used consisting of 24,300 elements and 106,309 nodes. This fine grid has been constructed by doubling the number of elements in all three directions within the conductor.

The finite element equations system has been solved by an incomplete shifted Cholesky preconditioned conjugate gradient solver. The iterations have been terminated once the squared norm of the residual vector became less than 10^{-12} times the squared norm of the right hand side of the equation.

The analysis of the problem has been carried out for both grids with the aid of edge elements to serve as reference. The erroneous fully gauged solutions have been computed, too. To illustrate the effects of the level of the ungauged region, partially gauged solutions have been further obtained with levels 1, 2 and 3. Finally, a totally ungauged analysis has also been performed with the coarse grid. To check the accuracy of the solutions, the current flowing through the surface shaded in Fig. 8 has been computed in each case.

The results have been summarized in Table 1 for the coarse grid and in Table 2 for the fine grid. The quantities shown are the number of degrees of freedom (DOF), the number of necessary conjugate gradient iterations (CG iterations) and the current flowing through the surface indicated in Fig. 8.

The wrong results obtained if a full gauge is used are clearly indicated by the currents computed by this formulation. Abandoning the gauge in level 1 regions strikingly improves the accuracy without appreciably increasing the number of necessary conjugate gradient iterations. The imaginary parts of the currents are, however, of poor precision. A partial gauge in regions of level 2 yields currents in very good agreement with the edge element results and it needs about 20 % less number of iterations. The improvement in accuracy brought about by using level 3 ungauged regions is negligible, but the solution time increases substantially, the convergence of the solver is poorer that in the case of edge elements. The totally unacceptable numerical behavior of the ungauged formulation is clearly visible in the last row of Table 1.

It seems that using level 2 regions without gauge in the vicinity of reentrant corners is the best compromise between accuracy and speed of solution.

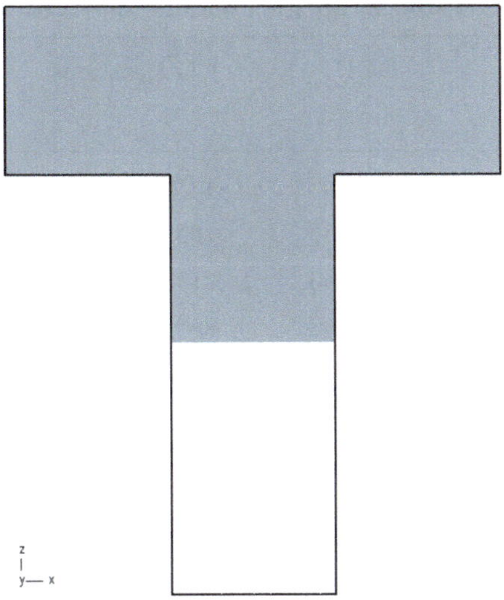

Fig. 8. The current given in Tables 1 and 2 has been computed through the surface shown shaded.

Table 1. Solution data for coarse mesh

Formulation	DOF	CG iterations	Current (A)
Edge elements	32,567	387	116.6-j10.97
Full gauge	32,690	78	44.51-j41.57
Partial gauge, level 1	32,690	97	114.5-j4.281
Partial gauge, level 2	32,690	315	116.6-j10.85
Partial gauge, level 3	32,690	52,491	116.7-j11.03
Without gauge	32,690	64,865	116.7-j11.04

Since this result is based on second order elements, levels 3 or 4 are probably needed if first order elements are used.

7 Conclusions

The error in the vicinity of reentrant corners caused by gauging the current vector potential can be eliminated by using the ungauged formulation in elements surrounding such corners. The resulting method is competitive with using edge elements.

Table 2. Solution data for fine mesh

Formulation	DOF	CG iterations	Current (A)
Edge elements	170,908	1,025	116.7-j11.01
Full gauge	171,353	165	46.00-j41.55
Partial gauge, level 1	171,353	215	114.0-j0.9793
Partial gauge, level 2	171,353	838	116.6-j10.90
Partial gauge, level 3	171,353	2,568	116.7-j11.02

References

1. Carpenter C.J. (1977) Comparison of alternative formulations of 3-dimensional magnetic-field and eddy-current problems at power frequencies. Proc. IEE 124:1026-1034
2. Bíró O., Preis K. (1990) Finite element analysis of 3-D eddy currents. IEEE Transactions on Magnetics 26:418-423
3. Bíró O. (1999) Edge element formulations of eddy current problems. Computer Methods in Applied Mechanics and Engineering 169:391-405

Finite Elements for the Time Harmonic Maxwell's Equations

Daniele Boffi

Università di Pavia, Dipartimento di Matematica "F. Casorati", via Ferrata 1, 27100 Pavia, ITALY

Summary. We review the time harmonic Maxwell's system and its approximation via the finite element method. The problem under consideration is strictly related to the so-called interior Maxwell's eigenproblem.

Standard nodal (Lagrangian) elements are known to provide useless results on general meshes. Special two-dimensional meshes have been shown to give good results, but the use of them is not recommended. The use of a penalty strategy with nodal elements has been proved to give wrong results for domains with singularities. Some special schemes, which make use of nodal elements, circumvent this problem; one of them is described in this paper.

On the other hand the so-called edge elements represent the natural choice. A new proof of convergence for a method based on edge elements is summarized.

1 The time harmonic Maxwell's system

We shall consider the Maxwell's system under the following form

$$\left(\varepsilon\frac{\partial}{\partial t}+\sigma\right)\mathcal{E}-\operatorname{curl}\mathcal{H}=\mathcal{J},\qquad \mu\frac{\partial}{\partial t}\mathcal{H}+\operatorname{curl}\mathcal{E}=0. \tag{1}$$

We shall include suitable boundary conditions later on. If we make the following time harmonic assumptions on the electric and magnetic field, respectively,

$$\mathcal{E}(x,t)=e^{-i\omega t}E(x),\qquad \mathcal{H}(x,t)=e^{-i\omega t}H(x),$$

then equations (1) take the form

$$-i\omega\left(\varepsilon+i\frac{\sigma}{\omega}\right)E-\operatorname{curl}H=J,\quad -i\omega\mu H+\operatorname{curl}H=0,$$

where we have also assumed $\mathcal{J}(x,t)=e^{-i\omega t}J(x)$. It is now a standard procedure to eliminate the magnetic phasor H, for instance, and to get a single equation for the electric field E, namely

$$\operatorname{curl}\left(\mu^{-1}\operatorname{curl}E\right)-\omega^2\left(\varepsilon+i\frac{\sigma}{\omega}\right)E=-i\omega J. \tag{2}$$

It is known that *for* $\sigma\neq 0$ problem (2), provided with suitable boundary conditions, is an elliptic one, so that any Galerkin approximation will be well posed and optimally convergent (see, for instance, [16, 17]). For this reason, we shall only consider the case $\sigma = 0$. In this case, with no loss of generality, we can consider E to be a real valued vectorfield.

For simplicity of notation, we shall denote our primal variable E by $\underline{u}$ and our right hand side $-i\omega J$ by $\underline{f}$, and complete equation (2) with a set of homogeneous standard boundary conditions on the tangential components of E (for more general boundary conditions see, for instance, [7]), so that our problem reads

$$\begin{cases} \operatorname{curl}\left(\mu^{-1}\operatorname{curl}\underline{u}\right) - \omega^2\varepsilon\underline{u} = \underline{f} & \text{in } \Omega \\ \underline{u}\times\underline{n} = 0 & \text{on } \partial\Omega. \end{cases} \tag{3}$$

Here Ω denotes a polyhedron. Some examples will be two-dimensional, in this case Ω will be a polygon and equation (3) will take the form

$$\begin{cases} \operatorname{curl}\left(\mu^{-1}\operatorname{rot}\underline{u}\right) - \omega^2\varepsilon\underline{u} = \underline{f} & \text{in } \Omega \\ \underline{u}\cdot\underline{t} = 0 & \text{on } \partial\Omega, \end{cases} \tag{4}$$

where $\operatorname{rot}(u, v) = v_x - u_y$, $\operatorname{curl}\phi = (\phi_y, -\phi_x)$ and $\underline{t}$ is the unit tangent vector to the boundary of Ω.

Notice that $\operatorname{div}\underline{f} = 0$ implies $\operatorname{div}(\varepsilon\underline{u}) = 0$, so that a divergence-free condition is hidden in these equations. This constraint is one of the major issues for the construction of any finite element approximation.

It is clear that a necessary condition for the well-posedness of problems (3) and (4) is ω not to be an interior Maxwell's eigenvalue. For the readers' convenience we recall that a real number ω and a real-valued function $\underline{u}$ are an interior Maxwell's eigenvalue and eigenfunction, respectively, if they solve the problem

$$\begin{cases} \operatorname{curl}\left(\mu^{-1}\operatorname{curl}\underline{u}\right) = \omega^2\varepsilon\underline{u} & \text{in } \Omega \\ \underline{u}\times\underline{n} = 0 & \text{on } \partial\Omega. \end{cases}$$

Similarly, one can define the eigenproblem for the two-dimensional case. It is well-known that ω not being an interior Maxwell's eigenvalue is also a sufficient condition for the well-posedness of problems (3) and (4), as it can be proven, for instance, by using the Fredholm's alternative.

2 Approximation of the time harmonic Maxwell's equations

We shall describe in detail the approximation to the three-dimensional problem (3), the same considerations holding for the two-dimensional problem (4) with the natural modifications.

In order to construct a finite element approximation of problem (3) we need a weak form, which can be obtained in a standard way with the introduction of the variational space $V = H_0(\operatorname{curl};\Omega)$ of the real valued vectorfields which are bounded in L^2 and whose curl is bounded in L^2; a function in V, moreover, has the trace of its tangential component vanishing on $\partial\Omega$. If we multiply equation (3) by a generic test function $\underline{v} \in V$, integrate on Ω and integrate by parts, by taking advantage of the boundary conditions, we are led to the following problem.

$$\begin{aligned}&\text{Find } \underline{u} \in V \text{ such that}\\ &\left(\mu^{-1}\operatorname{curl}\underline{u}, \operatorname{curl}\underline{v}\right) - \omega^2(\varepsilon\underline{u}, \underline{v}) = (\underline{f}, \underline{v}) \quad \forall \underline{v} \in V.\end{aligned} \tag{5}$$

A finite element approximation to problem (5) is given by a proper choice of a sequence of finite dimensional subspace $V_h \subset V$ and by the following sequence of discrete problems.

$$\begin{aligned}&\text{Find } \underline{u}_h \in V_h \text{ such that}\\ &\left(\mu^{-1}\operatorname{curl}\underline{u}_h, \operatorname{curl}\underline{v}\right) - \omega^2(\varepsilon\underline{u}_h, \underline{v}) = (\underline{f}, \underline{v}) \quad \forall \underline{v} \in V_h.\end{aligned} \tag{6}$$

It is immediately seen, in analogy to what we observed for the continuous problem, that a necessary condition for the solvability of problem (6) is that ω is not a *discrete* interior Maxwell's eigenvalue, i.e., ω does not solve the problem

$$\left(\mu^{-1}\operatorname{curl}\underline{u}_h, \operatorname{curl}\underline{v}\right) = \omega_h^2(\varepsilon\underline{u}_h, \underline{v}) \quad \forall \underline{v} \in V_h.$$

This remark raises the question whether the choice of V_h provides a good approximation of the interior Maxwell's eigenvalues. If this is not the case, in particular, we cannot discuss the convergence of the solution to (6) towards the solution to (5) because problem (6) might be not solvable even when problem (5) is well-posed. A close relationship between the approximation to the time-harmonic Maxwell's system and the approximation of the interior Maxwell's eigenvalues has been observed also in [12].

The good approximation of the interior Maxwell's eigenvalues has been deeply investigated during these last years, we shall review some known results and add some new remarks in the following two sections: in the first one we shall comment on the edge element approximation and in the second one we consider the nodal element approximation (with or without penalization)

3 Edge elements for Maxwell's eigenvalues

First of all, we recall the problem we are dealing with. When considering the approximation of the interior Maxwell's eigenvalues, it is a standard procedure to drop the divergence-free constraint and to consider the problem

$$\begin{aligned}&\text{find } \omega \in \mathbb{R} \text{ and } u \in V \text{ such that}\\ &\left(\mu^{-1}\operatorname{curl}\underline{u}, \operatorname{curl}\underline{v}\right) = \omega^2(\varepsilon\underline{u}, \underline{v}) \quad \forall \underline{v} \in V.\end{aligned} \tag{7}$$

The spectrum of problem (7) is given by the interior Maxwell's eigenvalues and by the eigenvalue $\omega = 0$, corresponding to the infinite-dimensional kernel associated with the gradients[1]. For this reason, a natural finite element approximation of the Maxwell's eigenproblem consists in taking a sequence of finite dimensional subspaces $V_h \subset V$ and in considering the following sequence of discrete eigenproblems.

$$\begin{aligned}&\text{Find } \omega_h \in \mathbb{R} \text{ and } u_h \in V_h \text{ suht that}\\&\left(\mu^{-1}\operatorname{curl}\underline{u}_h, \operatorname{curl}\underline{v}\right) = \omega_h^2(\varepsilon\underline{u}_h, \underline{v}) \quad \forall \underline{v} \in V_h.\end{aligned} \tag{8}$$

In practice, solving problem (8) corresponds to the solution of a generalized eigenvalue problem in matrix form.

The discrete eigenmodes of problem (8) should approximate the continuous solutions of problem (7) in the sense that we expect a number of eigenvalues approximating the value zero and the remaining ones approximating the interior Maxwell's eigenvalues. The former ones are usually called *spurious* solutions in the engineering literature. We follow this notation, even if we shall see in the remaining of the paper that a second kind of spurious solutions can be generated by a numerical scheme.

It is clear that a scheme can be used in practice only when the spurious solutions can be easily recognized among the physical ones. When V_h is made of edge elements, this property is definitely satisfied: a number of discrete eigenvalues vanishes (up to the machine precision) and the remaining ones approximate the continuous solutions with optimal order. The proof of this result is contained in the papers [7, 3, 4] and covers all families of edge elements which enjoy the commuting diagram property [18, 19, 12].

To conclude this section, we recall a useful definition of the resolvent operator associated with Maxwell's eigenproblem. To this aim, we make use of a suitable mixed problem introduced by Kikuchi in [14].

$$\begin{aligned}&\text{Given } \underline{g} \in L^2(\Omega)^3, \text{ find } (\underline{u}, p) \in H_0(\operatorname{curl};\Omega) \times H_0^1(\Omega) \text{ such that}\\&\begin{cases}(\mu^{-1}\operatorname{curl}\underline{u}, \operatorname{curl}\underline{v}) + (\varepsilon\underline{v}, \operatorname{grad} p) = (\varepsilon\underline{g}, \underline{v}) & \forall \underline{v} \in H_0(\operatorname{curl};\Omega)\\(\varepsilon\underline{u}, \operatorname{grad} q) = 0 & \forall q \in H_0^1(\Omega)\end{cases}\end{aligned} \tag{9}$$

Problem (9) is well posed as a mixed problem, the inf-sup condition being trivial and the ellipticity in the kernel being an easy consequence of the Helmholtz decomposition. We define the operator $T : L^2(\Omega)^3 \to L^2(\Omega)^3$ by setting $T\underline{g} = \underline{u}$ with the notation of formula (9). T is a compact operator and the eigenvalues of T correspond exactly to the interior Maxwell's eigenvalues.

Defining V_h as a space of edge elements and Q_h as the *natural* space of nodal elements such that $\operatorname{grad} Q_h \subset V_h$[2], one has that the following discrete

[1] If $\underline{u}$ is the gradient of a smooth function, then the left hand side of (7) vanishes while $\underline{u}$ might be not zero.

[2] Such a space exists for standard edge elements enjoying the commuting diagram property; for instance, for the lowest-order Nédélec element of the first family on tetrahedra, Q_h is the space of linear functions.

counterpart of (9) is well-posed.

$$\text{Find } (\underline{u}_h, p_h) \in V_h \times Q_h \text{ such that}$$
$$\begin{cases} (\mu^{-1}\operatorname{curl}\underline{u}_h, \operatorname{curl}\underline{v}) + (\varepsilon\underline{v}, \operatorname{grad} p_h) = (\varepsilon\underline{g}, \underline{v}) & \forall\underline{v} \in V_h \\ (\varepsilon\underline{u}_h, \operatorname{grad} q) = 0 & \forall q \in Q_h \end{cases} \tag{10}$$

The inf-sup condition for problem (10) is trivial, while the ellipticity in the kernel follows from the discrete compactness property introduced in [15] and proved in [4] for general families of edge elements.

With the help of problem (10) we can then define a discrete operator $T_h : L^2(\Omega)^3 \to V_h$ by $T_h\underline{g} = \underline{u}_h$. The operators T and T_h have been used in [8], where, in particular, it has been proved that T_h converges in norm to T, namely

$$||T\underline{g} - T_h\underline{g}||_{L^2} \leq \rho(h)||\underline{g}||_{L^2} \tag{11}$$

with $\rho(h)$ tending to zero as h goes to zero.

Remark 1. The uniform convergence (11) is not a trivial consequence of the results proven in [3, 4]. Note, in particular, that $\underline{g}$ is an arbitrary function in $L^2(\Omega)^3$ which need not be divergence free.

4 Nodal elements for Maxwell's eigenvalues

In this section we shall restrict ourselves to the two-dimensional case.

4.1 Unconstrained formulation

The use of standard nodal elements for the approximation (8) of problem (7) gives useless results. In Figure 1 the first eigenvalues computed with continuous piecewise linear elements are shown in a two-dimensional example with $\Omega =]0, \pi[\times]0, \pi[$. It is not possible to distinguish between the spurious eigenvalues and the three (marked with a star) values which are approximating the first (double) eigenvalue $\omega^2 = 1$ and the second one $\omega^2 = 2$.

On the other hand, some special two-dimensional meshes have been shown to give reasonable results. In Figure 2 a so called *consistent* mesh, introduced in [22], is shown which has been used for the computation of the eigenvalues plotted in Figure 3. Now the vanishing eigenvalues are well separated by the correct ones and the method seems to be robust. For this kind of consistent mesh there is no analysis of convergence, the key property described in [22] being that the gradients are well represented by the mesh (that is why the zero eigenvalues are obtained). This property is implied by the fact that on the mesh it is possible to construct a local basis for piecewise quadratic functions which are globally continuous together with their first derivatives. This property is also met by a second consistent mesh introduced by Powell [20] and plotted in Figure 4. Since gradients are well represented by the second

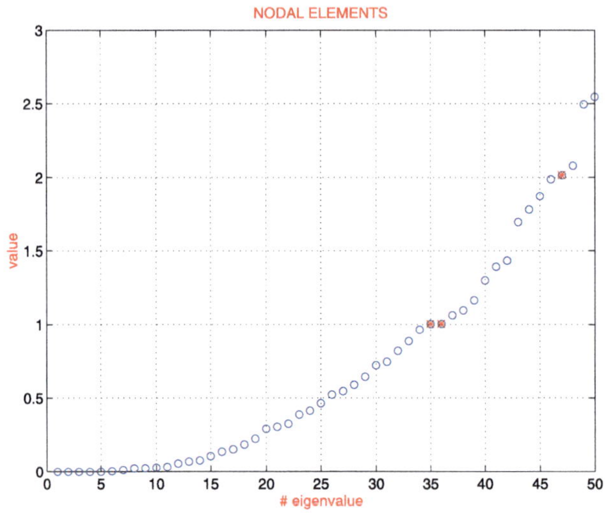

Fig. 1. Eigenvalues computed with piecewise linears

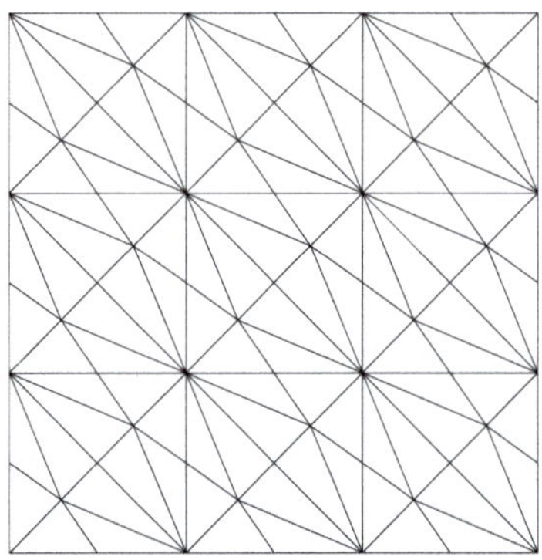

Fig. 2. A first consistent mesh

consistent mesh, a number of zero discrete eigenvalues is computed (up to the machine precision), but now a new kind of spurious eigenvalues appears. The results are summarized in Table 1, from which it is clear that now the spurious solution (which seems to converge to the value six) is much more dangerous than the previous ones. The corresponding eigenfunction is plotted in Figure 5; a checkerboard pattern can easily be recognized. We point out that other spurious solutions can be observed at a higher frequencies.

For this reason, we think that the use of consistent meshes should be avoided.

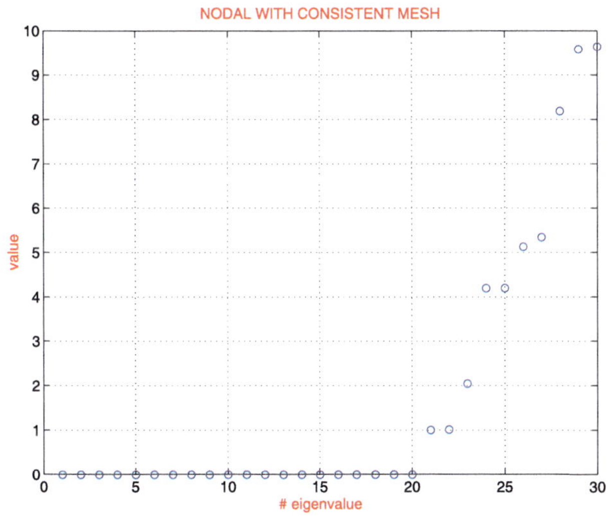

Fig. 3. Eigenvalues computed with the first consistent mesh

Fig. 4. A second consistent mesh

4.2 Penalty method

A penalty variational formulation for the interior Maxwell's eigenproblem reads:

$$
\begin{aligned}
&\text{find } \omega \in \mathbb{R} \text{ and } \underline{u} \in W = H_0(\mathrm{rot}; \Omega) \cap H(\mathrm{div}; \Omega) \text{ such that} \\
&(\mu^{-1} \operatorname{rot} \underline{u}, \operatorname{rot} \underline{v}) + s(\operatorname{div} \varepsilon \underline{u}, \operatorname{div} \varepsilon \underline{v}) = \omega^2 (\varepsilon \underline{u}, \underline{v}) \qquad \forall \underline{v} \in W.
\end{aligned}
\tag{12}
$$

The spectrum of (12) consists of two families of eigenvalues: the first one coincides with the Maxwell's eigenvalues and does not depend on s, the second one grows linearly with s and is associated with eigenfunctions $\underline{u}$ which have $\operatorname{rot} \underline{u} = 0$. For this reason, this method is an *exact* penalization and is sometimes referred to as a regularized method.

Table 1. Nodal approximation on crisscross mesh

exact	computed			
1	1.00428	1.00190	1.00107	1.00068
1	1.00428	1.00190	1.00107	1.00069
2	2.01711	2.00761	2.00428	2.00274
4	4.06804	4.03037	4.01710	4.01095
4	4.06804	4.03037	4.01710	4.01095
5	5.10634	5.04748	5.02674	5.01712
5	5.10634	5.04748	5.02674	5.01712
	5.92293	5.96578	5.98074	5.98767
8	8.27128	8.12151	8.06845	8.04383
9	9.34085	9.15309	9.08640	9.05537
9	9.34085	9.15309	9.08640	9.05537
d.o.f.	254	574	1022	1598
# zeros	63	143	255	399

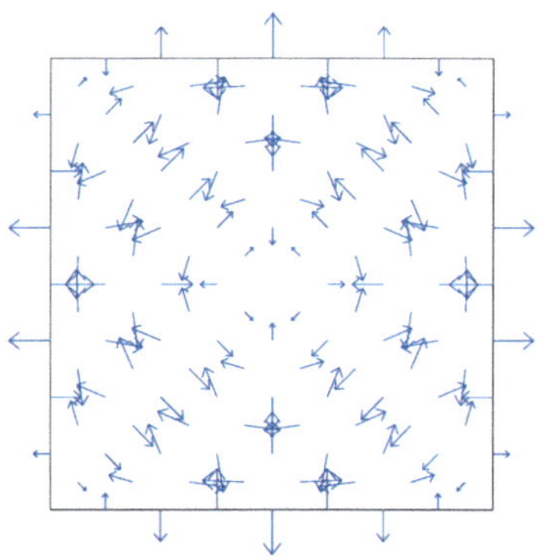

Fig. 5. The spurious eigenfunction

A discretization of (12) is obtained with the choice of a sequence of finite dimensional subspaces $W_h \subset W$ and with the solution of the following sequence of discrete problems.

$$\begin{aligned}&\text{Find } \omega_h \in \mathbb{R} \text{ and } \underline{u}_h \in W_h \text{ such that}\\&(\mu^{-1}\operatorname{rot}\underline{u}_h, \operatorname{rot}\underline{v}) + s(\operatorname{div}\varepsilon\underline{u}_h, \operatorname{div}\varepsilon\underline{v}) = \omega_h^2(\varepsilon\underline{u}_h, \underline{v}) \qquad \forall \underline{v} \in W_h.\end{aligned} \tag{13}$$

Unfortunately, the solutions to problems (13) in general do not converge towards those of problem (12) in presence of singularities (like the ones produced, for instance, by a nonconvex corner). This negative result has been proved and discussed in [11].

A reduced integration (or projection) method has been recently shown to give good results. It has been discussed and partly analyzed in [5, 6]; this scheme comes from a different problem arising in fluid-structure interaction (see [2, 13, 1]). The definition of this method (in its easies form, see [5] for a more general description) requires the introduction of the finite element space W_h of continuous piecewise biquadratic vectorfields and of the L^2-projection $\mathcal{P}_h$ onto the (discontinuous) piecewise linear functions. The method consists in introducing the projection $\mathcal{P}_h$ in the integrals on the left hand side of (13) as follows:

$$\begin{aligned}&\text{find } \omega_h \in \mathbb{R} \text{ and } \underline{u}_h \in W_h \text{ such that}\\&(\mathcal{P}_h\mu^{-1}\operatorname{rot}\underline{u}_h, \mathcal{P}_h \operatorname{rot}\underline{v}) + s(\mathcal{P}_h \operatorname{div}\varepsilon\underline{u}_h, \mathcal{P}_h \operatorname{div}\varepsilon\underline{v}) = \omega_h^2(\varepsilon\underline{u}_h, \underline{v}) \quad \forall \underline{v} \in W_h.\end{aligned} \tag{14}$$

In Figure 6 we plot the first singular eigenfunction computed with this method on an L-shaped domain. The good performance of this method can

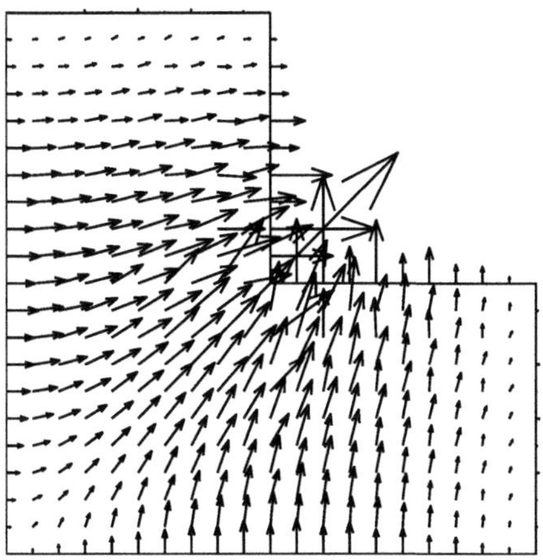

Fig. 6. The singular function computed with the reduced integration method

also be observed by looking at Figure 7 where the first eigenvalues computed on an L-shaped domain are plotted versus the penalty parameter s. It is clear that the behavior of the continuous problem is well reproduced.

Another special treatment of problem (13) can be found in [10]. We refer to that paper for the details.

5 Edge elements for the time harmonic Maxwell's equations

From the examples presented in the previous sections, the edge element strategy seems to be the natural choice for the approximation of problem (3).

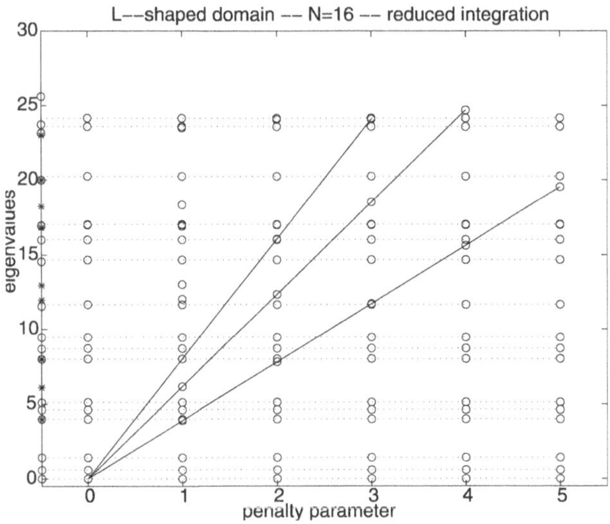

Fig. 7. Results of the reduced integration method

In this section we report on a new proof of convergence for the edge element approximation presented in [8]. The first proof of convergence, under stronger regularity assumptions, can be found in [17].

The proof makes use of an abstract theory by Brezzi, Rappaz and Raviart [9] for the approximation of nonlinear problems. Setting

$$\lambda = \omega^2, \quad G(\lambda, \underline{u}) = -\lambda \underline{u} - \varepsilon^{-1} \underline{f}$$

and with T as defined in Section 3, we can rewrite problem (3) under the form

$$\begin{cases} \underline{u} \in L^2(\Omega)^3, \\ \underline{u} + TG(\lambda, \underline{u}) = 0. \end{cases} \tag{15}$$

Given a sequence of finite dimensional subspaces $V_h \subset L^2(\Omega)^3$, the approximation of problem (15) reads

$$\begin{cases} \underline{u}_h \in V_h, \\ \underline{u}_h + T_h G(\lambda, \underline{u}_h) = 0. \end{cases} \tag{16}$$

We assume here and in the following that V_h is a space of edge elements satisfying the commuting diagram property (see [4] and [8] for more details).

Following [9] we introduce suitable conditions for the well-posedness of (15) and (16). We denote by Λ a compact interval of the real line which does not contain any Maxwell's eigenvalue and by Z a subspace of $L^2(\Omega)^3$ containing all the solutions $\underline{u}(\lambda)$ (for any possible datum $\underline{f}$) and introduce the following conditions.

A1 $TD_uG(\lambda, \underline{u}) : L^2(\Omega)^3 \to L^2(\Omega)^3$ is compact for any $\lambda \in \Lambda$;

A2 there exists a branch of nonsingular solutions such that $\lambda \mapsto \underline{u}(\lambda)$ is C^1 regular from Λ to $L^2(\Omega)^3$;

A3 $(\lambda, \underline{u}) \mapsto DG(\lambda, \underline{u})$ is Lipschitz continuous on the bounded subsets of $\Lambda \times L^2(\Omega)^3$.

Hypotheses A1–A3 are easily checked and imply the well-posedness of the continuous problem (15).

In order to show that problem (16) is solvable and that its solution converge to that of (15) we have to check the following conditions.

D1 There exists $\Pi_h : Z \to V_h$ such that
 - $||\Pi_h \underline{u}(\lambda_1) - \Pi_h \underline{u}(\lambda_2)||_{L^2} \leq C|\lambda_1 - \lambda_2| \quad \forall \lambda_1, \lambda_2 \in \Lambda$
 - $\sup_{\lambda \in \Lambda} ||\underline{u}(\lambda) - \Pi_h \underline{u}(\lambda)||_{L^2} \leq Ch^r$ for some $r > 0$;

D2 $||T - T_h||_{\mathcal{L}(L^2,L^2)} \to 0$ as h goes to zero;

D3 $\sup_{\lambda \in \Lambda} ||(T - T_h)G(\lambda, \underline{u}(\lambda))||_{L^2} \leq Ch^r$.

We now make the regularity assumption

$$Z \subseteq H^s(\Omega)^3, \quad s > 1/2.$$

This hypothesis can be weakened using the quasi-interpolant operators introduced bu Shöberl [21], see [8] for the details.

Conditions D1 and D3 are easily checked; they basically involve approximation properties for edge element spaces. On the other hand, condition D2, which is the most important, is exactly the uniform convergence stated in (11).

We are now in the position to state our main convergence result (here $k \leq 0$ denotes the order of used edge element).

Theorem 1. *For h small enough there exists a unique C^1 mapping $\lambda \mapsto \underline{u}_h(\lambda) \in E_h$ s.t.*

$$\underline{u}_h(\lambda) + T_h G(\lambda, \underline{u}_h(\lambda)) = 0 \quad \forall \lambda \in \Lambda$$

with the error estimate

$$||\underline{u}(\lambda) - \underline{u}_h(\lambda)||_V \leq Ch^\sigma \quad (\sigma = \min(s, k+1))$$

6 Acknowledgments

I would like to acknowledge that part of the work reported in this note has been carried out in collaboration with professor Lucia Gastaldi.

References

1. W. Bao, X. Wang, and K. J. Bathe. On the inf-sup condition of mixed finite element formulations for acoustic fluids. Math. Models and Methods Appl. Sci., in press.

2. K. J. Bathe, C. Nitikitpaiboon, and X. Wang. A mixed displacement-based finite element formulation for acoustic fluid-structure interaction. *Computers & Structures*, 56:225–237, 1995.
3. D. Boffi. Fortin operator and discrete compactness for edge elements. *Numer. Math.*, 87:229–246, 2000.
4. D. Boffi. A note on the de Rham complex and a discrete compactness property. *Appl. Math. Letters*, 14:33–38, 2001.
5. D. Boffi, C. Chinosi, and L. Gastaldi. Approximation of the grad div operator in nonconvex domains. *CMES Comput. Model. Eng. Sci.*, 1(2):31–43, 2000.
6. D. Boffi, M. Farina, and L. Gastaldi. On the approximation of maxwell's eigenproblem in general 2d domains. *Computers & Structures*, 79:1089–1096, 2001.
7. D. Boffi, P. Fernandes, L. Gastaldi, and I. Perugia. Computational models of electromagnetic resonators: analysis of edge element approximation. *SIAM J. Numer. Anal.*, 36:1264–1290, 1998.
8. D. Boffi and L. Gastaldi. Edge finite elements for the approximation of Maxwell resolvent operator. Submitted.
9. F. Brezzi, J. Rappaz, and P.A. Raviart. Finite dimensional approximation of nonlinear problems. Part i: Branches of nonsingular solutions. *Numer. Math.*, 36:1–25, 1980.
10. M. Costabel and M. Dauge. Weighted regularization of Maxwell equations in polyhedral domains. Technical Report 01-26, IRMAR, 2001.
11. Martin Costabel and Monique Dauge. Maxwell and Lamé eigenvalues on polyhedra. *Math. Methods Appl. Sci.*, 22(3):243–258, 1999.
12. L. Demkowicz and L. Vardapetyan. Modeling of electromagnetic absorption/scattering problems using *hp*-adaptive finite elements. *Comput. Methods Appl. Mech. Engrg.*, 152(1-2):103–124, 1998. Symposium on Advances in Computational Mechanics, Vol. 5 (Austin, TX, 1997).
13. L. Gastaldi. Mixed finite element methods in fluid structure systems. *Numer. Math.*, 74(2):153–176, 1996.
14. F. Kikuchi. Mixed and penalty formulations for finite element analysis of an eigenvalue problem in electromagnetism. In *Proceedings of the first world congress on computational mechanics (Austin, Tex., 1986)*, volume 64, pages 509–521, 1987.
15. F. Kikuchi. On a discrete compactness property for the Nédélec finite elements. *J. Fac. Sci., Univ. Tokyo, Sect. I A*, 36(3):479–490, 1989.
16. R. Leis. *Initial boundary value problems in Mathematical Physics.* Teubner, 1986.
17. P. Monk. A finite element method for approximating the time-harmonic Maxwell equations. *Numer. Math.*, 63(2):243–261, 1992.
18. J.-C. Nédélec. Mixed finite elements in $\mathbb{R}^3$. *Numer. Math.*, 35:315–341, 1980.
19. J.-C. Nédélec. A new family of mixed finite elements in $\mathbb{R}^3$. *Numer. Math.*, 50:57–81, 1986.
20. M. Powell. Piecewise quadratic surface fitting for contour plotting. In *Software for numerical mathematics*, pages 253–271. London Academic, 1974.
21. J. Schöberl. Commuting quasi-interpolation operators for mixed finite elements. In preparation.
22. S. Wong and Z. Cendes. Combined finite element-modal solution of three-dimensional eddy current problems. *IEEE Transactions on Magnetics*, 24:2685–2687, 1988.

Trace Theorems on Non-Smooth Boundaries for Functional Spaces Related to Maxwell Equations: an Overview

Annalisa Buffa

Istituto di Matematica Applicata e Tecnologie Informatiche.
Via Ferrata, 1 27100 Pavia, Italy

Summary. We study tangential vector fields on the boundary of a bounded Lipschitz domain in $\mathbb{R}^3$. Our attention is focused on the definition of suitable Hilbert spaces over a range of Sobolev regularity which we try to make as large as possible, and also on the construction of tangential differential operators. Hodge decompositions are proved to hold for some special choices of spaces which are of interest in the theory of Maxwell equations.

Introduction

In the present paper we collect results, observations and open problems as regards to a comprehensive functional theory for Maxwell equations in Lipschitz domains. Many results are known in this field and we refer e.g., to [1], [10], [11], [12], [13], [20], [21].

The main concern of our research is the construction of a suitable functional setting for non-homogeneous Dirichlet and Neumann problems for time-harmonic Maxwell equations, i.e.,

$$\begin{aligned} &\mathbf{curl}\,\mathbf{curl}\,\mathbf{u} - k^2\mathbf{u} = 0 \qquad \Omega \\ &\mathbf{u} \times \mathbf{n} = \mathbf{g} \ \text{ or } \ \mathbf{curl}\,\mathbf{u} \times \mathbf{n} = \mathbf{g}' \qquad \partial\Omega \end{aligned}$$

where Ω is a Lipschitz-continuous bounded domain, $\mathbf{n}$ denotes the outer normal to Ω, k the wave number, $\mathbf{u}$ either the magnetic or the electric field, and $\mathbf{g}$, $\mathbf{g}'$ need to be properly chosen.

More precisely, we characterize the space of tangential trace ($\mathbf{u} \mapsto \mathbf{u} \times \mathbf{n}$) for $\mathbf{H}(\mathbf{curl}\,, \Omega)$ as well as more and less regular fields under the assumption that Ω is a bounded domain with Lipschitz continuous boundary. This will be made precise in the next sections. This work is mainly inspired by [7] and [2] and we aim to extend (in a suitable way) the results contained in these papers. More precisely, we do not succeed in writing a completely general theory, but we present some extensions of the known results and we discuss some open problem.

We consider then polyhedral domains. The theory is deduced from the one developed for Lipschitz domains and the results presented in [4], [5] are reinterpreted under this point of view. In the case of piecewise regular domains, on one hand the theory should be easier, but on the other hand one expects to have more "explicit" informations. Here we characterize some of the spaces introduced for Lipschitz domains in terms of face by face regularity plus compatibility condition at the edges (i.e., "á la Grisvard" [15], [16]).

The trace theorems we present here have a direct impact on the application and, more precisely, they are important to properly formulate integral equations for Maxwell equations and to study their approximation by boundary elements. Some pioneering works in this direction are [6], [17], [3], [8].

1 Preliminaries

Before stating trace theorems for spaces related to Maxwell equations, we need to define some Sobolev spaces and some differential operator acting on them. We refer to [4, 5, 2] and to [7] for more detail.

1.1 Functional spaces

We denote by $\mathcal{D}(\bar{\Omega})^3$ the space of the $3D$ vector fields defined as $C^\infty_{comp}(\mathbb{R}^3)_{|\Omega}$.

Let $\Omega \subset \mathbb{R}^3$ be a bounded Lipschitz-continuous domain in $\mathbb{R}^3$. We denote by Γ its boundary, and the assumptions on Ω imply that Γ is locally subgraph of Lipschitz functions. Without loss of generality, we suppose that Γ is connected. When it is not the case, the theory presented in this paper can be applied separately at each connected component. We denote by $\mathbf{n}$ the outer unit normal vector to Ω. Moreover we denote by $H^s(\Omega)$, $\forall s \in \mathbb{R}^+$ and $H^t(\Gamma)$, $\forall t \in [-1,1]$ the standard real valued, Hilbertian Sobolev space defined on Ω, and Γ respectively (with the convention $H^0 = L^2$.) We denote by $H^{-s}(\Omega)$, $s \in (0,1/2]$ the dual space of $H^s(\Omega)$ with $L^2(\Omega)$ as pivot. Remark that in this way we are adopting the notation introduced by [19] which is in contrast with the one used in [15].

Although the electromagnetic fields are naturally complex valued vectors, here we consider real valued function spaces only for the sake of exposition. The results extend to complex valued function spaces with no change.

The duality pairing between $H^{-s}(\Gamma)$ and $H^s(\Gamma)$ is denoted by $\langle\cdot,\cdot\rangle_{s,\Gamma}$. We set:

$$\mathbf{H}^s(\Omega) := \left(H^s(\Omega)\right)^3, \tag{1}$$

$$\mathbf{H}^s(\mathbf{curl}\,,\Omega) = \{\mathbf{u} \in \mathbf{H}^s(\Omega) \mid \mathbf{curl}\,\mathbf{u} \in \mathbf{H}^s(\Omega)\}; \tag{2}$$

$$\mathbf{H}^s(\mathrm{div},\Omega) = \{\mathbf{u} \in \mathbf{H}^s(\Omega) \mid \mathrm{div}\mathbf{u} \in H^s(\Omega)\}; \tag{3}$$

$$\mathbf{L}^2_t(\Gamma) = \{\mathbf{v} \in \mathbf{L}^2(\Gamma) \mid \mathbf{n}\cdot\mathbf{v} = 0 \text{ on } \Gamma\}; \tag{4}$$

$$H^{-s}_\star(\Gamma) := \{\mathrm{u} \in H^{-s}(\Gamma) \mid \langle \mathrm{u}, 1\rangle_{s,\Gamma} = 0\} \quad (s \in [-1,1]) \tag{5}$$

$$\text{for } t > 1\,,\ H^t(\Gamma) := \{\mathrm{u}_{|\Gamma} \mid \mathrm{u} \in H^{t+1/2}(\Omega)\}. \tag{6}$$

We denote by $\|\cdot\|_{s,\mathbf{curl},\Omega}$ and $\|\cdot\|_{s,\mathrm{div},\Omega}$ the graph norms associated respectively to $\mathbf{H}^s(\mathbf{curl},\Omega)$ and $\mathbf{H}^s(\mathrm{div},\Omega)$.

The space $\mathbf{L}^2_t(\Gamma)$ is identified with the space of fields belonging to the tangent bundle $T\Gamma$ of Γ for almost every $\mathbf{x} \in \Gamma$ and which are square integrable.

The spaces $H^t(\Gamma)$, $t > 1$ have no intrinsic definition on the surface Γ. Nevertheless they are Hilbert spaces endowed with the norms:

$$||\lambda||_{t,\Gamma} := \inf_{u \in H^{t+1/2}(\Omega)} \{||u||_{t+1/2,\Omega} \text{ such that } u_{|\Gamma} = \lambda\}.$$

We denote by $H^{-t}(\Gamma)$ the dual space of $H^t(\Gamma)$ with $L^2(\Gamma)$ as pivot space. Finally, when Ω is a polyhedron these spaces can be characterized face by face. We refer to [4] for details.

Definition 1 *We define:*

- *the normal trace operator:* $\gamma_n : \mathcal{D}(\bar{\Omega})^3 \to \mathbf{L}^2_t(\Gamma)$, $\mathbf{u} \mapsto \mathbf{u} \cdot \mathbf{n}$;
- *the tangential components trace operator:* $\pi_\tau : \mathcal{D}(\bar{\Omega})^3 \to \mathbf{L}^2_t(\Gamma)$, $\mathbf{u} \mapsto \mathbf{u} - \gamma_n(\mathbf{u})\mathbf{n}$;
- *the "tangential trace" operator:* $\gamma_\tau : \mathcal{D}(\bar{\Omega})^3 \to \mathbf{L}^2_t(\Gamma)$, $\mathbf{u} \mapsto \mathbf{n} \times \mathbf{u}_{|\Gamma}$.

We denote by γ the standard trace operator acting on vectors: $\gamma : \mathbf{H}^1(\Omega) \to V$, $\gamma(\mathbf{u}) = \mathbf{u}_{|\Gamma}$. Let γ^{-1} be one of its right inverses. We will also use the notation π_τ (resp. γ_τ) for the composite operator $\pi_\tau \circ \gamma^{-1}$ (resp. $\gamma_\tau \circ \gamma^{-1}$) which acts only on traces. By density of $\mathcal{D}(\bar{\Omega})^3_{|\Gamma}$ into $\mathbf{L}^2(\Gamma)$, the operators π_τ and γ_τ can be extended to linear continuous operators in $\mathbf{L}^2(\Gamma)$.

We define:

Definition 2 *For any* $s > 0$, *let*

$$H^s_\times(\Gamma) := \gamma_\tau(\mathbf{H}^{s+1/2}(\Omega)) \quad H^s_T(\Gamma) := \pi_\tau(\mathbf{H}^{s+1/2}(\Omega)).$$

For $s = 0$, *we adopt the convention:* $H^0_\times(\Gamma) = H^0_T(\Gamma) = \mathbf{L}^2_t(\Gamma)$.

For $s > 0$, $H^s_\times(\Gamma)$ and $H^s_T(\Gamma)$ are Hilbert spaces endowed with norms that ensure the continuity of the operators γ_τ and π_τ, respectively. We set:

$$||\boldsymbol{\lambda}||_{s,\times} = \inf_{\mathbf{u} \in \mathbf{H}^{s+1/2}(\Omega)} \{||\mathbf{u}||_{s+1/2,\Omega} \mid \gamma_\tau(\mathbf{u}) = \boldsymbol{\lambda}\} \tag{7}$$

$$||\boldsymbol{\lambda}||_{s,T} = \inf_{\mathbf{u} \in \mathbf{H}^{s+1/2}(\Omega)} \{||\mathbf{u}||_{s+1/2,\Omega} \mid \pi_\tau(\mathbf{u}) = \boldsymbol{\lambda}\} \tag{8}$$

Note that $\pi_\tau : \mathbf{H}^{s+1/2}(\Omega) \to H^s_T(\Gamma)$ and $\gamma_\tau : \mathbf{H}^{s+1/2}(\Omega) \to H^s_\times(\Gamma)$ are isomorphisms by construction. The spaces $H^s_T(\Gamma)$, $H^s_\times(\Gamma)$ will be the bases of our construction. For any positive s, we denote by $H^s_T(\Gamma)'$, $H^s_\times(\Gamma)'$ the dual spaces of $H^s_T(\Gamma)$, $H^s_\times(\Gamma)$ respectively with $\mathbf{L}^2_t(\Gamma)$ as pivot. Note that $H^s_T(\Gamma)'$, $H^s_\times(\Gamma)'$ are Hilbert spaces endowed with their natural norms.

1.2 Tangential differential operators

In the following we need various differential operators defined on the surface Γ, which is a closed Lipschitz surface without boundary. The tangential functional spaces defined here above are suitable for their definition. The operators:

$$\nabla_\Gamma : H^1(\Gamma) \to \mathbf{L}^2_t(\Gamma), \qquad \mathbf{curl}_\Gamma : H^1(\Gamma) \to \mathbf{L}^2_t(\Gamma)$$

are defined on Γ in the usual way by a localization argument (see [22] or [7]). The adjoint operators of $-\nabla_\Gamma$ and $\mathbf{curl}_\Gamma$ are:

$$\mathrm{div}_\Gamma : \mathbf{L}^2_t(\Gamma) \to H^{-1}_\star(\Gamma), \qquad \mathrm{curl}_\Gamma : \mathbf{L}^2_t(\Gamma) \to H^{-1}_\star(\Gamma)$$

respectively, and they are linear and continuous for these choices of spaces. The operators ∇_Γ and curl_Γ can be restricted to more regular spaces.

Using the results in [7] and noting that for any regular $P \in \mathcal{D}(\bar{\Omega})$:

$$\pi_\tau(\nabla P) = \nabla_\Gamma(P_{|\Gamma}) \qquad \gamma_\tau(\nabla P) = \mathbf{curl}_\Gamma(P_{|\Gamma}), \tag{9}$$

we can easily deduce that for any $t \geq 1$, the operators

$$\nabla_\Gamma : H^t(\Gamma) \to H^{t-1}_T(\Gamma) \qquad \mathbf{curl}_\Gamma : H^t(\Gamma) \to H^{t-1}_\times(\Gamma). \tag{10}$$

are linear and continuous. Moreover, we easily have that $\nabla_\Gamma p = 0$ or $\mathbf{curl}_\Gamma p = 0$ if and only if $p = const.$

As a consequence, their adjoint operators $\mathrm{div}_\Gamma : H^{t-1}_T(\Gamma)' \to H^{-t}_\star(\Gamma)$ and $\mathrm{curl}_\Gamma : H^{t-1}_\times(\Gamma)' \to H^{-t}_\star(\Gamma)$ are linear and continuous operators.

Finally, we define the *Laplace-Beltrami* operator on the Lipschitz manifold Γ as $\Delta_\Gamma u = \mathrm{div}_\Gamma(\nabla_\Gamma u)$ for any $u \in H^1(\Gamma)$. One can prove [7] that $\Delta_\Gamma : H^1(\Gamma) \to H^{-1}_\star(\Gamma)$ is linear, continuous and admits a right inverse.

2 Green formulae

In order to have an insight on the functional spaces we have to deal with when treating the problem of traces for $\mathbf{H}(\mathbf{curl}, \Omega)$, we first introduce the related Green formulae.

Let $\mathbf{u} \in \mathcal{D}(\bar{\Omega})$, then

$$\int_\Omega \mathrm{div}(\mathbf{u}) = \int_\Gamma \gamma_n(\mathbf{u}). \tag{11}$$

Let now $\mathbf{v} \in \mathcal{D}(\bar{\Omega})^3$ and $v \in \mathcal{D}(\bar{\Omega})$. Since $\mathrm{div}(\mathbf{u}v) = \mathrm{div}(\mathbf{u})v + \mathbf{u} \cdot \nabla v$ and $\mathrm{div}(\mathbf{u} \times \mathbf{v}) = \mathbf{curl}\,\mathbf{u} \cdot \mathbf{v} - \mathbf{u} \cdot \mathbf{curl}\,\mathbf{v}$, and $(\mathbf{u} \times \mathbf{v}) \cdot \mathbf{n} = \gamma_\tau \mathbf{v} \cdot \mathbf{u}$, we deduce:

$$\int_\Omega \mathrm{div}(\mathbf{u})v + \mathbf{u} \cdot \nabla v = \int_\Gamma \gamma_n(\mathbf{u})\, v\ ; \tag{12}$$

$$\int_\Omega \mathbf{curl}\,\mathbf{u} \cdot \mathbf{v} - \mathbf{u} \cdot \mathbf{curl}\,\mathbf{v} = \int_\Gamma \gamma_\tau \mathbf{v} \cdot \mathbf{u}. \tag{13}$$

There are some consequences of (12, 13).

Theorem 2. *The operator γ_n extends to a linear and continuous operator from $\mathbf{H}^s(\mathrm{div},\Omega)$ to $H^{s-1/2}(\Gamma)$ for $-1/2 \le s < 1/2$.*

Proof. Straightforward using (12) and [18]. Remark that this result can be formulated in this way only thanks to our choice of notation, i.e., $H^{-s}(\Omega) = H^s(\Omega)'$, $s \in (0,1/2]$.

For positive values of s, let $H^s_n(\Gamma) = \gamma_n(H^{1/2+s}(\Omega))$. This is a Hilbert space which can be endowed with the norm that let γ_n be continuous. We set:

$$-1 \le s \le 0 \;:\; \mathbf{H}^s(\mathrm{div}_\Gamma,\Gamma) := \{\boldsymbol{\lambda} \in H^{-s}_T(\Gamma)' \mid \mathrm{div}_\Gamma \boldsymbol{\lambda} \in H^s_\star(\Gamma)\} \qquad (14)$$

$$s > 0 \;:\; \mathbf{H}^s(\mathrm{div}_\Gamma,\Gamma) := \{\boldsymbol{\lambda} \in H^s_\times(\Gamma) \mid \mathrm{div}_\Gamma \boldsymbol{\lambda} \in H^s_{n,\star}(\Gamma)\} \qquad (15)$$

where $H^s_{n,\star}(\Gamma)$ denote the space of zero mean value functions in $H^s_n(\Gamma)$.

Theorem 3. *The operator γ_τ extends to a linear and continuous mapping from $\mathbf{H}^s(\mathbf{curl},\Omega)$ to $\mathbf{H}^{s-1/2}(\mathrm{div}_\Gamma,\Gamma)$, for any $s \ge -1/2$, $s \ne 1/2$.*

Proof. For $s > 1/2$, the statement is obviously true thanks to the Definition 2 and the relation:

$$\forall \mathbf{u} \in H(\mathbf{curl},\Omega) \qquad \gamma_n(\mathbf{curl}\,\mathbf{u}) = \mathrm{div}_\Gamma(\gamma_\tau \mathbf{u}) \qquad (16)$$

which is an easy consequence of (12, 13). For $-1/2 \le s < 1/2$, let $\boldsymbol{\xi} \in H^{1/2-s}_T(\Gamma)$. By definition itself of $H^{1/2-s}_T(\Gamma)$, there exists a $\mathbf{u} \in \mathbf{H}^{1-s}(\Omega)$ such that $\pi_\tau(\mathbf{u}) = \boldsymbol{\xi}$. Let now $\mathbf{v} \in \mathcal{D}(\bar\Omega)^3$; using (13), we have:

$$\left|\int_\Gamma \pi_\tau(\mathbf{u})\cdot\gamma_\tau(\mathbf{v})\right| = \left|\int_\Omega \mathbf{curl}\,\mathbf{u}\cdot\mathbf{v} - \mathbf{curl}\,\mathbf{v}\cdot\mathbf{u}\right| \le \|\mathbf{u}\|_{1-s,\Omega}\|\mathbf{v}\|_{s,\mathbf{curl},\Omega}.$$

By density of $\mathcal{D}(\bar\Omega)^3$ in $\mathbf{H}^s(\mathbf{curl},\Omega)$ we deduce $\gamma_\tau : \mathbf{H}^s(\mathbf{curl},\Omega) \to H^{1/2-s}_T(\Gamma)'$ is continuous. Using then (16), we deduce that $\mathrm{div}_\Gamma(\gamma_\tau \mathbf{u}) \in H^{s-1/2}(\Gamma)$ as a consequence of Theorem 2.

Remark 2. For $s = 1/2$ the statement is false and a counterexample can be built using the construction proposed in [18] to show that the standard trace mapping $u \mapsto u_{|\Gamma}$ is "not" continuous from $H^{3/2}(\Omega)$ to $H^1(\Gamma)$. More precisely, the gradient of the $H^{3/2}(\Omega)$–function constructed in [18] furnishes a counterexample for the statement of Theorem 3 in the case $s = 1/2$.

The next task is to study the existence of suitable right inverses for the trace mappings γ_n and γ_τ.

Theorem 4. *For any s, $-1/2 < s \le 1/2$, there exists a lifting operator $\mathcal{L}_n : H^{s-1/2}(\Gamma) \to \mathbf{H}^s(\mathrm{div},\Omega)$ such that $\gamma_n(\mathcal{L}_n(u)) = u\ \forall u \in H^{s-1/2}(\Gamma)$ and* $\mathrm{div}\mathcal{L}(u) = \int_\Gamma u$.

Proof. For $0 \leq s < 1/2$ it is standard [14]. For $s = 1/2$ it is a consequence of the results proved in [18]. For $-1/2 < s < 0$, it is an application of the transposition argument [19].

Theorem 5. *There exists a linear and continuous extension operator*

$$T^s : \mathbf{H}^{s-1/2}(\mathrm{div}_\Gamma, \Gamma) \to \mathbf{H}^s(\mathbf{curl}\,, \Omega)$$

for any $-1/2 < s \leq 1/2$ *such that* $\gamma_\tau T^s(\boldsymbol{\lambda}) = \boldsymbol{\lambda}$.

Proof. In [25], Tartar proved that $\gamma_\tau : \mathbf{H}(\mathbf{curl}\,, \Omega) \to \mathbf{H}^{-1/2}(\mathrm{div}_\Gamma, \Gamma)$ admits a linear right inverse. It is not hard to see that his proof holds true for any s, $s \in (-1/2, 1/2]$.

This theorem has several consequences.

- We define the bilinear form $b(\cdot,\cdot) : \mathbf{L}^2_t(\Gamma) \times \mathbf{L}^2_t(\Gamma) \to \mathbb{R}$,

$$b(\boldsymbol{\lambda}, \boldsymbol{\xi}) = \int_\Gamma \boldsymbol{\lambda} \cdot \boldsymbol{\xi} \times \mathbf{n}.$$

  Theorem 5 together with the first green formula (13) implies that the bilinear form b defines a duality between $\mathbf{H}^s(\mathrm{div}_\Gamma, \Gamma)$ and $\mathbf{H}^{-1-s}(\mathrm{div}_\Gamma, \Gamma)$ for $s \in (-1, 0)$; i.e., b is continuous from $\mathbf{H}^s(\mathrm{div}_\Gamma, \Gamma) \times \mathbf{H}^{-1-s}(\mathrm{div}_\Gamma, \Gamma)$ to $\mathbb{R}$, and $\mathbf{v} \mapsto b(\mathbf{v}, \cdot)$ maps $\mathbf{H}^{-s}(\mathrm{div}_\Gamma, \Gamma)$ onto its dual.
- The differential operator $\mathbf{curl}_\Gamma : H^1_\star(\Gamma) \to \mathbf{L}^2_t(\Gamma)$ can be extended by using the formula (9) (see [3]):

$$0 < s \leq 1\,,\ \mathbf{curl}_\Gamma : H^s_\star(\Gamma) \to H_T^{-(s-1)}(\Gamma)'\,; \qquad (17)$$

  The question : "Does the $\mathbf{curl}_\Gamma$ have closed range in $H_T^{-(s-1)}(\Gamma)'$?" remains open for the moment and the answer can not be easily deduced from Theorem 5.
  Note that the following diagram commutes for $0 < s < 1$:

$$\begin{array}{ccc} H^{s+1/2}(\Omega) & \xrightarrow{\ \nabla\ } & \mathbf{H}^{s-1/2}(\mathbf{curl}\,, \Omega) \\ \big\downarrow \gamma & & \big\downarrow \gamma_\tau \\ H^s(\Gamma) & \xrightarrow{\ \mathbf{curl}_\Gamma\ } & \mathbf{H}^{s-1}(\mathrm{div}_\Gamma, \Gamma). \end{array}$$

- Many choices of extension operator are possible. The next theorem explains one of these.

  Theorem 6. *There exists a linear and continuous extension operator* $\mathcal{L}_\times$: $\mathbf{H}^{s-1/2}(\mathrm{div}_\Gamma, \Gamma) \to \mathbf{H}^s(\mathbf{curl}\,, \Omega)$ *for any* $0 \leq s \leq 1/2$ *such that* $\gamma_\tau \mathcal{L}_\times(\boldsymbol{\lambda}) = \boldsymbol{\lambda}$ *and* $\mathrm{div}(\mathcal{L}_\times \mathbf{u}) = 0$ *in* Ω.

Proof. For $0 \leq s \leq 1/2$, let $\boldsymbol{\lambda} \in \mathbf{H}^{s-1/2}(\mathrm{div}_\Gamma, \Gamma)$. We can solve the problem: Find $\mathbf{u} \in \mathbf{H}(\mathbf{curl}\,, \Omega)$ such that

$$\mathbf{curl}\,\mathbf{curl}\,\mathbf{u} + \mathbf{u} = 0\,, \quad \gamma_\tau(\mathbf{u}) = \boldsymbol{\lambda}.$$

The result is then a consequence of the results stated in [9].

Remark 3. Using Theorem 6 together with [1, Theorem 3.5], and since Γ is assumed to be connected, we know that actually any vector $\boldsymbol{\lambda} \in \mathbf{H}^s(\mathrm{div}_\Gamma, \Gamma)$ can be extended in the form of a **curl**. This observation is important since it makes sure that both Dirichlet trace operator ($\mathbf{u} \mapsto \gamma_\tau(\mathbf{u})$) and the Neumann trace operator ($\mathbf{u} \mapsto \gamma_\tau(\mathbf{curl}\,\mathbf{u})$) have the same range (for suitable, and different, choices of the spaces where they act on).

3 Hodge decompositions

We now focus the attention on the structure of the spaces $\mathbf{H}^s(\mathrm{div}_\Gamma, \Gamma)$ and to this aim, we introduce suitable Hodge decompositions of these spaces. This section is mainly inspired by [7], [2] and [3].

Following [3], for any $s \in [-1, 1]$, we define the operator $\mathcal{G}^s \colon \mathbf{H}^s(\mathrm{div}_\Gamma, \Gamma) \to \mathbf{H}^s(\mathrm{div}_\Gamma, \Gamma)$ as follows: For any $\mathbf{u} \in \mathbf{H}^s(\mathrm{div}_\Gamma, \Gamma)$, $\mathcal{G}^s(\mathbf{u}) = \nabla_\Gamma p$ where $p \in H^1_\star(\Gamma)$ is the solution of the problem:

$$\int_\Gamma \nabla_\Gamma p \cdot \nabla_\Gamma q = \int_\Gamma \mathrm{div}_\Gamma \mathbf{u}\, q \qquad q \in H^1_\star(\Gamma).$$

Since $\mathrm{div}_\Gamma \mathcal{G}^s(\mathbf{u}) = \mathrm{div}_\Gamma \mathbf{u}$, the operator $\mathcal{G}^s$ verifies the following:

$$\ker \mathcal{G}^s = \{\mathbf{u} \in \mathbf{H}^s(\mathrm{div}_\Gamma, \Gamma) \mid \mathrm{div}_\Gamma \mathbf{u} = 0\} \quad \mathcal{G}^s(\mathbf{u} - \mathcal{G}^s(\mathbf{u})) = 0.$$

Thus the operator $\mathcal{G}^s$ is a projection and it generates a splitting for the space $\mathbf{H}^s(\mathrm{div}_\Gamma, \Gamma)$. Namely let

$$\mathcal{H}^s(\Gamma) = \{p \in H^1_\star(\Gamma) \mid \Delta_\Gamma p \in H^s_\star(\Gamma)\}\,, \; s \in [-1, 0];$$

we have:

$$s \in [-1, 0] \quad \mathbf{H}^s(\mathrm{div}_\Gamma, \Gamma) = \nabla_\Gamma \mathcal{H}^s(\Gamma) \oplus \ker\{\mathrm{div}_\Gamma\} \cap \mathbf{H}^s(\mathrm{div}_\Gamma, \Gamma). \tag{18}$$

The next step is the characterization of $\ker\{\mathrm{div}_\Gamma\} \cap \mathbf{H}^s(\mathrm{div}_\Gamma, \Gamma)$ and for this we take advantage of the results in [2]. He first need to introduce some notations: Let $\gamma_\tau^{\Omega_e}$ denote the tangential trace operator acting on functions defined on the exterior domain $\Omega_e := \mathbb{R}^3 \setminus \bar{\Omega}$. We set:

$$\mathbb{H}_1 := \{\mathbf{u} \in \mathbf{H}(\mathbf{curl}\,, \Omega) \cap \mathbf{H}(\mathrm{div}, \Omega)\ \mathbf{curl}\,\mathbf{u} = 0\,,\ \mathrm{div}\mathbf{u} = 0\,,\ \mathbf{u} \cdot \mathbf{n}_{|\Gamma} = 0\};$$
$$\mathbb{H}_2 := \{\mathbf{u} \in \mathbf{H}(\mathbf{curl}\,, \Omega_e) \cap \mathbf{H}(\mathrm{div}, \Omega_e)\ \mathbf{curl}\,\mathbf{u} = 0\,,\ \mathrm{div}\mathbf{u} = 0\,,\ \mathbf{u} \cdot \mathbf{n}_{|\Gamma} = 0\}.$$

It is well known that the spaces $\mathbb{H}_1$ and $\mathbb{H}_2$ have finite dimensions [23], [10] or [1], and moreover $dim\{\mathbb{H}_1\} = dim\{\mathbb{H}_2\}$ (see [24, theorem 16, p. 296]).

We set:

$$\mathbf{H}_1 = \gamma_\tau \mathbb{H}_1 \qquad \mathbf{H}_2 = \gamma_\tau^{\Omega_e} \mathbb{H}_2$$

and we state the following proposition:

Proposition 1 *We have that* $\mathbf{H}_1,\ \mathbf{H}_2 \subseteq \mathbf{L}^2_t(\Gamma)$ *and the following decomposition holds:*

$$\ker\{\mathrm{div}_\Gamma\} \cap \mathbf{L}^2_t(\Gamma) = \mathbf{curl}_\Gamma H^1(\Gamma) \oplus \mathbf{H}_1 \oplus \mathbf{H}_2 \tag{19}$$

where $\oplus$ *denotes the direct sum with respect to the* $\mathbf{L}^2_t(\Gamma)$ *scalar product.*

Proof. First of all:

$$\mathbb{H}_1 \subset \{\mathbf{u} \in \mathbf{L}^2(\Omega) \,:\, \mathbf{curl}\,\mathbf{u} \in \mathbf{L}^2(\Omega)\,,\ \mathrm{div}\mathbf{u} \in \mathbf{L}^2(\Omega)\,,\ \mathbf{u}\cdot\mathbf{n}_{|\Gamma} \in L^2(\Gamma)\}$$
$$\mathbb{H}_2 \subset \{\mathbf{u} \in \mathbf{L}^2(\Omega_e) \,:\, \mathbf{curl}\,\mathbf{u} \in \mathbf{L}^2(\Omega_e)\,,\ \mathrm{div}\mathbf{u} \in \mathbf{L}^2(\Omega_e)\,,\ \mathbf{u}\cdot\mathbf{n}_{|\Gamma} \in L^2(\partial\Omega_e)\}.$$

Using standard regularity results for Maxwell equations [9], we deduce

$$\mathbf{H}_1\,,\ \mathbf{H}_2 \subseteq \mathbf{L}^2_t(\Gamma).$$

Then the proof of the statement proposed in [2] extends with no change to the case of Lipschitz domains. From [2], we know also that (19) can be rewritten as

$$\ker\{\mathrm{div}_\Gamma\} \cap \mathbf{L}^2_t(\Gamma) = \mathbf{curl}_\Gamma H^1(\Gamma) \overset{\perp}{\oplus} \mathbb{H}$$

where $\mathbb{H} := \{\boldsymbol{\lambda} \in \mathbf{L}^2_t(\Gamma) \mid \mathrm{div}_\Gamma\boldsymbol{\lambda} = 0\,,\ \mathrm{curl}_\Gamma\boldsymbol{\lambda} = 0\}$ and the orthogonality is in the sense of $\mathbf{L}^2_t(\Gamma)$.

We can now prove the following:

Proposition 2 *For any* s, $-1 < s \le 0$ *we have:*

$$\ker\{\mathrm{div}_\Gamma\} \cap \mathbf{H}^s(\mathrm{div}_\Gamma, \Gamma) = \mathbf{curl}_\Gamma H^{s+1}(\Gamma) \oplus \mathbb{H}.$$

Proof. This result is basically a consequence of Theorem 5 and the proof can be obtained adapting the proof Theorem 2 in [2].

Note that a consequence of this theorem is the following:

Corollary 1. *The operator* $\mathbf{curl}_\Gamma : H^s(\Gamma) \to H_T^{-(s-1)}(\Gamma)'$ *has closed range for any* s, $0 < s < 1$. *Moreover, since* $\ker\{\mathbf{curl}_\Gamma\} = \mathbb{R}$, *we have:*

$$||p||_{s,\Gamma} \le C||\mathbf{curl}_\Gamma p||_{(s-1),T} \qquad \forall p \in H^s_\star(\Gamma).$$

Note that by symmetry the same holds true for the gradient operator: $\nabla_\Gamma : H^s(\Gamma) \to H_\times^{-(s-1)}(\Gamma)'$ has closed range for any s, $0 < s < 1$ and it holds:

$$||p||_{s,\Gamma} \le C||\nabla_\Gamma p||_{(s-1),T} \qquad \forall p \in H^s_\star(\Gamma).$$

Summarizing we have proved the following theorem:

Theorem 7. *For any $s \in [-1,0]$, we have:*

$$\mathbf{H}^s(\mathrm{div}_\Gamma, \Gamma) = \nabla_\Gamma \mathcal{H}^s(\Gamma) \oplus \ker\{\mathrm{div}_\Gamma\} \cap \mathbf{H}^s(\mathrm{div}_\Gamma, \Gamma).$$

Moreover, for $s \in (-1,0)$, $\ker\{\mathrm{div}_\Gamma\} \cap \mathbf{H}^s(\mathrm{div}_\Gamma, \Gamma)$ can be characterized as follows:

$$\ker\{\mathrm{div}_\Gamma\} \cap \mathbf{H}^s(\mathrm{div}_\Gamma, \Gamma) = \mathbf{curl}_\Gamma H^{s+1}(\Gamma) \oplus \mathbb{H}.$$

Note that, the duality induced by b reads:

$$\begin{aligned} &\mathbf{u} \in \mathbf{H}^s(\mathrm{div}_\Gamma, \Gamma)\,, \quad \mathbf{u} = \nabla_\Gamma p_u + \mathbf{curl}_\Gamma q_u + \mathbf{h}_u \\ &\mathbf{v} \in \mathbf{H}^{-1-s}(\mathrm{div}_\Gamma, \Gamma)\,, \quad \mathbf{u} = \nabla_\Gamma p_v + \mathbf{curl}_\Gamma q_v + \mathbf{h}_v \\ &\quad b(\mathbf{u}, \mathbf{v}) = -\langle \Delta_\Gamma p_u, q_v \rangle_{-s,\Gamma} + \langle \Delta_\Gamma p_v, q_u \rangle_{-s,\Gamma} + \int_\Gamma \mathbf{h}_u \cdot \mathbf{h}_v \times \mathbf{n}. \end{aligned}$$

4 Polyhedral domains

When Ω is a polyhedron, the spaces $H^s_\times(\Gamma)$ and $H^s_T(\Gamma)$ could be fully characterized in an intrinsic way on the surface Γ for some values of s. As far as we know, the theory is not complete in this context and we report here some recent results in this framework. This section is largely inspired by [4], [5]. To this end, we introduce some notation. We denote by Γ_j, $j = 1, .., N_\Gamma$ the boundary faces of the polyhedron Ω and by $e_{ij} = \bar{\Gamma}_j \cap \bar{\Gamma}_i$ (for some i, j) the set of edges. Let $\boldsymbol{\tau}_{ij}$ be a unit vector parallel to e_{ij} and $\mathbf{n}_j = \mathbf{n}_{|\Gamma_j}$; $\boldsymbol{\tau}_i := \boldsymbol{\tau}_{ij} \wedge \mathbf{n}_i$. The couple $(\boldsymbol{\tau}_i, \boldsymbol{\tau}_{ij})$ is an orthonormal basis of the plane generated by Γ_i (resp. Γ_j); $(\boldsymbol{\tau}_i, \boldsymbol{\tau}_{ij}, \mathbf{n}_i)$ is an orthonormal basis of $\mathbb{R}^3$. Finally, we denote by $\mathcal{I}_j$ the set of indices i such that Γ_i shares an edge (namely e_{ij}) with Γ_j.

For any $\varphi \in L^2(\Gamma)$ we adopt the notation $\varphi_j = \varphi_{|\Gamma_j}$. This notation is used whenever the restriction to a face is considered, that is as regards to any functional space in which the restriction to a face is meaningful.

We set $\mathbf{H}^s_-(\Gamma) := \{\boldsymbol{\varphi} \in \mathbf{L}^2_t(\Gamma)$ such that $\boldsymbol{\varphi}_j \in H^s(\Gamma_j)^2\}$, $s > 0$. First of all, we have the following:

Theorem 8. *For $0 < s < 1/2$,*

$$H^s_T(\Gamma) = H^s_\times(\Gamma) = \mathbf{H}^s_-(\Gamma).$$

For any $\boldsymbol{\varphi} \in \mathbf{H}^{\frac{1}{2}}_-(\Gamma)$, we define:

$$\mathcal{N}^{||}_{ij}(\boldsymbol{\varphi}) := \int_{\Gamma_i} \int_{\Gamma_j} \frac{|\boldsymbol{\varphi}_i(\mathbf{x}) \cdot \boldsymbol{\tau}_{ij} - \boldsymbol{\varphi}_j(\mathbf{y}) \cdot \boldsymbol{\tau}_{ij}|^2}{||\mathbf{x} - \mathbf{y}||^3} \, d\sigma(\mathbf{x}) d\sigma(\mathbf{y}) \quad \forall\, i \in \mathcal{I}_j \ \forall\, j$$

$$\mathcal{N}^{\perp}_{ij}(\boldsymbol{\varphi}) := \int_{\Gamma_i} \int_{\Gamma_j} \frac{|\boldsymbol{\varphi}_i(\mathbf{x}) \cdot \boldsymbol{\tau}_{i} - \boldsymbol{\varphi}_j(\mathbf{y}) \cdot \boldsymbol{\tau}_{j}|^2}{||\mathbf{x} - \mathbf{y}||^3} \, d\sigma(\mathbf{x}) d\sigma(\mathbf{y}) \quad \forall\, i \in \mathcal{I}_j \ \forall\, j$$

and we adopt the notation $\boldsymbol{\varphi}_i \cdot \boldsymbol{\tau}_{ij} \stackrel{\frac{1}{2}}{=} \boldsymbol{\varphi}_j \cdot \boldsymbol{\tau}_{ij}$ at e_{ij}, $i \in \mathcal{I}_j$ (resp. $\boldsymbol{\varphi}_i \cdot \boldsymbol{\tau}_i \stackrel{\frac{1}{2}}{=} \boldsymbol{\varphi}_j \cdot \boldsymbol{\tau}_j$ at e_{ij}) if and only if $\mathcal{N}_{ij}^{\|}(\boldsymbol{\varphi})$ (resp. $\mathcal{N}_{ij}^{\perp}(\boldsymbol{\varphi})$) is finite.

The proof of the following lemma can be found in [5].

Lemma 1. *Let Ω be a polyhedron. The spaces $H_T^{1/2}(\Gamma)$ and $H_\times^{1/2}(\Gamma)$ can be characterized in the following way:*

$$\begin{aligned} \mathbf{H}_T^{\frac{1}{2}}(\Gamma) &:= \left\{ \boldsymbol{\psi} \in \mathbf{H}_-^{\frac{1}{2}}(\Gamma) \mid \boldsymbol{\psi}_i \cdot \boldsymbol{\tau}_{ij} \stackrel{\frac{1}{2}}{=} \boldsymbol{\psi}_j \cdot \boldsymbol{\tau}_{ij} \text{ at } e_{ij}\ \forall i \in \mathcal{I}_j,\ \forall j \right\}. \\ \mathbf{H}_\times^{\frac{1}{2}}(\Gamma) &:= \left\{ \boldsymbol{\psi} \in \mathbf{H}_-^{\frac{1}{2}}(\Gamma) \mid \boldsymbol{\psi}_i \cdot \boldsymbol{\tau}_i \stackrel{\frac{1}{2}}{=} \boldsymbol{\psi}_j \cdot \boldsymbol{\tau}_j \text{ at } e_{ij}\ \forall i \in \mathcal{I}_j,\ \forall j \right\} \end{aligned} \tag{20}$$

As far as more regular fields are concerned we can develop a theory which is somehow parallel to the one proposed in the case of Lipschitz domains. For any $t > 1$, we define the space:

$$H^t(\Gamma) = \{u \in H^1(\Gamma) \mid u_j \in H^t(\Gamma_j)\} \tag{21}$$

endowed with its natural norm

$$\|u\|_{t,\Gamma} := \left(\|u\|_{1,\Gamma}^2 + \sum_{j=1}^{N_\Gamma} \|u_j\|_{t,\Gamma_j}^2 \right)^{\frac{1}{2}}.$$

We define:

$$\begin{aligned} \mathbf{H}_-^s(\Gamma) &= \{\varphi \in \mathbf{L}_t^2(\Gamma) \mid \varphi_j \in H^s(\Gamma_j)^2\} \quad (s \geq 0)\ ; \\ \mathbf{H}_\|^s(\Gamma) &= \{\varphi \in \mathbf{H}_-^s(\Gamma) \mid \varphi_i \cdot \boldsymbol{\tau}_{ij} = \varphi_j \cdot \boldsymbol{\tau}_{ij} \text{ at } e_{ij}\} \quad (s > \tfrac{1}{2})\ ; \\ \mathbf{H}_\perp^s(\Gamma) &= \{\varphi \in \mathbf{H}_-^s(\Gamma) \mid \varphi_i \cdot \boldsymbol{\tau}_i = \varphi_j \cdot \boldsymbol{\tau}_i \text{ at } e_{ij}\} \quad (s > \tfrac{1}{2})\ . \end{aligned} \tag{22}$$

The space $\mathbf{H}_-^s(\Gamma)$ is endowed with its natural norm

$$\|\varphi\|_{s,-,\Gamma} := \left(\sum_{j=1}^{N_\Gamma} \|\varphi_j\|_{s,\Gamma_j}^2 \right)^{\frac{1}{2}}.$$

The spaces $\mathbf{H}_\|^s(\Gamma)$ and $\mathbf{H}_\perp^s(\Gamma)$ are closed subspaces of $\mathbf{H}_-^s(\Gamma)$ for any $s > \frac{1}{2}$.

Finally it is easy to see that, for any $s \geq \frac{1}{2}$, the operators

$$\nabla_\Gamma : H^{s+1}(\Gamma) \to \mathbf{H}_\|^s(\Gamma)\ ; \qquad \mathbf{curl}_\Gamma : H^{s+1}(\Gamma) \to \mathbf{H}_\perp^s(\Gamma)$$

are linear and continuous.

Remark 4. The equalities $H_\times^s(\Gamma) \equiv \mathbf{H}_\perp^s(\Gamma)$ and $H_T^s(\Gamma) \equiv \mathbf{H}_\|^s(\Gamma)$ for $s > 1/2$ are not obvious. Moreover, the definition (21) seems natural for polyhedra, but cannot be extended to the general case of Lipschitz surfaces. In particular, in the case $s = 3/2$, in [4] it is shown that the two definitions (6) and (21) give the same space both algebraically and topologically.

References

1. C. Amrouche, C. Bernardi, M Dauge, and V. Girault. Vector potentials in three-dimensional non-smooth domains. *Math. Meth. Appl. Sci.*, 21:823–864, 1998.
2. A. Buffa. Hodge decompositions on the boundary of a polyhedron: the multi connected case. *Math. Meth. Model. Appl. Sci.*, 11(9):1491-1504, 2001.
3. A. Buffa and S.A. Christiansen. The electric field integral equation on Lipschitz screens: definition and numerical approximation. *Numer. Mathem.* , 2002 (in press)
4. A. Buffa and P. Ciarlet, Jr. On traces for functional spaces related to Maxwell's equations. Part I: An integration by parts formula in Lipschitz polyhedra. *Math. Meth. Appl. Sci.*, 21(1):9–30, 2001.
5. A. Buffa and P. Ciarlet, Jr. On traces for functional spaces related to Maxwell's equations. Part II: Hodge decompositions on the boundary of Lipschitz polyhedra and applications. *Math. Meth. Appl. Sci.*, 21(1):31–48, 2001.
6. A. Buffa, M. Costabel, and C. Schwab. Boundary element methods for Maxwell equations in non-smooth domains. *Numer. Mathem.*, (electronic) DOI 10.1007/s002110100372, 2001.
7. A. Buffa, M. Costabel, and D. Sheen. On traces for $\mathbf{H}(\mathbf{curl}, \Omega)$ for Lipschitz domains. *J. Math. Anal. Appl.*, 2002 (in press)
8. A. Buffa, R. Hipmair, T. von Petersdorff, and Ch. Schwab. Boundary element methods for Maxwell equations in Lipschitz domains. *Numer. Mathem.*, 2002. (in press).
9. M. Costabel. A remark on the regularity of solutions of Maxwell's equations on Lipschitz domains. *Math. Meth. Applied Sci.*, 12:365–368, 1990.
10. P. Fernandes and G. Gilardi. Magnetostatic and electrostatic problems in inhomogeneous anisotropic media with irregular boundary and mixed boundary conditions. *Math. Mod. Meth. Appl. Sci.*, 7:957–991, 1997.
11. P. Fernandes and I. Perugia. Vector potential formulation for magnetostatics and modelling of permanent magnets. *IMA J. of Appl. Mathem.*, 66:293–318, 2001.
12. N. Filonov. *Système de Maxwell dans des domaines singuliers.* PhD thesis, Université de Bordeaux 1, 1996.
13. N. Filonov. Principal singularities of the magnetic field in resonators with boundary of given smoothness. *St. Petersburg Math. J.*, 9(2):379–390, 1998.
14. V. Girault and P.-A. Raviart. *Finite element methods for Navier-Stokes equations.* Sringer-Verlag, Berlin, 1986.
15. P. Grisvard. *Elliptic problems in nonsmooth domains*, volume 24 of *Monographs and studies in Mathematics.* Pitman, London, 1985.
16. P. Grisvard. *Singularities in boundary value problems*, volume RMA 22. Masson, Paris, 1992.
17. R. Hiptmair. Symmetric coupling for eddy current problems. Technical Report 148, Sonderforschungsbereich 382, University of Tübigen, March 2000. Submitted to SIAM J. Numer. Anal.
18. D. Jerison and C.E. Kenig. The inhomogeneous dirichlet problem in lipschitz domains. *J. Funct. Anal.*, 130:161–219, 1995.
19. J.-L. Lions and E. Magenes. *Problèmes aux limites non homogènes et applications.* Dunod, Paris, 1968.

20. Alan McIntosh and Marius Mitrea. Clifford algebras and Maxwell's equations in Lipschitz domains. *Math. Methods Appl. Sci.*, 22(18):1599–1620, 1999.
21. M. Mitrea. Generalized Dirac operators on nonsmooth manifolds and Maxwell's equations. *J. Fourier Anal. Appl.*, 7(3):207–256, 2001.
22. J. Necăs. *Les méthodes directes en théorie des équations elliptiques.* Masson, Paris, 1967.
23. R. Picard. On the boundary value problem of electro- and magnetostatics. *Proc. Royal Soc. Edinburgh*, (92 A):165–174, 1982.
24. E. H. Spanier. *Algebraic Topology.* McGraw Hill Book Company, 1966.
25. L. Tartar. On the characterization of traces of a sobolev space used for Maxwell's equation. In *Proceedings of a meeting held in Bordeaux, in honour of Michel Artola*, November 1997.

Applications of the Mortar Element Method to 3D Electromagnetic Moving Structures

Annalisa Buffa[1], Yvon Maday[2], and Francesca Rapetti[3]

[1] Istituto di Analisi Numerica del C.N.R., Via Ferrata 1, 27100 Pavia, Italy
[2] Laboratoire J.-L. Lions, UMR 7598 CNRS, Université Paris 6, boîte courrier 187, 4 place Jussieu, 75252 Paris cedex 5, France
[3] Laboratoire de Mathématiques J. A. Dieudonné, UMR 6621 CNRS, Université de Nice Sophia-Antipolis, Parc Valrose, 06108 Nice cedex 2, France

Summary. This paper deals with the modelling, the analysis and a numerical approach for the simulation of the dynamical behavior of a three-dimensional coupled magneto-mechanical system such as a damping machine. The model is based on the electric formulation of the eddy currents problem for the electromagnetic part and on the motion equation of a rotating rigid body for the mechanical part.

For the approximation, the magnetic system is discretized in space by means of edge elements and the sliding mesh mortar element method is used to account for the rotation. In time, a one step Euler method is used, implicit for the magnetic and velocity equations and explicit for the rotation angle. The coupled differential system can then be solved with an explicit procedure.

Here, we analyse the well-posedness of the continuous problem and give some details on its discretization.

1 Introduction

The full simulation of electromagnetic devices involves the solution of systems of linear or non–linear partial and ordinary differential equations. There is a well–known interaction among the electromagnetic field distribution, the heating and the dynamics of the device. Although the model of each separated phenomenon can be chosen linear, the coupling is, in general, non–linear. Few analysis and/or numerical methods are available in this context and they strongly depend on the application. We refer, e.g., to [17]-[18] for the analysis of a coupled electromagnetic-heating system and to [14] for the simulation of a magneto-mechanical system.

In this paper we are concerned with the modeling, the analysis and the simulation of a damping machine as the one presented in Figure 1.

The forces resulting from the magnetic field make the structure move. The variation in the configuration of the structure modifies the distribution of the magnetic field and consequently of the induced forces. Therefore, the interaction between magnetic and mechanical phenomena cannot be simulated independently and, in this paper, we propose a method to discretize the coupled problem. As an example we study a system composed of two solid parts:

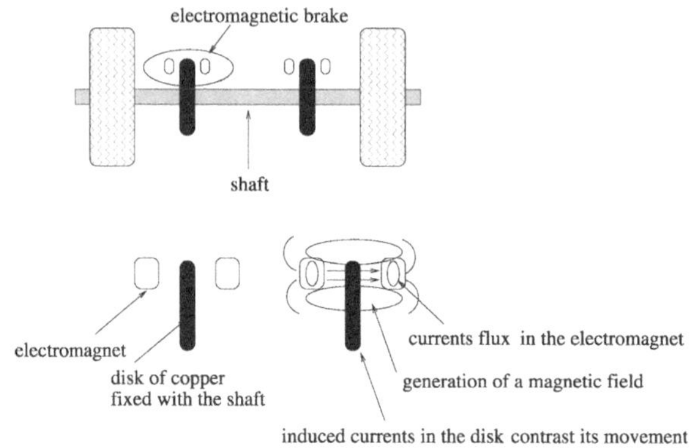

Fig. 1. Simplified example of an electromagnetic brake. Conducting disks are installed on the axes of the vehicle and electromagnets are placed around them such that the disks move in the gap of the electromagnets. When the mechanical brakes are applied, a current is passed through the electromagnets and the braking effects of the mechanical and magnetic brakes are added together. We note, however, that the braking effect assumes a non-zero speed for the disks. For this reason, electromagnetic brakes can not be used to completely stop the vehicle, but only to slow it down.

the stator, which stands still, and the rotor, which can turn around a given rotation axis.

For the electromagnetic part, we consider a three-dimensional model resulting from the following assumptions. The displacement currents are neglected with respect to the conducting ones: we have to solve a degenerated parabolic problem. The magneto-mechanical interaction is here analyzed when the rotor moves: we work in the time-dependent domain.

Concerning the spatial system of coordinates, we choose to work in Lagrangian variables in order to avoid the presence of a convective term in the equations. Among the possible variables to describe the involved phenomena, we select the *modified magnetic vector potential.*

A similar problem has already been presented, with no rigorous mathematical analysis, in [14] where the moving band technique has been used to take into account the rotor movement in two dimensions. In [9] we have introduced a sliding mesh technique for a two-dimensional problem based on the mortar element method. We extend here this spatial discretization for a three-dimensional simulation.

The coupling is obtained by means of Lagrange multipliers, suitably chosen to ensure optimal properties on the discrete problem, and the problem is set in the constrained space (the Lagrange multipliers are eliminated). This non-conforming non-overlapping domain decomposition technique which allows for independent meshes in adjacent subdomains, is now known in the literature as the *mortar element method.* It has been first introduced in [6] and studied recently in the Maxwell's equations framework (see [2, 15] for its mathematical analysis and [10, 19] for its first application to magnetostatics and magnetodynamics in three dimensions). According to the authors' knowledge, it is the first time that such a technique is analysed to deal with a coupled magneto-mechanical problem in three dimensions.

This approach leads naturally to a sliding mesh method which has several advantages with respect to other approaches. Among them, the fact that remeshing and interpolation procedures are avoided as well as the fact that no heavy constraints are imposed among the time step δt, the spatial parameter h and the rotation angle at every time step, as in the paper [13].

For the mechanical part, the motion of the rotor (a rigid rotating body) is the solution of a second order ordinary differential equation. Its coefficients depend on the mechanical features of the system such as the momentum of inertia of the rotor, the friction coefficient and, more importantly here, the global magnetic torque acting on the rotor axis due to the induced electromagnetic forces.

To analyze the magneto-mechanical system we have to solve simultaneously the electromagnetic and the mechanical equations. It is then necessary to evaluate the global magnetic torque acting on the moving part of the structure through the numerical computation of the magnetic field. The algorithm we propose here is based on an "explicit" coupling procedure that consists in solving alternatively the magnetic equations and the mechanical ones. At each time step, the magnetic force obtained from the field solution is inserted in the mechanical equation to compute the displacement. The latter is imposed to the moving part for the next step of the magnetic field calculation. A procedure to check the convergence of either the force or the displacement is necessary. This decoupling algorithm is proved to be stable and convergent in a two-dimensional situation [9]. We propose here some limited analysis and refer to a forthcoming paper for more about theory and numerical results.

Concerning the structure of the paper, Section 1 is devoted to the derivation of a model problem both for the magnetic equation and the mechanical one. The coupled problem is also stated. In Section 2 the well–posedeness of the problem is analyzed together with the regularity of the solution of the model problem. In Section 3 we then propose the discretization of the involved equations (linear PDEs and ODEs).

2 The model of a damping machine

The mathematical model describing the eddy current problem in the conductors at low frequencies is given by the quasi-stationary Maxwell's equations. One may eliminate the electric field **E** or the current density **J** and set up a formulation in terms of the magnetic field **H** [7]. Here we consider the alternative which consists in eliminating **H** and the magnetic induction **B** by means of a *modified magnetic vector potential* **A**. In this formulation, however, we have to ensure the uniqueness of the potential in the non-conducting parts. This can be done by imposing explicitely a gauge condition. We can also adopt an approach similar to the one presented in [19]: for a current density $\mathbf{J} = \sigma\,\mathbf{E} + \mathbf{J}_s$, we introduce a vector $\mathbf{T}$ such that the source current $\mathbf{J}_s = \text{curl}\;\mathbf{T}$, then starting with a vector in the space orthogonal to ker(curl), the Conjugate Gradient algorithm applied to the final algebraic system generates, at each iteration, a solution which is again in the space orthogonal to ker(curl). This guarantees the potential's uniqueness; moreover, the approach is still valid in a domain decomposition framework.

Let Ω be an open set in $\mathbb{R}^3$ containing Ω_c a conducting and Ω_{nc} a non-conducting part. We introduce the vector potential $\mathbf{A}$ such that $\mathbf{B} = \text{curl}\,\mathbf{A}$ in Ω, $\mathbf{A} = -\int_0^t \mathbf{E}(t')\,dt'$ in Ω_c and whose tangential component $(\mathbf{A})_{\tau,\partial\Omega} = \mathbf{0}$ on $\partial\Omega$. We assume that the magnetic permeability μ and the electric conductivity σ are linear, bounded, scalar functions of the space variable $\mathbf{x}$; then, denoting $\nu = 1/\mu$, the equation we solve reads

$$\sigma\,\frac{\partial \mathbf{A}}{\partial t} + \text{ curl } (\nu \text{ curl } \mathbf{A}) = \mathbf{J}_s\,. \tag{1}$$

Equation (1) is expressed in Lagrangian variables. This equation admits a unique solution in the conducting regions (i.e. where $\sigma > 0$), with Dirichlet, Dirichlet-Neumann, Neumann boundary conditions; in the region where σ is zero, the solution $\mathbf{A}$ is defined up to the gradient of a scalar function but the magnetic induction $\mathbf{B} = \text{curl}\,\mathbf{A}$ is unique.

To introduce the mechanical part, we need to decompose Ω into its moving and fixed parts, say Ω_1 and Ω_2 respectively. The two subdomains are separated by an interface Γ.

The presence of a magnetic field in the physical system described above generates an induced electromagnetic force which acts as a torque on the moving part, Ω_1. In particular the magnetic field time variation induces an electric field and so dissipative currents $\sigma\,\mathbf{E}$. In general, the electromagnetic force $\mathbf{F}$ is given by the generalized Laplace form [16]

$$\mathbf{F} = (\sigma\,\mathbf{E} + \mathbf{J}_s) \times \mu\mathbf{H} \;-\; \frac{1}{2}|\mathbf{H}|^2\,\text{grad}\,\mu \;+ \mathbf{f}',$$

where $\mathbf{f}'$ denotes the internal forces. Since simulations are led with μ constant everywhere in Ω and magnetostriction phenomena are neglected ($\mathbf{f}' = \mathbf{0}$), the expression of the force density in terms of the magnetic vector potential becomes

$$\mathbf{F} = (-\sigma\,\frac{\partial \mathbf{A}}{\partial t} + \mathbf{J}_s) \times \text{curl}\,\mathbf{A}$$

and it produces a torque which is parallel to the cylinder axis

$$\mathbf{T}_m = \int_{\Omega_1} \mathbf{r} \times \mathbf{F}\,d\Omega$$

where $\mathbf{r}$ is the position vector.

When a torque is present, the moving part of the system turns around its axis with the following law:

$$\begin{cases} J\,\dfrac{d\omega}{dt} + k\,\omega = \mathrm{T}_m & \forall t \in [0,T], \\ \omega(0) = \omega_0, \end{cases} \qquad \begin{cases} \omega = \dfrac{d\theta}{dt} & \forall t \in [0,T], \\ \theta(0) = \theta_0, \end{cases} \tag{2}$$

where ω is the angular speed and θ the rotation angle, J is the momentum of inertia of the rotor, k is the friction constant ($k > 0$), and T_m is the magnetic torque intensity.

Now, we can state the coupled magneto-mechanical problem. We denote by $r_t : \Omega_1 \to \Omega_1$ the rotation operator which turns the moving part around its axis of an angle $\theta = \theta(t)$ solution of (2). We adopt the notation $\Omega_1(t) = r_t\,\Omega_1(0)$ where $\Omega_1(0)$ is of course the initial configuration of Ω_1. Moreover, from now on, we denote the magnetic vector potential $\mathbf{A}$ by $\mathbf{u}$ or, when useful, by the couple $(\mathbf{u}_1, \mathbf{u}_2)$ where $\mathbf{u}_1$ and $\mathbf{u}_2$ denote the restrictions of $\mathbf{u}$ to Ω_1 and Ω_2 respectively. We shall need also a rotation operator R_t for vectors: we define $R_t\,\mathbf{u}_1(\mathbf{x},t) = r_t[\mathbf{u}_1(r_{-t}\mathbf{x},t)]$ and $\mathcal{R}_t\,\mathbf{u} = (R_t\,\mathbf{u}_1, \mathbf{u}_2)$.

In the following we denote by $\mathcal{C}$ and $\mathcal{I}$ the conductor and the insulator respectively at time $t = 0$, i.e. $\mathcal{C} = \Omega_c(0)$ and $\mathcal{I} = \Omega_{nc}(0)$. Note that $\mathcal{C} \subset \Omega_1$. For the sake of simplicity we assume that the insulator is connected and simply connected.

Finally the magnetic potential depends on the angular speed of the moving part: this means that Maxwell equations and the structure equations are coupled. Then the behavior of the system is described by the following non-linear system of partial differential equations:

$$\begin{array}{lll}
(E_1) & \sigma\dfrac{\partial \mathbf{u}_1}{\partial t} + \operatorname{curl}(\nu\operatorname{curl}\mathbf{u}_1) = \mathbf{J}_s & \Omega_1(0)\times]0,T[\\
(E_2) & \operatorname{curl}(\nu\operatorname{curl}\mathbf{u}_2) = \mathbf{J}_s & \Omega_2\times]0,T[\\
(G) & \operatorname{div}(\mathcal{R}_t\mathbf{u}) = 0 & \mathcal{I}\times]0,T[\\
(E_3) & \mathbf{T}_m = \int_{\Omega_1}\mathbf{r}\times\big[\big(-\sigma\dfrac{\partial \mathbf{u}_1}{\partial t} + \mathbf{J}_s\big)\times\operatorname{curl}\mathbf{u}_1\big]\,d\Omega & \\
(E_4) & J\,\dfrac{d\omega}{dt} + k\,\omega = \mathrm{T}_m \quad , \quad \omega = \dfrac{d\theta}{dt} &]0,T[\\
(IC_1) & (R_t(\mathbf{u}_1))_{\tau,\Gamma}(\mathbf{x},t) = (\mathbf{u}_2)_{\tau,\Gamma}(\mathbf{x},t) & \Gamma\times]0,T[\\
(IC_2) & (R_t(\nu\operatorname{curl}\mathbf{u}_1))_{\tau,\Gamma}(\mathbf{x},t) = (\nu\operatorname{curl}\mathbf{u}_2)_{\tau,\Gamma}(\mathbf{x},t) & \Gamma\times]0,T[\\
(IC_3) & \omega(0) = 0 \quad , \quad \theta(0) = \theta_0 & \\
(BC) & (\mathbf{u}_2)_{\tau,\partial\Omega} = \mathbf{0} & \partial\Omega\times]0,T[\\
(0C) & \mathbf{u}(\mathbf{x},0) = 0 & C\times\{0\}
\end{array} \tag{3}$$

where T is the final time, the partial differential equation (E_i) is in the sense of distributions in $\Omega_i\times]0,T[$ (i=1,2). The transmission conditions (IC_1)

and (IC_2) describe, respectively, the continuity of $\mathbf{E}_{\tau,\Gamma}$ (and consequently of $\mathbf{B}_{n,\Gamma}$) and the continuity of $\mathbf{H}_{\tau,\Gamma}$ across Γ. We remark that the condition (G) implies that the normal component of $\mathcal{R}_t\mathbf{u}$ is continuous across the insulator boundary.

3 Lagrangian versus Eulerian formulation

We make now same remarks around the terminology "Lagrangian" and "Eulerian" (hoping not to confuse the reader more than necessary). Only in this section, we denote by ∂_t the partial derivative with respect to the time variable t. Maxwell equations when expressed on the material manifold, and when one neglects displacement currents, read as:

$$\partial_t \mathbf{B} + \operatorname{curl} \mathbf{E} = \mathbf{0}, \qquad \operatorname{curl} \mathbf{H} = \mathbf{J}. \tag{4}$$

One recognizes Faraday's law and Ampère's theorem in equation (4). These laws are always valid, whatever the movement of matter in space. The goal of this section is to verify that eddy current equations (4) are fully *covariant* or "form invariant", in the rotating case. Adopting the terminology of [8], we "push-forward" equations (4) by means of the rotation operator R_t.

For the magnetic vector potential $\mathbf{A}$ we introduce $\mathbf{a}$ such that $\mathbf{A} = R_t\mathbf{a}$. We denote $\partial_t\mathbf{A}$ the Eulerian temporal derivative of $\mathbf{A}$ and $d_t\mathbf{A}$ the Lagrangian one: the two derivatives are related through

$$d_t\mathbf{A} = \partial_t\mathbf{A} + (\mathbf{V}\cdot \operatorname{grad})\mathbf{A}$$

where $\mathbf{V}$ is the velocity associated to the rotation movement. We denote by $\mathbf{x} = (x, y, z)$ the position of a point P with respect to the origin O of the cartesian reference system. The tangential velocity at the point P is $\mathbf{V} = \bar{\omega}\times\mathbf{x}$ where $\bar{\omega}$ is the vector $(0, 0, \omega)$. So we have

$$\mathbf{V} = (y\,\omega,\, -x\,\omega,\, 0)\,, \qquad \operatorname{div} \mathbf{V} = 0 \qquad (\omega = \omega(t))\,,$$

together with

$$r_t = \begin{pmatrix} \cos\theta(t) & -\sin\theta(t) & 0 \\ \sin\theta(t) & \cos\theta(t) & 0 \\ 0 & 0 & 1 \end{pmatrix}, \quad (r_t)^{-1} = r_{-t} = \begin{pmatrix} \cos\theta(t) & \sin\theta(t) & 0 \\ -\sin\theta(t) & \cos\theta(t) & 0 \\ 0 & 0 & 1 \end{pmatrix},$$

$$d_t r_t = -\omega \begin{pmatrix} \sin\theta(t) & \cos\theta(t) & 0 \\ -\cos\theta(t) & \sin\theta(t) & 0 \\ 0 & 0 & 0 \end{pmatrix}, \qquad r_t^{-1} d_t r_t = -\omega \begin{pmatrix} 0 & 1 & 0 \\ -1 & 0 & 0 \\ 0 & 0 & 0 \end{pmatrix},$$

and

$$(r_t^{-1} d_t r_t)\mathbf{u} = (-u_y\,\omega,\, u_x\,\omega,\, 0) = -\mathbf{u}\times\bar{\omega}\,.$$

From the above identities on r_t and related operators r_{-t}, $d_t r_t$, $r_t^{-1} d_t r_t$, we have the ones for R_t and R_{-t}, $d_t R_t$, $R_t^{-1} d_t R_t$. We need the vector relation

$$\mathbf{V} \times (\operatorname{curl} \mathbf{a}) = \operatorname{grad} (\mathbf{V} \cdot \mathbf{a}) - (\mathbf{V} \cdot \operatorname{grad})\mathbf{a} - (\operatorname{grad} \mathbf{V}) \cdot \mathbf{a}\,,$$

with

$$(\operatorname{grad} \mathbf{V}) \cdot \mathbf{a} = \mathbf{a} \times \bar{\omega}\,,$$

and

$$\operatorname{grad} (\mathbf{V} \cdot \mathbf{a}) = \mathbf{V} \operatorname{div} \mathbf{a} + \mathbf{a} \operatorname{div} \mathbf{V} = \mathbf{0}$$

for the properties of $\mathbf{V}$ and of the vector potential $\mathbf{a}$. Note that the Hodge operators curl, divergence and gradient commutes with R_t, implying that $R_t(\operatorname{curl} \cdot) = \operatorname{curl}(R_t \cdot)$, for example. From equation $\sigma\, \partial_t \mathbf{A} + \operatorname{curl} (\nu \operatorname{curl} \mathbf{A}) = \mathbf{J}_s$ we get

$$\begin{aligned}
&\sigma\, [d_t(R_t\mathbf{a}) - (\mathbf{V} \cdot \operatorname{grad})(R_t\mathbf{a})] + \operatorname{curl} (\nu \operatorname{curl} (R_t\mathbf{a})) = R_t\mathbf{j}_s\,,\\
&\sigma\, [(d_t\, R_t)\mathbf{a} + R_t(\partial_t\mathbf{a}) - (\mathbf{V} \cdot \operatorname{grad})(R_t\mathbf{a})] + R_t(\operatorname{curl} (\nu \operatorname{curl} \mathbf{a})) = R_t\mathbf{j}_s\,,\\
&\sigma\, (R_t^{-1} d_t R_t)\mathbf{a} + \sigma\, \partial_t \mathbf{a} - \sigma\, (\mathbf{V} \cdot \operatorname{grad})(\mathbf{a}) + \operatorname{curl} (\nu \operatorname{curl} \mathbf{a}) \;= \mathbf{j}_s\,,\\
&\sigma\, (R_t^{-1} d_t R_t)\mathbf{a} + \sigma\, \partial_t \mathbf{a} + \sigma\, (\mathbf{V} \times \operatorname{curl} \mathbf{a} + \mathbf{a} \times \bar{\omega}) + \operatorname{curl} (\nu \operatorname{curl} \mathbf{a}) \;= \mathbf{j}_s\,,\\
&\sigma\, (\partial_t \mathbf{a} + \,\mathbf{V} \times \operatorname{curl} \mathbf{a}) + \operatorname{curl} (\nu \operatorname{curl} \mathbf{a}) \;= \mathbf{j}_s\,.
\end{aligned}$$

We then have that $\partial_t \mathbf{A} = R_t(\partial_t \mathbf{a} + \mathbf{V} \times \operatorname{curl} \mathbf{a})$.

Let $\mathbf{b}$, $\mathbf{h}$, $\mathbf{j}$ such that $\mathbf{B} = R_t\mathbf{b}$, $\mathbf{H} = R_t\mathbf{h}$, $\mathbf{J} = R_t\mathbf{j}$. From the commutativity of the Hodge operators curl, divergence and gradient with R_t, one has $\operatorname{curl} \mathbf{h} = \mathbf{j}$ as in (4).

Faraday's law is not so straightforward since it is *not true* that R_t and ∂_t commute when R_t depends on time as in this case. Let us go into details. We denote by $\partial_t\mathbf{B}$ the Eulerian temporal derivative of $\mathbf{B}$ and by $d_t\mathbf{B}$ the Lagrangian one: the two derivatives are related through

$$d_t\mathbf{B} = \partial_t\mathbf{B} + (\mathbf{V} \cdot \operatorname{grad})\mathbf{B}\,.$$

We recall that

$$\operatorname{curl}(\mathbf{V} \times \mathbf{b}) = (\mathbf{b} \cdot \operatorname{grad})\mathbf{V} - \mathbf{b}\operatorname{div} \mathbf{V} - (\mathbf{V} \cdot \operatorname{grad})\mathbf{b} + \mathbf{V}\operatorname{div} \mathbf{b} \qquad (5)$$

with

$$\begin{aligned}
&\operatorname{div} \mathbf{b} = 0 \qquad (\operatorname{div} \mathbf{B} = 0 \quad \text{and} \quad \operatorname{div}(R_t\cdot) = R_t(\operatorname{div}\cdot))\,,\\
&(\mathbf{b} \cdot \operatorname{grad})\mathbf{V} = -\mathbf{b} \times \bar{\omega}\,.
\end{aligned}$$

So, from $\partial_t\, \mathbf{B} + \operatorname{curl} \mathbf{E} = \mathbf{0}$ we get

$$\begin{aligned}
&d_t(R_t\mathbf{b}) - (\mathbf{V} \cdot \operatorname{grad})(R_t\mathbf{b}) + \operatorname{curl} R_t R_t^{-1}\mathbf{E} = \mathbf{0}\,,\\
&(d_t R_t)\mathbf{b} + R_t(\partial_t\mathbf{b}) - (\mathbf{V} \cdot \operatorname{grad})(R_t\mathbf{b}) + R_t\operatorname{curl} R_t^{-1}\mathbf{E} = \mathbf{0}\,,\\
&R_t^{-1}(d_t R_t)\mathbf{b} + \partial_t\mathbf{b} - (\mathbf{V} \cdot \operatorname{grad})\mathbf{b} + \operatorname{curl} R_t^{-1}\mathbf{E} = \mathbf{0}\,,\\
&R_t^{-1}(d_t R_t)\mathbf{b} + \partial_t\mathbf{b} - (\mathbf{b} \cdot \operatorname{grad})\mathbf{V} + \operatorname{curl} R_t^{-1}\mathbf{E} + \operatorname{curl} (\mathbf{V} \times \mathbf{b}) = \mathbf{0}\,,\\
&\partial_t\mathbf{b} + \operatorname{curl} (R_t^{-1}\mathbf{E} + \mathbf{V} \times \mathbf{b}) = \mathbf{0}\,.
\end{aligned}$$

We *define* the electric field $\mathbf{e} = (R_t^{-1}\mathbf{E} + \mathbf{V} \times \mathbf{b})$. So, in Lagrangian variables we have $\partial_t \mathbf{b} + \operatorname{curl} \mathbf{e} = \mathbf{0}$. Note that the relation between $\mathbf{a}$ and $\mathbf{A}$ is the same that exists between $\mathbf{e}$ and $\mathbf{E}$. In the following we use the relation

$$\partial_t R_t \mathbf{u}(\mathbf{x}, t) = R_t(\partial_t \mathbf{u} + \mathbf{V} \times \operatorname{curl} \mathbf{u}) \,. \tag{6}$$

4 Well–posedness of the continuous problem (3)

We are going to use a fixed point theorem of Schauder type in order to prove that (3) admits at least one solution. To this aim, we first fix an angular speed $\omega \in C^0([0,T])$ and we analyze the existence and regularity of the magnetic potential $\mathbf{u}$ resulting from (E_1), (E_2), (IC_1), (IC_2), (BC) and $(0C)$. Above all, we need a variational formulation of this problem and we introduce the notations:

$$\begin{aligned}
H(\operatorname{curl}, \Omega_k) &= \{\mathbf{v} \in \mathbf{L}^2(\Omega_k) \,|\, \operatorname{curl} \mathbf{v} \in \mathbf{L}^2(\Omega_k)\}; \\
||\mathbf{v}||^2_{H(\operatorname{curl}, \Omega_k)} &= ||\mathbf{v}||^2_{\mathbf{L}^2(\Omega_k)} + ||\operatorname{curl} \mathbf{v}||^2_{\mathbf{L}^2(\Omega_k)}, \quad \forall \mathbf{v} \in H(\operatorname{curl}, \Omega_k); \\
\mathcal{H}(\Omega) &= H(\operatorname{curl}, \Omega_1) \times H_{0,\partial\Omega}(\operatorname{curl}, \Omega_2); \\
H_0(\operatorname{curl}, S) &= \{\mathbf{v} \in \mathcal{H}(\Omega) \,|\, \operatorname{curl} \mathbf{v} \in \mathbf{L}^2(\Omega) \text{ and } \mathbf{v}_{|\Omega\setminus S} = 0\}, \quad S = \Omega, \mathcal{C}, \mathcal{I}. \\
H(\operatorname{div}^0, \mathcal{I}) &= \{\mathbf{u} \in \mathbf{L}^2(\Omega) \,|\, \operatorname{div} \mathbf{u} = 0 \text{ on } \mathcal{I}\}; \\
X &= H_0(\operatorname{curl}, \Omega) \cap H(\operatorname{div}^0, \mathcal{I}).
\end{aligned}$$

Other standard Bochner-type spaces will be used and we refer to [12] for their definitions. In the following for any $\mathbf{u} \in \mathbf{L}^2(\Omega)$, we denote by $\mathbf{u}_i$ its restriction to Ω_i, $i = 1, 2$. We denote by $(\cdot,\cdot)_{\Omega_i}$, $i = 1,2$ the scalar product in $\mathbf{L}^2(\Omega_i)$.

We note that X is a Hilbert space endowed with the following norm:

$$||\mathbf{u}||^2_\star := ||\sqrt{\sigma}\,\mathbf{u}||^2_{\mathbf{L}^2(\mathcal{C})} + \sum_{i=1,2} ||\sqrt{\nu}\operatorname{curl} \mathbf{u}_i||^2_{\mathbf{L}^2(\Omega_i)}. \tag{7}$$

The fact that (7) is a semi-norm, is an easy consequence of the results in [1]. Actually it is also a norm due to two facts: X is a closed subspace of $\mathcal{H}(\Omega)$ and if $||\mathbf{u}||_\star = 0$, then $\mathbf{u}_{|\mathcal{C}} = \mathbf{0}$ which implies $\mathbf{u} \in H_0(\operatorname{curl}, \mathcal{I})$. Since also $\operatorname{div}(\mathbf{u}) = 0$ on $\mathcal{I}$ and the insulator is supposed to be simply connected, we immediately deduce that $\mathbf{u} \equiv \mathbf{0}$. Now, in order to obtain a variational equation in space, we need to introduce "spatial" test functions. The natural choice is:

$$\mathcal{X}_t := \mathcal{R}_{-t}\{X\}. \tag{8}$$

Analogously we can define the space $L^p(0,T;\mathcal{X}_t)$, $p \in \{2, \infty\}$ as:

$$L^p(0,T;\mathcal{X}_t) := \{\mathbf{v} \in L^p(0,T;\mathbf{L}^2(\Omega)) \mid \mathbf{v} = \mathcal{R}_{-t}(\mathbf{u}),\ \mathbf{u} \in L^p(0,T;X)\}.$$

These are Banach spaces endowed with the norms:

$$||\mathbf{v}||_{L^2(0,T;\mathcal{X}_t)} := \left(\int_0^T ||\mathbf{v}(\cdot,\cdot)||_\star^2\, dt\right)^{1/2}, \; ||\mathbf{v}||_{L^\infty(0,T;\mathcal{X}_t)} = \sup_{t\in]0,T[} ||\mathbf{v}(\cdot,\cdot)||_\star\, dt.$$

Let now $\mathbf{u} \in L^2(0,T;\mathcal{X}_t)$, by definition we have $\mathbf{u}_1 = R_t\mathbf{v}_1$, $\mathbf{v}_1 = \mathbf{v}_{|\Omega_1}$,$\mathbf{v} \in L^2(0,T;X)$; then, using (6), where $\mathbf{V}$ is the rotation speed of Ω_1, we obtain:

$$\frac{\partial \mathbf{u}_1}{\partial t} = R_t(\frac{\partial \mathbf{v}_1}{\partial t} + \mathbf{V} \times \mathrm{curl}\mathbf{v}_1)$$

which implies $\frac{\partial \mathbf{u}_1}{\partial t} \in H^{-1}(0,T;\mathbf{L}^2(\Omega_1))$.

The variational formulation of the considered problem reads:

$$\begin{aligned} &\text{Find } \; \mathbf{u}(\cdot,\cdot) \in L^2(0,T;\mathcal{X}_t) \cap C^0(0,T;L^2(\mathcal{C})) \text{ such that} \\ &\left(\sigma \tfrac{\partial \mathbf{u}}{\partial t}, \mathbf{v}\right)_{\Omega_1} + a(\mathbf{u},\mathbf{v}) = \textstyle\int_\Omega \mathbf{J}_s \cdot \mathbf{v} \quad \forall \mathbf{v} \in \mathcal{X}_t\,, \end{aligned} \tag{9}$$

where, $\forall\, \mathbf{u},\, \mathbf{v} \in \mathcal{X}_t$,

$$a(\mathbf{u},\mathbf{v}) = \sum_{k=1}^{2} a_k(\mathbf{u}_k,\mathbf{v}_k) = \sum_{k=1}^{2} \int_{\Omega_k} \nu\, \mathrm{curl}\mathbf{u}_k \cdot \mathrm{curl}\mathbf{v}_k\,. \tag{10}$$

In order to give a meaning to equation (9), we need to be sure that it can be interpreted in the distributional sense in time. In particular, we simply need to prove that the quantity $\left(\sigma \frac{\partial \mathbf{u}}{\partial t}, \mathbf{v}\right)_{\Omega_1}$ is a distribution in time. For $\mathbf{v} \in \mathcal{X}_t$, we have that $\mathbf{v} = \mathcal{R}_{-t}\xi$, $\xi \in H_0(\mathrm{curl},\Omega)$, and by formula (6) we get:

$$\frac{\partial \mathbf{v}}{\partial t} = -\mathcal{R}_{-t}(\mathbf{V} \times \mathrm{curl}\xi).$$

We obtain then, for any $\varphi \in \mathcal{D}(]0,T[)$:

$$\begin{aligned} \langle \left(\sigma \tfrac{\partial \mathbf{u}}{\partial t}, \mathbf{v}\right)_{\Omega_1}, \varphi\rangle &= -\int_0^T \left(\sigma\, \mathbf{u}, \frac{\partial(\mathbf{v}\,\varphi)}{\partial t}\right)_{\Omega_1} \\ &= -\int_0^T \left[\left(\sigma\, \mathbf{u}, \mathbf{v}\frac{\partial \varphi}{\partial t}\right)_{\Omega_1} - (\sigma\, \mathbf{u}, -\varphi\, \mathcal{R}_{-t}(\mathbf{V} \times \mathrm{curl}\,\xi))_{\Omega_1}\right] \end{aligned} \tag{11}$$

where the duality pairing in the left hand side is the one between $\mathcal{D}(]0,T[)$ and $\mathcal{D}'(]0,T[)$ and, in the right hand side, the integral is well defined and bounded by the norms of $\mathbf{u}$ and $\mathbf{v}$ associated to their own spaces and the norm of φ in $C^1(]0,T[)$.

Moreover, for sufficiently regular vector fields $\mathbf{u}$ the standard integration by parts formula holds true and we have:

$$2\langle \left(\sigma \frac{\partial \mathbf{u}}{\partial t}, \mathbf{u}\right)_{\Omega_1}, \varphi\rangle = \langle \frac{\partial (\sigma\, \mathbf{u}, \mathbf{u})_{\Omega_1}}{\partial t}, \varphi\rangle = -\int_0^T (\sigma\, \mathbf{u}, \mathbf{u})_{\Omega_1} \frac{\partial \varphi}{\partial t}. \tag{12}$$

As a consequence, for solution $\mathbf{u}$ of (9) regular enough to ensure that (9) holds in $L^1(]0,T[)$, we have:

$$\frac{1}{2}\frac{\partial\,(\sigma\ \mathbf{u},\mathbf{u})_{\Omega_1}}{\partial t} + a(\mathbf{u},\mathbf{u}) = (\mathbf{J}_s,\mathbf{u})_{\Omega_2} \tag{13}$$

Integrating in time, and using the fact that $||\cdot||_\star$ is a norm on X, we have:

$$||\sqrt{\sigma}\,\mathbf{u}||^2_{\mathbf{L}^2(C)} + \frac{1}{2}\int_0^T a(\mathbf{u},\mathbf{u}) \leq C\int_0^T ||\mathbf{J}_s||^2_{\mathbf{L}^2(\Omega_2)}\,. \tag{14}$$

Proposition 1. *Let $\omega \in C^0([0,T])$ be a fixed angular speed of the rotating part Ω_1 and $\mathbf{J}_s \in L^2(0,T;\mathbf{L}^2(\Omega))$ be an imposed density current. Problem (9) admits a solution $\mathbf{u}$ that verifies (E_1), (E_2), (G), (IC_1), (IC_2), (BC) and $(0C)$ in the sense of distributions.*

Proof. To prove the existence of a solution $\mathbf{u}$ to problem (9), we use the Faedo-Galerkin method (see [12], vol. 5, for details). We solve the system (9) for a suitable sequence of nested finite-dimensional spaces $\{\mathcal{X}_N^t\}_{N\geq 1}$ such that:

$$\mathcal{X}_N^t \subset \mathcal{X}_{N+1}^t \subset \mathcal{X}_t \quad \forall N \geq 1 \quad \text{and} \quad \bigcup_{N\geq 1} \mathcal{X}_N^t \quad \text{dense in} \quad \mathcal{X}_t\,; \tag{15}$$

then by using the stability estimate (14), we prove the existence of the solution.

The procedure is quite standard in the case of a completely conducting system while in the other case we have to pay attention to the fact that the equations we are dealing with are not strictly parabolic.

Thanks to the definition (8), in what follows we construct a suitable sequence of approximation spaces in X and then, by applying the operator $\mathcal{R}_t$ we obtain $\{\mathcal{X}_N^t\}_N$.

For convenience we set:

$$\begin{aligned} V &= H_0(\text{curl},\mathcal{I}) \cap H(\text{div}^0,\mathcal{I}) \\ W &= (H_0(\text{curl},\mathcal{I}) \cap H(\text{div}^0,\mathcal{I}))^\perp \end{aligned}$$

where $^\perp$ stands for L^2-orthogonal projection. We then have:

$$X = V \overset{\perp}{\oplus} W\,; \qquad \mathbf{u} = \mathbf{u}_I + \mathbf{u}_C\,,\ \mathbf{u}_I \in V, \mathbf{u}_C \in W$$

the symbol $\overset{\perp}{\oplus}$ stands for the L^2-orthogonal direct sum, where, of course, the functions in V are extended by zero outside the insulator $\mathcal{I}$. Since the two spaces V and W are separable Hilbert spaces, it is always possible to find two sequences of nested finite-dimensional spaces such that

$$\begin{aligned} &V_N \subset V_{N+1} \subset V \quad , \qquad \textstyle\bigcup_{N\geq 1} V_N \quad \text{dense in} \quad V \\ &W_N \subset W_{N+1} \subset W \quad , \qquad \textstyle\bigcup_{N\geq 1} W_N \quad \text{dense in} \quad W \end{aligned}$$

Finally we define $\mathcal{X}_N^t := \mathcal{R}_{-t}V_N \overset{\perp}{\oplus} \mathcal{R}_{-t}W_N$. It is not hard to see that this space verifies (15). Note that the movement is exactly taken into account in the construction of these nested finite-dimensional spaces.

We have now to study the existence of a solution for the variational problem (9) when replacing $\mathcal{X}_t$ with $\mathcal{X}_N^t$.

Let $\Phi = \{\phi_i(\mathbf{x},t)\}_{1\leq i\leq N}$ be a basis for the space $\mathcal{X}_N^t$ counting first the basis functions associated to the subspace $\mathcal{R}_{-t}V_N$. For any $\mathbf{u}_N \in \mathcal{X}_N^t$, we denote again by $\mathbf{u}_N$ the vector of coefficients associated to $\mathbf{u}_N$ through the basis Φ. For any $\mathbf{u}_N$, $\mathbf{v}_N \in \mathcal{X}_N^t$, the integral $\int_\Omega \sigma \mathbf{u}_N \cdot \mathbf{v}_N$ is equal to $(M(t)\mathbf{v}_N, \mathbf{u}_N)$ where $M(t)$ is the usually called *mass matrix*.

Due to the fact that $\sigma = 0$ at $\mathcal{I}$, we clearly have that the matrix $M(t)$ has the form (thanks to the chosen ordering):

$$M(t) = \begin{pmatrix} 0 & 0 \\ 0 & M_W(t) \end{pmatrix}$$

where $M_W(t)$ is the *mass matrix* associated only to the space $\mathcal{R}_{-t}W_N$.

It is easy to show that the matrix $M_W(t)$ is non-singular: for every given $\mathbf{w}_N \in \mathcal{R}_{-t}W_N$ such that $\mathbf{w}_N^T(t)\, M_W(t)\, \mathbf{w}_N(t) = 0$ we want to show that $\mathbf{w}_N = \mathbf{0}$. We surely have $\mathbf{w}_N = \mathbf{0}$ over $\mathcal{C}$ and this means in particular that $(\mathbf{w}_N)_{\tau,\partial\mathcal{C})} = \mathbf{0}$. Since $(\mathbf{w}_N)_{\tau,\partial\Omega} = \mathbf{0}$ and $\mathbf{w}_N \in \mathcal{R}_{-t}W_N$, we deduce that $\mathbf{w}_N = \mathbf{0}$ also over $\mathcal{I}$ and so that $\mathbf{w}_N$ is identically $\mathbf{0}$. The problem (9) on the finite-dimensional spaces $\mathcal{X}_N^t$ reads:

$$\text{Find } \mathbf{u}_N \in L^2(0,T;\mathcal{X}_N^t) \cap C^0(0,T;L^2(\mathcal{C})) \text{ such that :}$$

$$\left(\sigma \frac{\partial \mathbf{u}_N}{\partial t}, \mathbf{v}_N\right)_{\mathcal{C}} + a(\mathbf{u}_N, \mathbf{v}_N) = (\mathbf{J}_s, \mathbf{v}_N)_{\Omega_2}\,, \qquad \forall \mathbf{v}_N \in \mathcal{X}_N^t \tag{16}$$

and it can be easily written in matrix form and using our decomposition. The system reads

$$\frac{d}{dt}\left[\begin{pmatrix} 0 & 0 \\ 0 & M_W \end{pmatrix}\begin{pmatrix} \mathbf{u}_V \\ \mathbf{u}_W \end{pmatrix}\right] + \begin{pmatrix} A_{VV} & A_{VW} \\ A_{WV} & A_{WW} \end{pmatrix}\begin{pmatrix} \mathbf{u}_V \\ \mathbf{u}_W \end{pmatrix} = \begin{pmatrix} \mathbf{J}_V \\ \mathbf{J}_W \end{pmatrix}.$$

We can obtain $\mathbf{u}_V$ as a function of $\mathbf{u}_W$ from the first set of equations since, thanks to our construction, the sub-matrix A_{VV} is surely non-singular at every time t and $\mathbf{u}_W$ from the second one by solving a first order ordinary differential equation.

By using the a-priori stability estimate (14) applied on the sequence $\mathbf{u}_N$, there exists a subsequence of $\mathbf{u}_N$ (called again $\mathbf{u}_N$ for convenience), such that $\mathbf{u}_N \rightharpoonup \mathbf{u}$ in $L^2(0,T;\mathcal{X}_t)$. Let now $\mathbf{v}$ be any function in $\mathcal{X}_t$ and $\mathbf{v}_N \in \mathcal{X}_N^t$ be a sequence strongly convergent to $\mathbf{v}$ in $\mathcal{X}_t$. If we let N go to ∞, we obtain (9).

We pass now to study the regularity of the solution $\mathbf{u}$ of (9). We have:

Theorem 9. *Let $\omega \in C^0([0,T])$ be a fixed angular speed for the rotating part Ω_1, $\nu \in W^{1,\infty}(\Omega_1)$ and $\boldsymbol{J}_s \in H^1(0,T;\mathbf{L}^2(\Omega))$. Then, the solution $\boldsymbol{u}$ of problem (9) belongs to $H^1(0,T;L^2(\mathcal{C})) \cap L^\infty(0,T;\mathcal{X}_t)$. Moreover, $\boldsymbol{u}$ is also unique.*

Proof. As far as uniqueness is concerned, it is a direct consequence of (14).

We use the same Faedo-Galerkin approximation spaces (and also the same notation) as in the previous proof. The solution $\mathbf{u}_N$ of (16) can be written as $\mathbf{u}_N(x,t) = \sum_{i=1}^N \mathbf{u}_i(t)\phi_i(\mathbf{x},t)$ where $\phi_i(\mathbf{x},t)$ are the basis functions for $\mathcal{X}_N^t$. We suppose without loss of generality that $\{\phi_i(\mathbf{x},t)\}_i$ are regular in $\Omega_1 \cup \Omega_2$. Now, we choose the test function $\mathbf{v}_N(x,t) = \sum_{i=1}^N \frac{d\,\mathbf{u}_i}{dt}(t)\phi_i(x,t)$ in (16).

By (6), we have:

$$\mathbf{v}_N(x,t) = \begin{cases} \dfrac{\partial \mathbf{u}_N}{\partial t} - \mathbf{V} \times \operatorname{curl} \mathbf{u}_N \text{ on } \Omega_1 \\ \dfrac{\partial \mathbf{u}_N}{\partial t} \qquad\qquad\qquad \text{ on } \Omega_2 \,. \end{cases}$$

The function $\mathbf{v}_N$ is admissible by construction; plugging it into (16) we get

$$\begin{aligned} \int_{\mathcal{C}} \sigma \left| \frac{\partial \mathbf{u}_N}{\partial t} \right|^2 + a(\mathbf{u}_N, \frac{\partial \mathbf{u}_N}{\partial t}) &= \int_{\Omega_2} \mathbf{J}_s \cdot \left(\frac{\partial \mathbf{u}_N}{\partial t} - \mathbf{V} \times \mathrm{curl}\mathbf{u}_N \right) \\ &+ \int_{\Omega_1} \nu \mathrm{curl}\mathbf{u}_N \cdot \mathrm{curl}(\mathbf{V} \times \mathrm{curl}\mathbf{u}_N) + \int_{\mathcal{C}} \sigma \frac{\partial \mathbf{u}_N}{\partial t} \cdot \mathbf{V} \times \mathrm{curl}\mathbf{u}_N . \end{aligned} \tag{17}$$

We estimate the three terms in the right hand side starting from the last one:

$$\int_{\mathcal{C}} \sigma \frac{\partial \mathbf{u}_N}{\partial t} \cdot \mathbf{V} \times \mathrm{curl}\mathbf{u}_N \le C ||\mathrm{curl}\mathbf{u}_N||^2_{L^2(\mathcal{C})} + \frac{1}{4} \int_{\Omega_1} \sigma \left| \frac{\partial \mathbf{u}_N}{\partial t} \right|^2 . \tag{18}$$

To estimate the second term we use the fact that $\mathrm{div}(\mathcal{R}_t \mathbf{u}_N) = 0$ implies that $\mathrm{div}(\mathbf{u}_N) = 0$ on Ω_1. Using the chain rule (5) and reminding that $\mathrm{div}\, \mathbf{V} = 0$ and that, by integrating by parts, $\int_{\Omega_1} \nu (\mathbf{V} \cdot \mathrm{grad})(\mathrm{curl}\, \mathbf{u}_N)^2$ is zero, we get:

$$\begin{aligned} &\int_{\Omega_1} \nu \mathrm{curl}\mathbf{u}_N \cdot \mathrm{curl}(\mathbf{V} \times \mathrm{curl}\mathbf{u}_N) \\ &\quad = \int_{\Omega_1} \nu \mathrm{curl}\mathbf{u}_N \cdot (\mathrm{curl}\mathbf{u}_N \cdot \mathrm{grad})\mathbf{V} \le C ||\omega||_{L^\infty} ||\mathrm{curl}\mathbf{u}_N||^2_{L^2(\Omega_1)} \end{aligned} \tag{19}$$

Finally, about the first one, we integrate it in time and we obtain the equality:

$$\begin{aligned} \int_{\Omega} \int_0^t \mathbf{J}_s \cdot \left(\frac{\partial \mathbf{u}_N}{\partial t} - \mathbf{V} \times \mathrm{curl}\mathbf{u}_N \right) &= - \int_{\Omega_2} \int_0^t \frac{\partial \mathbf{J}_s}{\partial t} \cdot \mathbf{u}_N \\ &+ \int_{\Omega_2} \mathbf{J}_s(\cdot,t) \cdot \mathbf{u}_N(\cdot,t) - \int_{\Omega_2} \int_0^t \mathbf{J}_s \cdot \mathbf{V} \times \mathrm{curl}\mathbf{u}_N . \end{aligned} \tag{20}$$

Then, we have that

$$\text{the right hand side of (20)} \le C\|\mathbf{J}_s\|_{H^1(0,T;\mathbf{L}^2(\Omega))}\|\mathbf{u}_N(\cdot,t)\|_{L^\infty(0,t;\mathcal{X}_t)}. \quad (21)$$

By using these estimates in (17), and integrating in time, we obtain:

$$\|\mathbf{u}_N\|^2_{L^\infty(0,T;\mathcal{X}_t)} + \int_0^T \int_{\mathcal{C}} \sigma \left|\frac{\partial \mathbf{u}_N}{\partial t}\right|^2 \le C\|\mathbf{J}_s\|^2_{H^1(0,T;\mathbf{L}^2(\Omega))}.$$

There is then a subsequence of $\mathbf{u}_N$ (that we call again $\mathbf{u}_N$ for convenience) which converge weakly also in $H^1(0,T;\mathbf{L}^2(\mathcal{C}))\cap L^\infty(0,T;\mathcal{X}_t)$. By uniqueness, the weak limit is again $\mathbf{u}$ and this ends the proof.

Since the existence of an angular speed verifying (E_4) and (IC_3) for a given torque with intensity in $L^1(]0,T[)$ is standard, we pass directly to the analysis of the coupled problem.

Theorem 10. *Let $\boldsymbol{J}_s \in H^1(0,T,L^2(\Omega_1))$. The system of equations (3) admits at least one solution $(\boldsymbol{u}, \boldsymbol{T}_m, \omega) \in H^1(0,T;\mathbf{L}^2(\mathcal{C})) \cap L^\infty(0,T;\mathcal{X}_t) \times L^2(]0,T[)^3 \times H^1(]0,T[)$. For J large enough the solution is also unique.*

Proof. Let $T \in \mathbb{R}^+$ and $G : C^0([0,T]) \to C^0([0,T])$ be the open feedback operator, that associates to every angular speed $\bar{\omega} \in C^0([0,T])$, the angular speed ω calculated by means of (E_4) in (3) when the torque $\mathbf{T}_m$ is computed by means of the solution $\mathbf{u}$ of $(E_1), (E_2)$ associated with the speed $\bar{\omega}$. Using the results obtained in Proposition (1), system (E_1), (E_2), (G), (IC_1), (IC_2), (BC), $(0C)$ admits a unique solution $\mathbf{u} \in L^2(0,T;\mathcal{X}_t)\cap C^0(0,T;\mathbf{L}^2(\mathcal{C}))$. Now, thanks to the regularity results of Theorem 9, in the "open feedback" system, for any $\bar{\omega} \in C^0([0,T])$, the integral in the equation (E_3) of (3) is meaningful and moreover the resulting torque $\mathbf{T}_m = \mathbf{T}_m(t)$ belongs to $L^2(]0,T[)$. Using finally (E_4) and (IC_3), we find the new angular speed ω which turns out to belong to $H^1([0,T])$. Moreover the following stability holds:

$$\|\omega\|_{H^1([0,T])} \le C(J)\|\bar{\omega}\|_{C^0([0,T])}.$$

Since the embedding $H^1([0,T]) \hookrightarrow C^0([0,T])$ is compact, the operator G is also compact. By applying the Schauder fixed-point theorem [20], we deduce that G has at least one fixed point which corresponds to a solution of (3).

The constant $C(J)$ tends to zero as J tends to infinity, as results from (E_4). We deduce that, when J is large enough, the operator G is also contractive and, by applying the Banach fixed-point theorem, we have that such a fixed point is also unique.

5 Discretization of (3)

For what concerns the discretization of (3), we adopt the technique presented in [9, 19]. Once discretized both domains Ω_k by two independently created meshes of tetrahedra $\mathcal{T}_{k,h}$ (h is the maximum size of all mesh tetrahedra), we introduce two edge element spaces $X_{k,h}$ as in [10] and $T_h := \{(\mathbf{v}_{2,h})_{\tau,\Gamma} \mid \mathbf{v}_{2,h} \in X_{2,h}\}$ the space of tangential components on Γ of functions in $X_{2,h}$. As now a standard procedure for non-conforming domain decomposition methods, the mortar element method leads to impose the transmission condition (TC_1) in a weak form by means of a suitable space of Lagrange multipliers M_h. We choose M_h as a proper subset of T_h, in such a way that dim $(M_h) =$ dim $(T_h \cap H_0(\text{curl}, \Gamma))$ as in [10]. The approximation space is

$$\mathcal{H}_h^t = \{ (\mathbf{u}_{1,h}, \mathbf{u}_{2,h}) \in X_{1,h} \times X_{2,h} \mid \int_\Gamma [(R_t(\mathbf{u}_{1,h})_{\tau,\Gamma}) - (\mathbf{u}_{2,h})_{\tau,\Gamma}] \cdot \varphi_h \, d\Gamma = 0 \,, \forall \varphi_h \in M_h \}.$$

In time, we use finite difference schemes, explicit for the rotation angle θ and implicit for the rotation speed ω, on a uniform partition of the time interval $[0,T]$. Let δt be the time step and N be an integer such that $T = N\,\delta t$. The solution $\mathbf{u}_h^n$ at time $t = t^n$ as well as the rotation operator r_t are computed by means of the "calculated speed" and "angle" and not of the exact ones since these ones are unknowns of the problem. We then define $\tilde{\mathbf{u}}_h^n$ and $\tilde{r}_t$ these perturbed quantities. The approximation space is discretized in time as well and we denote by $\tilde{\mathcal{H}}_h^{t^n}$ the space associated to the rotation $\tilde{r}_n$ of the computed angle θ_h^n. To shorten the notations, we consider

$$t^n = n\,\delta t \,, \; \tilde{r}_{t^n} = \tilde{r}_n \,, \; \tilde{\mathbf{u}}_h^n \in \tilde{\mathcal{H}}_h^{t^n} = \tilde{\mathcal{H}}_h^n \,.$$

Now, at each time step, say $t = t^n$, we are given with the angle θ_h^n and the magnetic vector potential $\tilde{\mathbf{u}}_h^{n-1}$ at the previous time step. The magnetic vector potential is computed by means of the fully discrete equation:

$$\begin{aligned} &\text{find } \tilde{\mathbf{u}}_h^n \in \tilde{\mathcal{H}}_h^n \text{ such that :} \\ &\int_C \sigma\, \partial_{\delta t} \tilde{\mathbf{u}}_h^n\, \tilde{\mathbf{v}}_h^n + a(\tilde{\mathbf{u}}_h^n, \tilde{\mathbf{v}}_h^n) = (\mathbf{J}_z, \tilde{\mathbf{v}}_h^n)_{\Omega_1} \quad \forall \tilde{v}_h^n \in \tilde{\mathcal{H}}_h^n, \end{aligned} \tag{22}$$

where $\partial_{\delta t}\tilde{\mathbf{u}}_h^n$ stands for $\dfrac{\tilde{\mathbf{u}}_h^n - \tilde{\mathbf{u}}_h^{n-1}}{\delta t}$. Note that the gauge condition (G) is taken into account by solving (22) by a Conjugate Gradient procedure. This technique has been first used in the magnetostatic case in [10] where it is shown to have good performances by means of suitable numerical tests.

The torque vector at time t^n can be computed by means of known quantities as:

$$(\mathbf{T}_m)_h^n = \int_{\Omega_1} \mathbf{r} \times \left[\left(-\sigma \frac{\partial \tilde{\mathbf{u}}_h^n}{\partial t} + \mathbf{J}_s\right) \times \text{curl}\, \tilde{\mathbf{u}}_h^n\right]. \tag{23}$$

Its intensity $(\mathrm{T}_m)_h^n$ is used to compute the actual speed at $t = t^n$, by the equation obtained applying the implicit Euler method to the equation (E_4) in (3):

$$\text{Find } \omega_h^n \text{ such that} \qquad J\frac{\omega_h^n - \omega_h^{n-1}}{\delta t} + k\omega_h^n = (\mathrm{T}_m)_h^n. \tag{24}$$

Finally, once we have the angular speed, the new angle is computed by the equation:

$$\theta_h^{n+1} = \theta_h^n + \delta t\, \omega_h^n. \tag{25}$$

Now the feedback in Figure 2 is closed and, starting again, the procedure allows the computation of the new value of the magnetic vector potential $\mathbf{u}_h^{n+1}$ at the further time step by induction.

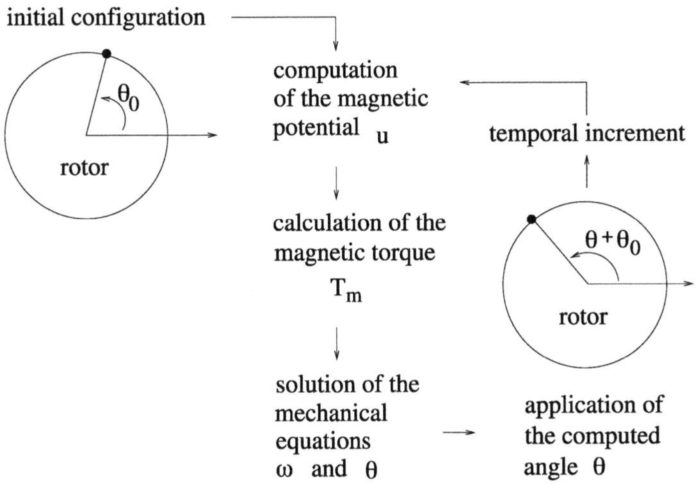

Fig. 2. Flowchart of the "explicit" coupling procedure. The procedure ends when the rotor reaches the equilibrium position, corresponding to a zero-value for the acting magnetic torque T_m.

Numerical simulations for a coupled three-dimesional magneto-mechanical system are in progress and we refer the interested reader to [9] for numerical results concerning the two-dimensional coupled case. Existing results for three-dimensional systems are presented in [10] for a magnetostatic problem and in [19] for a magnetodynamic problem, the latter without mechanical coupling.

Acknowledgements

The authors wish to warmly thank Professor Alain Bossavit for his helpfull remarks on the Maxwell equations modelling and "what lies underneath".

References

1. Amrouche C., Bernardi C., Dauge M., Girault V. (1998) Vector potentials in three-dimensional non-smooth domains, Math. Meth. Appl. Sci., **21**, 823–864.

2. Ben Belgacem F., Buffa A., Maday Y. (2001) The mortar element method for Maxwell equations: first results, SIAM J. Num. Anal., **39**, 880–901.
3. Ben Belgacem F., Maday Y. (1994) Non–conforming spectral element methodology tuned to parallel implementation, Comput. Meth. in Appl. Mech. and Engng., **116**, 59–67.
4. Bergh J., Löfström J. (1976) Interpolation Spaces, An introduction, Springer Verlag.
5. Bernardi C. (1989) Optimal finite element interpolation of curved domains, SIAM J. Num. An., **26**, 1212–1240.
6. Bernardi C., Maday Y., Patera A. (1994) A new nonconforming approach to domain decomposition: the mortar element method, in Nonlinear Partial Differential Equations and Their Applications, H. Brezis and J.L. Lions, eds. Pitman, 13–51.
7. Bossavit A. (1992) Edge-element computation of the force field in deformable bodies, IEEE Trans. Magn., **28**(2), 1263-1266.
8. Bossavit A. (1992) On local computation of electromagnetic force field in deformables bodies, Int. J. Appl. Electromagnetics in Materials, **2**(4), 333-343.
9. Bouillault F., Buffa A., Maday Y., Rapetti F. (2002) Simulation of a magneto-mechanical damping machine: analysis, discretization, results, Computer Methods in Appl. Mech. and Engng., **191**(23-24), 2587-2610.
10. Bouillault F., Buffa A., Maday Y., Rapetti F. (2001) The mortar edge element method in three dimensions: applications to magnetostatics, SIAM J. on Scient. Comp., in press.
11. Ciarlet P.G. (1978) The Finite Element Method for Elliptic Problems, North-Holland, Amsterdam.
12. Dautray R., Lions J.-L. (1987) Analyse mathématique et calcul numérique pour les sciences et les techniques, **1**, Modéles physiques, Masson.
13. Emson C. R. I., Riley C. P., Walsh D. A., Ueda K., Kumano T. (1998) Modelling eddy currents induced in rotating systems, IEEE Trans. on Magn., **28**, 2593–2596.
14. Gaspalou B., Colamartino F., Marchand C., Ren Z. (1995) Simulation of an electromagnetic actuator by a coupled magneto-mechanical modelling, Int. J. for Comp. and Math. in Electric and Electronic Eng., **14**, 203–206.
15. Hoppe R.H.W. (1999) Mortar edge element method in $\mathbb{R}^3$, East-West J. Num. Math., **7**, 159–173.
16. Jackson J. D. (1952) Classical Electrodynamics, New York, Wiley.
17. Parietti C., Rappaz J. (1998) A quasi–static two–dimensional induction heating problem. I. Modeling and analysis, Math. Methods Models Appl. Sci., **8**, 1003–1021.
18. Parietti C., Rappaz J. (1999) A quasi–static two–dimensional induction heating. II. Numerical analysis, Math. Methods Models Appl. Sci., **9**, 1333–1350.
19. Rapetti F., Maday Y., Bouillault F., Razek A. (2002) Eddy current calculations in three-dimensional moving structures, IEEE Trans. on Magn., **38**(2), 613–616.
20. Zeidler E. (1991) Nonlinear Functional Analysis and its Applications I - Fixed Point Theorems, Springer-Verlag, New-York.

Numerical Stability of Collocation Schemes for Time Domain Boundary Integral Equations

Penny Davies[1] and Dugald Duncan[2]

[1] Department of Mathematics, University of Strathclyde, 26 Richmond St, Glasgow, G1 1XH, UK
[2] Department of Mathematics, Heriot-Watt University, Riccarton, Edinburgh, EH14 4AS, UK

Summary. Time domain boundary integral formulations of transient scattering problems involve retarded potential integral equations (RPIEs). Collocation schemes for RPIEs are often unstable, having errors which oscillate and grow exponentially with time. We describe how Fourier analysis can be used to analyse the stability of uniform grid schemes and to show that the instabilities are often very different from those observed in PDE approximations. We also present a new stable collocation scheme for a scalar RPIE, and show that it converges.

1 Introduction

The physical problem that we are interested in addressing is to find the field scattered by a conductor Γ when it is struck by an incident electromagnetic pulse.

If Γ is a smooth, closed, perfectly conducting surface bounding the finite domain (open connected region) $\Omega^- \subset \mathbb{R}^3$, and $\Omega^+ = \mathbb{R}^3 \setminus \{\Omega^- \cup \Gamma\}$, then the exterior scattering problem is: given an incident electric field $\mathbf{e}$ on the surface Γ, find scattered electromagnetic fields $\mathbf{E}^+(\mathbf{x},t)$ and $\mathbf{B}^+(\mathbf{x},t)$ in the region $\Omega^+ \times \mathbb{R}$. Both these and the scattered interior fields $\mathbf{E}^-$ and $\mathbf{B}^-$ in $\Omega^- \times \mathbb{R}$ satisfy Maxwell's equations:

$$\left.\begin{aligned} \mathbf{E}_t^\pm - \operatorname{curl}\mathbf{B}^\pm &= \mathbf{0} \\ \mathbf{B}_t^\pm + \operatorname{curl}\mathbf{E}^\pm &= \mathbf{0} \end{aligned}\right\} \quad \text{in } \Omega^\pm \times \mathbb{R} \tag{1}$$

and the boundary condition

$$\mathbf{n} \times \mathbf{E}^\pm\big|_\Gamma + \mathbf{n} \times \mathbf{e} = \mathbf{0} \quad \text{on } \Gamma \times \mathbb{R}\,, \tag{2}$$

where the subscript $_t$ denotes the partial derivative $\partial/\partial t$ and $\mathbf{n}(\mathbf{x})$ is the outward unit normal to Γ at the point $\mathbf{x}$. We suppose all fields to be identically zero before time $t = 0$, i.e. that

$$\mathbf{E}^\pm \equiv \mathbf{0}, \quad \mathbf{B}^\pm \equiv \mathbf{0}, \quad \mathbf{e} \equiv \mathbf{0} \quad \text{for all } t \le 0.$$

The fields can be represented in terms of vector and scalar potentials $\mathbf{A}$ and ϕ as

$$\left.\begin{aligned} \mathbf{E}^{\pm} &= -\mathbf{A}_t - \operatorname{grad}\phi \\ \mathbf{B}^{\pm} &= \operatorname{curl}\mathbf{A} \end{aligned}\right\} \quad \text{in } \Omega^{\pm}\times\mathbb{R} \tag{3}$$

(see e.g. [15, 25]), where $\mathbf{A}$ and ϕ are respectively retarded potential integrals of the surface current $\mathbf{J}$ and charge ρ:

$$\left.\begin{aligned} \mathbf{A}(\mathbf{x},t) &= \frac{1}{4\pi}\int_{\Gamma}\frac{\mathbf{J}(\mathbf{x}',t-|\mathbf{x}-\mathbf{x}'|)}{|\mathbf{x}-\mathbf{x}'|}\,dS'_{\Gamma} \\ \phi(\mathbf{x},t) &= \frac{1}{4\pi}\int_{\Gamma}\frac{\rho(\mathbf{x}',t-|\mathbf{x}-\mathbf{x}'|)}{|\mathbf{x}-\mathbf{x}'|}\,dS'_{\Gamma}\ . \end{aligned}\right\} \tag{4}$$

As described by Rynne [25], if $\mathbf{J}$ and ρ satisfy the surface continuity equation

$$\rho_t + \operatorname{div}_{\Gamma}\mathbf{J} \tag{5}$$

on $\Gamma\times\mathbb{R}$ (where $\operatorname{div}_{\Gamma}$ denotes the surface divergence), then the fields (3) satisfy Maxwell's equations (1); and the boundary condition (2) yields the electric field integral equation (EFIE)

$$\mathbf{n}\times(\mathbf{A}_t + \operatorname{grad}\phi) = \mathbf{n}\times\mathbf{e} \quad \text{on } \Gamma\times\mathbb{R}, \tag{6}$$

which from (4) is an integro–differential equation involving $\mathbf{J}$ and ρ. If (5)–(6) can be solved for $(\mathbf{J},\rho)$, the solution $\mathbf{E}^{+}$ of the exterior scattering problem can then be found from (3) via the potentials (4).

The problem of determining $(\mathbf{J},\rho)$ from $\mathbf{e}$ is considered by Bachelot and Pujols [1, 20], who use an approach originally applied to acoustic scattering by Bamberger and Ha-Duong [2]. This involves taking the Laplace transform of the hyperbolic PDE system (1) and then studying the associated boundary integral formulation. In [25] Rynne instead uses the theory of uniformly characteristic hyperbolic systems to analyse the problem. He shows that (in a suitable function space setting) the equations (5) and (6) are well posed in the sense that for a given function $\mathbf{e}$ there exists a unique solution $(\mathbf{J},\rho)$, and this solution depends continuously on $\mathbf{e}$. This approach gives better regularity results for $(\mathbf{J},\rho)$ (dropping only one order of differentiability) and also enables Rynne to prove that $(\mathbf{J},\rho)\to 0$ in an appropriate sense as $t\to\infty$ (taking Laplace transforms typically yields only that the current and charge are bounded by $Ce^{\sigma t}$ for positive constants C and σ).

Various numerical methods for computing $(\mathbf{J},\rho)$ have been reported in the literature. Bachelot and Pujols [1, 20] describe a variational method based on the coercivity of a bilinear form corresponding to a full Galerkin approximation in time and space (similar to that used for acoustics by Bamberger and Ha-Duong [2]). Collocation schemes for the problem have been described by many authors including Rao et al [22] and Rynne [24]. However, numerical instabilities are frequently observed for collocation approximations (see e.g. [7] and the references therein), even in the computationally simpler case for which Γ is a flat plate [26].

Fourier analysis of collocation approximations for the flat plate problem [5, 6, 8] indicates that the most likely cause of numerical instability is the inaccurate approximation and solution of retarded potential integral equations (RPIEs) like (4). Hence from now on we shall restrict attention to computing the solution of the scalar RPIE

$$a(\mathbf{x},t) = \frac{1}{4\pi}\int_{\Gamma} \frac{u(\mathbf{x}',t-|\mathbf{x}-\mathbf{x}'|)}{|\mathbf{x}-\mathbf{x}'|}\, d\mathbf{x}' \quad \text{for } \mathbf{x}\in\Gamma, \quad t\in(0,T) \qquad (7)$$

for u when a is given on $\Gamma\times(0,T)$ for fixed $T>0$, and

$$u\equiv 0, \quad a\equiv 0 \quad \text{for all } t\leq 0. \qquad (8)$$

This is the single layer potential equation for acoustic scattering from Γ [19, Sect. 2.3].

Existence, uniqueness and well–posedness results for (7) are given in [2, 13, 19]. A similar argument to that used by Lubich [19, Sect. 2.3] in the case that Γ is a smooth, closed surface (based on results of Bamberger and Ha-Duong [2, Prop. 3]) can be used to deduce the following Proposition from results of Ha-Duong [13] when Γ is a flat plate. We use the notation

$$H_*^m(0,T) = \left\{ f\,|_{(0,T)} : f\in H^m(\mathbb{R}) \text{ with } f\equiv 0 \text{ on } (-\infty,0) \right\} ,$$

where for real m, $H^m(\mathbb{R})$ denotes the usual Sobolev space of order m.

Proposition 1 *(Ha-Duong [13, Thm. 3], Lubich [19, Sect. 2.3])*
For temporally smooth data $a(\cdot,t)\in H^{1/2}(\Gamma)$ which vanish near $t=0$, the RPIE (7) has a unique smooth solution $u(\cdot,t)\in H^{-1/2}(\Gamma)$. Moreover there exists a constant C depending only on T such that

$$||u||_{H_*^m(0,T;H^{-1/2}(\Gamma))} \leq C\,||a||_{H_*^{m+1}(0,T;H^{1/2}(\Gamma))} \quad (m\in\mathbb{R})\ .$$

In this article we give stability and convergence results for collocation approximations of (7) on a flat plate. The algorithms are described in the next section, and Fourier methods are used to define and analyse their stability in Sect. 3. Section 4 contains the statement and (brief) proof of convergence for one of the schemes. We conclude with a discussion of our results.

2 Algorithms

In this section we describe collocation approximation schemes for solving the RPIE (7). Because we shall use Fourier methods to analyse the stability of these schemes (in Sect. 3) we assume Γ to be a square flat surface and discretise it uniformly in space into squares of side h. We also use a uniform discretisation in time, with time-step Δt and define the mesh ratio to be $r=\Delta t/h$. An arbitrary spatial node point is denoted as $\mathbf{x}_\beta$, and t^s represents $s\,\Delta t$ for $s\in\mathbb{R}$.

Various choices need to be made in order to approximate the RPIE (7):

– how should u be approximated?
– what collocation points (i.e. $\mathbf{x}$ and t) should be used?
– how should the resulting integral be evaluated or approximated?

Approximation. We suppose that u is expanded in terms of low–order piecewise–smooth spatial and temporal basis functions (i.e. either piecewise constant "pulse" or piecewise linear "hat" functions):

$$u(\mathbf{x},t) \approx U(\mathbf{x},t) = \sum_{\beta,s} U_\beta^s \,\varphi_\beta(\mathbf{x})\,\psi_s(t)\,. \tag{9}$$

If the $\varphi_\beta(\mathbf{x})$ are piecewise constant (resp. linear), then the nodes $\mathbf{x}_\beta$ are taken to be at the centre (resp. corners) of the square space mesh. Similarly, if the $\psi_s(t)$ are piecewise constant (resp. linear), then $s = m + 1/2$ (resp. $s = m$) for integer m.

Collocation. Temporal collocation points are always taken to be $t = t^n$ for integer n, whilst the collocation points in space are taken to be the nodes $\mathbf{x}_\beta$ defined above.

Integral. The integrals to be evaluated have the form

$$\int_\Gamma \frac{\varphi_\beta(\mathbf{x}')\,\psi_s(t^n - |\mathbf{x}_\alpha - \mathbf{x}'|)}{|\mathbf{x}_\alpha - \mathbf{x}'|}\,d\mathbf{x}'\,. \tag{10}$$

As demonstrated in the next section, the integral approximation used has a big effect on the scheme's stability.

A general collocation approximation of (7) has the form

$$a(\mathbf{x}_\alpha,t^n) = \frac{1}{4\pi}\sum_{\beta,s} U_\beta^s \int_\Gamma \frac{\varphi_\beta(\mathbf{x}')\,\psi_s(t^n - |\mathbf{x}_\alpha - \mathbf{x}'|)}{|\mathbf{x}_\alpha - \mathbf{x}'|}\,d\mathbf{x}'\,.$$

Because of the causality assumption (8) this can be written as a convolution sum (for more details see e.g. [5, 6])

$$\underline{a}^n = \sum_{0\le s\le n} \mathbb{Q}^s\,\underline{U}^{n-s}, \tag{11}$$

where $\underline{a}^s$ and $\underline{U}^s$ are vectors containing the values of the functions a and U at mesh-points on Γ at time-level $t^s < T$, and where each $\mathbb{Q}^s$ is a (sparse) square matrix. This can be inverted at each time-step to yield the solution U^s, where $s = n$ (resp. $s = n - 1/2$) if the temporal basis functions are piecewise linear (resp. constant), provided the matrix $\mathbb{Q}^0$ (resp. $\mathbb{Q}^{1/2}$) is nonsingular.

Evaluating the right hand side of (11) takes $O(N_S^2)$ floating point operations (flops) per time-step n, where N_S is the number of spatial degrees of freedom used in the approximation (9). Thus, computing the solution up to time $T = N_T\,\Delta t$ takes $O(N_T\,N_S^2)$ flops – so the solution algorithm is highly computationally intensive. The fully variational approach for solving

(7) also leads to a convolution sum like (11) [2, 14]. However, it is much more complicated (and costly) to evaluate the entries in the matrices $\mathbb{Q}^s$ for these schemes, since they involve five dimensional integrals over $\Gamma \times \Gamma \times (0, T)$.

Recently Michielssen and co-workers [10, 11, 18] have introduced "fast methods" for time dependent boundary integral equations (BIEs) such as (7) that reduce the operation count to $O(N_T\, N_S^{3/2} \log N_S)$ (for a two-level scheme), or $O(N_T\, N_S \log^2 N_S)$ (multi-level). Although complicated to implement, these make the BIE approach for time-dependent scattering problems viable compared to methods based on solving PDEs in 3D space.

3 Stability

Unfortunately the numerical solution of (11) is often found to be exponentially unstable (i.e. to grow exponentially with n) in practice. A good, comprehensive discussion of such instances of instability is given by Jones in [15, Section 7] (see also [5, 6, 8, 21, 24, 26]). The unstable modes can sometimes be filtered out by averaging the solution in time or space [5, 8, 23, 26], but it may not be possible to remove them entirely [7].

The instabilities exhibited by RPIE schemes behave differently from those observed in time-stepping PDE approximations. RPIE schemes appear to be *more* unstable at lower values of the mesh ratio r, and schemes can be completely stable on a coarse spatial mesh but become unstable as the mesh is refined. This is illustrated in Fig. 1, which shows the behaviour of the solution for an approximation scheme that uses piecewise constant basis functions in space, piecewise linears in time, and the midpoint quadrature rule (i.e. the scheme proposed in [26]) when a is a transient pulse. It is completely stable on an $N \times N$ spatial grid when $N = 10$, but exponentially unstable for higher values of N. The mesh ratio used in the calculation is $r = 0.65$; at higher mesh ratios ($r > 0.75$) the scheme appears to be stable for all N. More details are given in [8].

Here we follow [5] and show how Fourier analysis can be used to characterise the stability of schemes like (11) when Γ is an infinite flat plate. We also show how a comparison with the continuous Fourier transform of (7) can give useful insight. We conclude the section with stability results for different approximation schemes.

3.1 Stability Analysis when Γ is an Infinite Flat Plate

When the temporal basis functions ψ used in (9) are piecewise linear the convolution sum (11) can be written as

$$\underline{a}^n = \sum_{m=0}^{n} \mathbb{Q}^m \, \underline{U}^{n-m} \,. \tag{12}$$

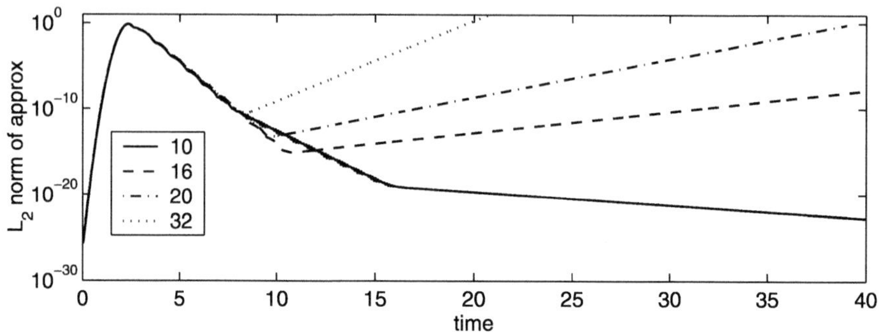

Fig. 1. L_2 norm over Γ of the approximate solution U on an $N \times N$ grid for $N =$ 10, 16, 20, 32. See text for details

For simplicity we shall assume this form for the approximation from now on (an identical analysis can be carried out for other choices of ψ [9]). The sum is written this way for convenience, but note that the term $\underline{U}^0$ is zero by causality (8).

Taking the discrete Fourier transform (DFT) of (12) at frequency ω yields

$$\tilde{a}^n(\omega) = \sum_{m=0}^{n} q_m(h\omega)\, \tilde{U}^{n-m}(\omega) \,, \tag{13}$$

where for any mesh function $\underline{v} = (v_{j,k})$ the DFT is defined by

$$\tilde{v}(\omega) = h^2 \sum_{j,k} v_{j,k}\, e^{-i(x_j, y_k)\cdot\omega} \qquad \text{for } \omega \in S_h = [-\pi/h, \pi/h]^2 \,.$$

The q_m correspond to the DFT of rows of the infinite version of $\mathbb{Q}^m$, thought of as shift operators (see [5] for more details).

As shown in [5], it is sufficient to consider stability of the homogeneous problem

$$q_0\, \tilde{U}^n = -\sum_{m=1}^{n} q_m\, \tilde{U}^{n-m} \,,$$

where the initial data $\tilde{U}^0$ are now non-zero. This can be rewritten as

$$\tilde{U}^n(\omega) = p_n(h\omega)\, \tilde{U}^0(\omega)$$

where the amplification factors p_n are found from

$$p_0(h\omega) = 1 \,, \quad p_n(h\omega) = \frac{-1}{q_0(h\omega)} \left(\sum_{m=0}^{n} q_{n-m}(h\omega)\, p_m(h\omega) \right) , \; n \geq 1 \,. \tag{14}$$

The scheme (12) is said to be **stable** if the norm of the solution $\underline{U}^n$ of the homogeneous version (i.e. with $\underline{a}^n = \underline{0}$ for all n and non-zero initial data

$\underline{U}^0$) is bounded independently of n and h for all $t^n < T$. It is straightforward to show that stability corresponds to the existence of a constant C such that $|p_n(\xi)| \le C$ for all n and all $\xi \in [-\pi, \pi]^2$ (details are given in [5]).

This is a useful stability test, since evaluating the coefficients p_n over N_f discrete frequencies is a relatively straightforward computation (compared with calculating the solution of (12)), taking only $O(N_f\, N_T^2)$ flops. We shall give stability results for specific algorithms in Sect. 3.2, but first it is instructive to compare the transformed approximate equation (13) with the continuous Fourier transform (CFT) of the RPIE (7). This yields the first kind Volterra convolution equation

$$\widehat{a}(\omega, t) = \frac{1}{2} \int_0^t J_0(\omega R)\, \widehat{u}(\omega, t - R)\, dR \,, \tag{15}$$

after some manipulation, where $\boldsymbol{\omega}$ is the Fourier frequency, $\omega = |\boldsymbol{\omega}|$, and J_0 is the first kind Bessel function of order zero. Using a piecewise linear approximation ψ in time for u in (7) corresponds to using the trapezoidal rule to solve (15) once spatial approximation errors are neglected. Similarly, a piecewise constant temporal approximation for u corresponds to the midpoint rule for (15).

These two quadrature rules are both known to give stable schemes for (15), although the leading error term for the trapezoidal rule solution is oscillatory [12, 16, 17]. It is also known [12, 17] that all higher order Newton–Cotes quadrature rules give rise to unstable approximations of (15). Hence one must be careful in constructing approximations of (7) that use temporal basis functions of higher degree, incase they give rise to the same instabilities. We therefore restrict our attention to low (zero or first) order temporal basis functions ψ in (9). An unfortunate consequence of this is that either the integrand of (10) or its first derivative will be discontinuous along curves within the spatial elements (as shown in Fig. 2), making quadrature over the spatial elements very slow to converge.

Comparing the approximate transform equation (13) with the trapezoidal rule approximation of (15) suggests that the coefficients $q_m(h\omega)$ approximate $\Delta t\, J_0(\omega t^m)/2$. Fig. 3 shows that this is indeed the case for two simple schemes, with a good match tending to give a more stable scheme.

3.2 Stability Results

As previously noted, the approximation scheme that uses piecewise constant basis functions in space, piecewise linears in time, and the midpoint quadrature rule is unstable when the mesh ratio is less than about 0.75. Rynne and Smith [26] show that this scheme can be stabilised by averaging it in time (this filters out high frequency instabilities [6, 8]). This is not entirely satisfactory because electromagnetic scattering problems involve more complicated RPIEs and hence are harder to stabilise [7]. We believe that a minimum

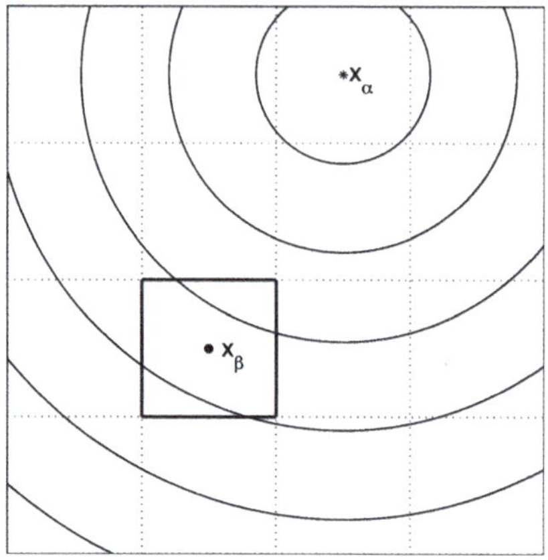

Fig. 2. Intersection of support of spatial (square) and temporal (annular) basis functions for the integrand of (10) when $r = 0.65$

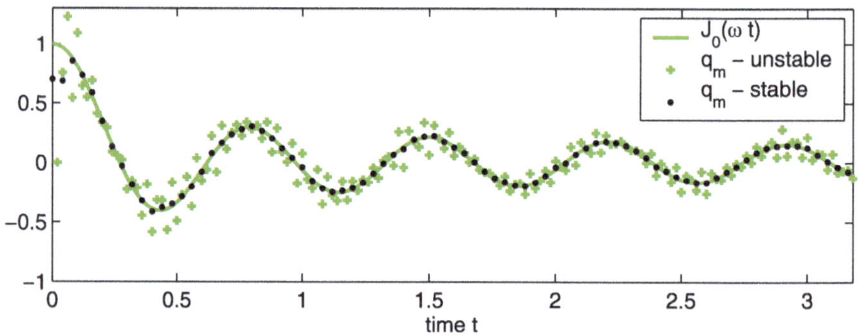

Fig. 3. Comparison of normalised coefficients $2q_m(h\omega)/\Delta t$ (dots) with $J_0(\omega t^m)$ (grey solid) at frequency $\omega = (0.1\pi, 0.1\pi)$. The grey dots correspond to an unstable scheme and the black dots to a stable scheme

requirement for a scalar RPIE scheme to be useful is that it should be stable when applied on an infinite flat plate without recourse to any filtering. Stability over a wide range of mesh ratios is very important, since practical calculations over general surfaces may involve space mesh elements of vastly different sizes.

Using a low order composite quadrature rule for the integrals (10) with $\mu^2 \geq 1$ quadrature points in each spatial element does not give a reliably stable scheme and results are summarised in Fig. 4. We have also investigated the stability of schemes that remove the singularity in the integrals (10) by using local polar coordinates, replacing them by integrals

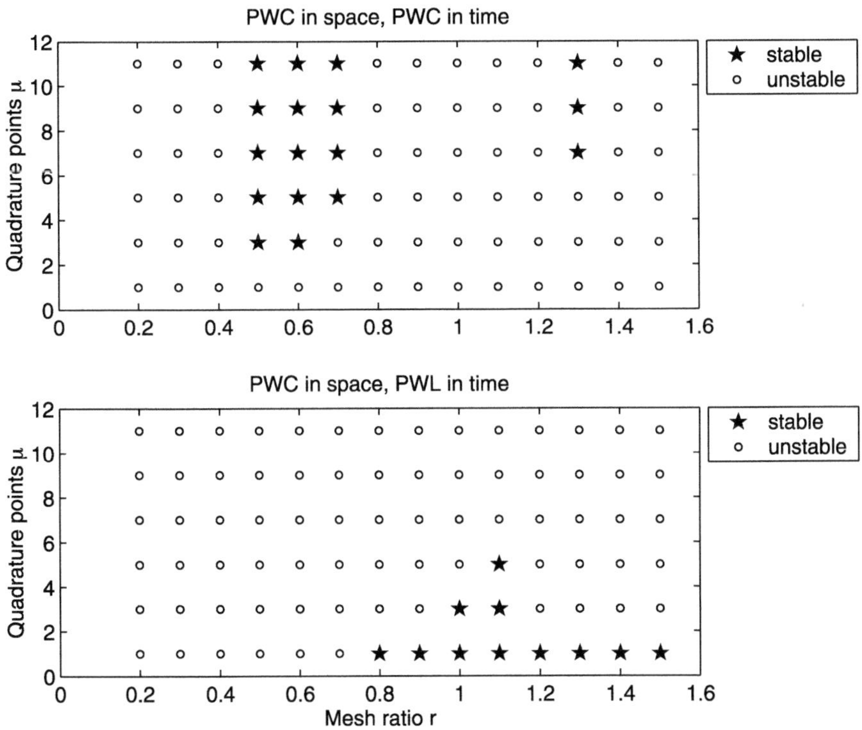

Fig. 4. Stability summary for schemes that are piecewise constant in space and piecewise constant and linear in time with μ^2 quadrature points per space element

$$\int_{I_\alpha} \varphi_\beta(\mathbf{x}_\alpha + R\,\mathbf{e}_\theta)\,\psi_s(t^n - R)\,dR\,d\theta\ , \tag{16}$$

over an appropriate region I_α, where $\mathbf{e}_\theta = (\cos\theta, \sin\theta)$. These integrals are then approximated using the trapezoidal rule in R and "exact" (i.e. arbitrarily accurate) integration in θ. The disadvantage is that the transformed region of integration has a complicated shape (that depends on $\mathbf{x}_\alpha$) and so this scheme is harder to implement in practice. However it is easy to calculate the stability coefficients p_n and the 'polar' scheme that uses piecewise linear basis functions φ and ψ in time and space appears to be **stable for all values of the mesh ratio** r. A sample mesh plot of $p_n(\xi)$ against ξ is shown in Fig. 5. We demonstrate the convergence of this scheme in the next section.

We have seen above that stability of collocation RPIE schemes depends crucially on the way in which the integrals (10) are evaluated. Ha-Duong [14, Sect. 4] claims that the variational approach for (7) yields schemes that are much more stable than collocation methods. However none of the papers [1, 2, 13, 14, 20] that describe and analyse variational BIE approximations for time–dependent scattering problems give explicit details on how the integrals

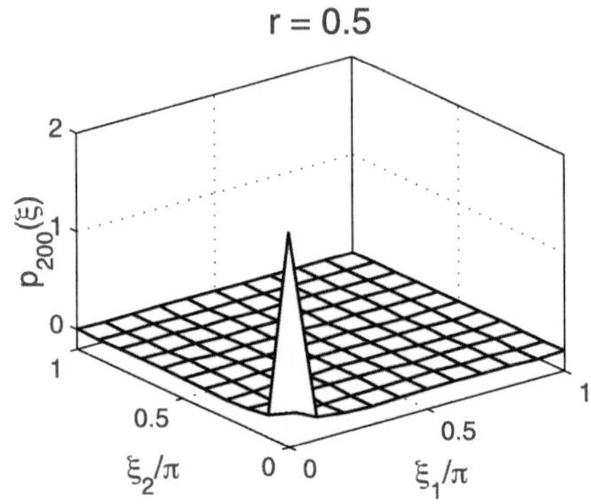

Fig. 5. Mesh plot of $p_n(\xi)$ against ξ when $n = 200$ and $r = 0.5$ for the 'polar' approximation scheme

are evaluated. Numerical tests by Bartels [3] indicate that the stability of variational methods is also dependent on the type of integral approximation used, with some common quadrature rules giving rise to unstable schemes.

During his presentation at this conference Nédélec stated that the best variational methods for the boundary integral formulation of the electromagnetic scattering problem use Stokes' theorem to convert integrals over the support of space and time basis functions in $\Gamma \times \Gamma \times (0, T)$ to integrals over the boundaries of these sub-regions. The boundary integrals are then evaluated arbitrarily accurately using high order quadrature, although organisation of the computations is apparently very complicated. A similar approach is shown in [9] to yield stable collocation schemes for (7) (for arbitrary mesh ratio) when the basis functions φ_β are piecewise linear in space and the ψ_s are piecewise constant or linear in time. In this case Stokes' theorem is used to transform surface integrals over patches of Γ (corresponding to the intersections of the annuli and squares in the simple flat plate geometry illustrated in Fig. 2) to line integrals around the patch boundaries.

4 Convergence

We now prove a convergence result for the stable 'polar' scheme for (7) described in the previous section. Recall that the spatial and temporal basis functions φ and ψ for this scheme are both piecewise linear and that integrals (16) are calculated using the trapezoidal rule in R and "exact" integration in θ. The proof relies on the CFT of (7) being a convolution equation and we use techniques due originally to Jones [16] to obtain pointwise bounds for the Fourier transform of the approximation error. We then use classical estimates derived by Bramble and Hilbert [4] and Thomée [27] to bound the discrete norm of the error as $h \to 0$. Space constraints prevent giving more than a brief outline of the proof, but we note that the steps are similar to those described for schemes based on exact integration in [9], where full details may be found.

Because the analysis involves Fourier transforms we again take Γ to be an infinite flat plate, and make the following assumptions.

Hypotheses. *Suppose:*

(H1) that $a \in H_*^{m+1}\left(0,T;H_0^{s+1/2}(\Gamma)\right)$ *for sufficiently high* m *and* s*, where the subscript* $_0$ *denotes that* a *has compact support on* Γ*, and let* $u \in H_*^m\left(0,T;H_0^{s-1/2}(\Gamma)\right)$ *be the corresponding solution of (7);*

(H2) that the 'polar' scheme for (7) has approximate solution U_β^n *and is stable at mesh ratio* r*.*

Note that we explicitly need to make assumption (H2) because stability for this scheme has only been verified numerically and not proved rigorously.

We say that the scheme is **convergent** if $||\underline{U}^n - \underline{u}^n||_h \to 0$ as $h \to 0$ whenever $t^n < T$, where $\underline{u}^n$ and $\underline{U}^n$ are respectively the values of the exact and approximate solutions on node points of Γ at time t^n, and the discrete norm of $||\cdot||$ is defined by

$$||\underline{v}||_h = \left(h^2 \sum_{\mathbf{x}_\beta} |g(\mathbf{x}_\beta)|^2\right)^{1/2} .$$

The discrete norm is equal to the norm of the DFT by Parseval's identity, i.e.

$$||\underline{v}||_h = ||\tilde{v}||_F \equiv \left(\frac{1}{4\pi^2}\int_{S_h} |\tilde{v}(\omega)|^2\, d\omega\right)^{1/2} .$$

Using this and the triangle inequality gives

$$||\underline{U}^n - \underline{u}^n||_h \le ||\tilde{U}^n - \widehat{u}^n||_F + ||\widehat{u}^n - \tilde{u}^n||_F . \tag{17}$$

It follows from (H1) and a theorem due to Bramble and Hilbert [4, Thm. 5] that the second term on the right of (17) is bounded by $h^{s-1/2}$ times the $H^{s-1/2}$ norm of $u(\cdot, t^n)$. Hence

$$||\underline{U}^n - \underline{u}^n||_h \le ||\tilde{U}^n - \widehat{u}^n||_F + C\, h^{s-1/2}$$

and it remains to bound $||\tilde{U}^n - \widehat{u}^n||_F$. We split this into a number of steps. The first result is to make the relationship between the coefficients q_m and the first kind Bessel functions more precise.

Lemma 1 *There exists a constant* C *independent of* m*,* h *and* ω *such that*

$$\left| J_0(\omega t^m) - \frac{2q_m(h\omega)}{\Delta t}\right| \le C(h\omega)^2 \quad \textit{for all } m \ge 1 \textit{ and all } \omega \in S_h . \tag{18}$$

The proof of this relies on calculating the CFT of the approximate RPIE and then using the Poisson sum formula [4, Thm. 6]. Details are similar to those given in [9].

Comparing the CFT (15) of the exact RPIE with the scheme's DFT (13) gives

$$\sum_{m=1}^{n} q_{n-m}(h\omega)\,\varepsilon^m(\omega) = \tilde{a}^n(\omega) - \widehat{a}^n(\omega) + e^n(\omega) \tag{19}$$

where $\varepsilon^m = \tilde{U}^m - \widehat{u}^m$ is the term we wish to bound and the error term is

$$e^n = \frac{1}{2}\int_0^{t^n} J_0(\omega(t^n - R))\,\widehat{u}(\omega, R)\,dR - \sum_{m=0}^{n} q_{n-m}(h\omega)\,\widehat{u}^m(\omega)\;.$$

We bound ε^n in two ways; the first bound is valid for all ω in S_h and the second when $h\omega$ is small.

Bound for all $\omega \in S_h$

Inverting the sum (19), which is a $Z-$transform convolution, gives

$$\varepsilon^n = q_0^{-1} \sum_{m=1}^{n} p_{n-m}\left(\tilde{a}^m - \widehat{a}^m + e^m\right)$$

(the p_m's are defined in (14)). The coefficient $q_0 = \Delta t/4$ at any frequency and hence it follows from the stability assumption (H2) that

$$|\varepsilon^n| \le \frac{C}{h} \sum_{m=1}^{n} \left(|\tilde{a}^m - \widehat{a}^m| + |e^m|\right)$$

for some constant C. The bound (18) and standard results for the trapezoidal rule approximation of integrals then yield

$$|\varepsilon^n| \le C\left(\omega^2 Z_0(\omega) + \omega Z_1(\omega) + Z_2(\omega) + h^{-1}\sum_{m=1}^{n} |\tilde{a}^m - \widehat{a}^m|\right) \tag{20}$$

for all $\omega \in S_h$, where

$$Z_k(\omega) = \max_{t\in[0,T]} \left|\frac{\partial^k}{\partial t^k}\,\widehat{u}(\omega, t)\right|.$$

Bound for small $h\omega$

This follows from an analysis similar to those used by Jones [16] for the trapezoidal rule approximation of a first kind convolution Volterra integral equation. First we define β^n to be the solution of (19) without the terms $\tilde{a}^m - \widehat{a}^m$, i.e.

$$\sum_{m=1}^{n} q_{n-m}\,\beta^m = e^n\,. \tag{21}$$

Arguments similar to those used above yield the bound

$$|\varepsilon^n| \le |\beta^n| + Ch^{-1}\sum_{m=1}^{n} |\tilde{a}^m - \hat{a}^m| \tag{22}$$

for all $\omega \in S_h$. It remains to bound β^n when $h\omega$ is small.

Taking the second forward difference of (21) in n and using the triangle inequality gives

$$|\beta^{n+2}| \le \frac{|e^{n+2} - 2e^{n+1} + e^n|}{q_0} + \sum_{m=1}^{n+1} \frac{|q_{n-m+2} - 2q_{n-m+1} + q_{n-m}|}{q_0}\,|\beta^m| \tag{23}$$

and it follows from (18) and properties of first kind Bessel functions that

$$\frac{|q_{l+2} - 2q_{l+1} + q_l|}{q_0} \le \begin{cases} C(h\omega)^2 & \text{if } l \ge 1 \text{ or } l = -1 \\ 1 + C(h\omega)^2 & \text{if } l = 0. \end{cases}$$

for some constant C. Trapezoidal rule error estimates can be used to bound the first term in (23), resulting in the inequality

$$|\beta^{n+2}| \le |\beta^n| + \delta \sum_{m=1}^{n+1} |\beta^m| + \eta_0(\omega)$$

when h is sufficiently small, where $\delta = Ch^2(1+\omega)^2$, $\eta_0 = Ch^3(1+\omega)\eta$ and

$$\eta(\omega) = (1+\omega)\left(\omega^2 Z_0 + \omega Z_1 + Z_2\right) + Z_3 + Z_4\,.$$

Again arguments due to Jones [16] imply that if the sequence $\{g_n\}$ is defined by $g_0 = |\beta^1|$ and

$$g_n = g_{n-1} + 2\delta \sum_{m=0}^{n} g_m + \eta_0(\omega)$$

for $n \ge 1$, then $|\beta^{2m}| \le |\beta^{2m+1}| \le g_m$ for all $m \ge 0$. Taking the forward difference of this in n yields the two–step linear difference equation

$$(1-2\delta)\,g_{n+1} - 2\,g_n + g_{n-1} = 0$$

for $n \ge 1$. This has the solution $g_n = A\lambda_+^n + B\lambda_-^n$ where $\lambda_\pm = 1/(1 \pm \sqrt{2\delta})$ and $|A| \le |B| \le Ch^2\eta(\omega)$. The quantities $\lambda_\pm$ are bounded by $0 < \lambda_+ \le 1$ and

$$\lambda_-^n \le (1 + 2\sqrt{\delta})^n \le e^{\gamma(1+\omega)}$$

for some constant γ when $h(1+\omega)$ is sufficiently small. Hence g_n satisfies the bound

$$|g_n| \le Ch^2 e^{\gamma\,\omega}\,\eta(\omega)\,,$$

and together with (22) this implies that

$$|\varepsilon^n| \le C\left\{h^2 e^{\gamma\,\omega}\,\eta(\omega) + h^{-1}\sum_{m=1}^{n}|\tilde{a}^m - \widehat{a}^m|\right\} \tag{24}$$

for constants C and γ if $h(1+\omega)$ is sufficiently small. In particular, this estimate is valid when $\omega \in L_h = \{\omega : \omega \le |\ln h|/\gamma\}$ and h is small enough.

We now use the two bounds (20) (valid for all $\omega \in S_h$) and (24) (valid for $\omega \in L_h$ with small h) to obtain an estimate of the discrete Fourier norm of the error ε^n. The integral used in the Fourier norm $\|\varepsilon^n\|_F$ is split into a "low" frequency section $\omega \in L_h$ where inequality (24) is used, and a "high" frequency section $\omega \in S_h \backslash L_h$ where inequality (20) is used. The result is

$$\|\varepsilon^n\|_F \le h^{-1}\sum_{m=1}^{n}\|\tilde{a}^m - \widehat{a}^m\|_F + C\left(\int_{L_h}\left|h^2 e^{\gamma\,\omega}\eta(\omega)\right|^2\,d\omega\right)^{1/2} +$$
$$C\left(\int_{S_h\backslash L_h}\left|\omega^2 Z_0 + \omega Z_1 + Z_2\right|^2\,d\omega\right)^{1/2}.$$

We examine each term on the right hand side separately. Hypothesis (H1) and the Bramble–Hilbert theorem [4, Thm. 5] gives

$$h^{-1}\sum_{m=1}^{n}\|\tilde{a}^m - \widehat{a}^m\|_F \le Ch^{s-3/2}\max_{0\le t\le T}\|a(\cdot,t)\|_{s+1/2}\,.$$

The integral over low frequencies satisfies

$$\int_{L_h}\left|h^2 e^{\gamma\,\omega}\eta(\omega)\right|^2\,d\omega \le h^2\int_{L_h}|\eta(\omega)|^2\,d\omega$$

if s is sufficiently high to guarantee $\eta(\omega) \in L^2(\mathbb{R}^2)$, since $\omega \in L_h$ implies that $h^2 e^{\gamma\,\omega} \le h$. Following the arguments used by Thomeé [27], the high frequency integral satisfies

$$\int_{S_h\backslash L_h}|\eta_1(\omega)|^2\,d\omega \le \int_{S_h\backslash L_h}\left|\frac{\gamma\,\omega}{|\ln h|}\eta_1(\omega)\right|^2\,d\omega \le \frac{C}{|\ln h|^2}\,,$$

when $\omega\eta_1(\omega) \in L^2(\mathbb{R}^2)$ for $\eta_1 = \omega^2 Z_0 + \omega Z_1 + Z_2$. Combining the three bounds above yields the following result.

Theorem 11. *Under hypotheses (H1) and (H2) with m, $s \ge 4$, the global error for the 'polar' approximation scheme satisfies the bound*

$$\|\underline{U}^n - \underline{u}^n\|_h \le C/|\ln h|$$

as $h \to 0$ whenever $t^n < T$, where C is a constant.

Although we have demonstrated convergence, it is at a rate slower than any positive power of h! The reason is the exponential term $e^{\gamma\omega}$ in the small $h\omega$ bound. Unfortunately any standard method of bounding ε^n leads to a result like (24), but of course it may be possible to obtain a better small $h\omega$ bound in a different way. If u is sufficiently smooth and its Fourier transform decays fast enough when ω is large (like $e^{-\gamma\omega}$ or faster), then arguments similar to those used in the proof of Thm. 1 imply that the convergence rate is $O(h^2)$. Numerical tests on a finite plate indicate that the actual error is $O(h^\sigma)$ for σ somewhere between 1 and 2. Although not directly comparable with the convergence result, it is reasonable to expect that the extra problem of dealing with the boundaries of the plate adds to the error rather than reducing it.

5 Summary

In this article we have demonstrated the stability and convergence of a collocation scheme for the scalar RPIE (7) and attempted to give some insight as to why such time–marching schemes often exhibit numerical instabilities. A more detailed discussion of other stable and convergent collocation schemes can be found in [9].

The variational approach may well be more "natural" for RPIEs, and certainly allows a more sophisticated and general analysis of stability and convergence to be carried out [1, 2, 13, 14, 20]. However, we note that collocation schemes are much easier to implement than numerical schemes based on a variational approach, which involve the computation of five dimensional integrals over $\Gamma \times \Gamma \times (0, T)$, rather than just over Γ for collocation.

References

1. A. Bachelot and A. Pujols. Time dependent integral equations for the Maxwell system. *C. R. Acad. Sci. Paris Ser. I Math.*, 314:639–644, 1992.
2. A. Bamberger and T. Ha Duong. Formulation variationnelle espace–temps pour le calcul par potentiel retardé de la diffraction d'une onde acoustique (i). *Math. Meth. Appl. Sci.*, 8:405–435, 1986.
3. S. Bartels. Numerical analysis of retarded potential integral equations of electromagnetism, 1998. M.Sc. Thesis, Heriot-Watt University.
4. J. H. Bramble and S. R. Hilbert. Estimation of linear functionals on Sobolev spaces with applications to Fourier transforms and spline interpolation. *SIAM J. Numer. Anal.*, 7:112–124, 1970.
5. P. J. Davies. Numerical stability and convergence of approximations of retarded potential integral equations. *SIAM J. Numer. Anal.*, 31:856–875, 1994.
6. P. J. Davies. Stability of time-marching numerical schemes for the electric field integral equation. *J. Electromag. Waves & Appl*, 8:85–114, 1994.

7. P. J. Davies. A stability analysis of a time marching scheme for the general surface electric field integral equation. *Applied Numerical Mathematics*, 27:33–57, 1998.
8. P. J. Davies and D. B. Duncan. Averaging techniques for time marching schemes for retarded potential integral equations. *Applied Numerical Mathematics*, 23:291–310, 1997.
9. P. J. Davies and D. B. Duncan. Stability and convergence of collocation schemes for retarded potential integral equations. In preparation.
10. A. A. Ergin, B. Shanker, and E. Michielssen. The plane–wave time–domain algorithm for the fast analysis of transient wave phenomena. *IEEE Ant. Prop. Magazine*, 41(4):39–52, 1999.
11. A. A. Ergin, B. Shanker, and E. Michielssen. Fast analysis of transient acoustic wave scattering from rigid bodies using the multilevel plane wave time domain algorithm. *J. Acoust. Soc. Am.*, 107(3):1168–1178, 2000.
12. C. J. Gladwin and R. Jeltsch. Stability of quadrature rule methods for first kind Volterra integral equations. *BIT*, 14:144–151, 1974.
13. T. Ha Duong. On the transient acoustic scattering by a flat object. *Japan J. Appl. Math.*, 7:489–513, 1990.
14. T. Ha-Duong. On boundary integral equations associated to scattering problems of transient waves. *Z. Angew. Math. Mech.*, 76 (suppl. 2):261–264, 1996.
15. D. S. Jones. *Methods in Electromagnetic Wave Propagation.* Clarendon Press, Oxford, second edition, 1994.
16. J. G. Jones. On the numerical solution of convolution integral equations and systems of such equations. *Math. Comp.*, 15:131–142, 1961.
17. P. Linz. *Analytical and Numerical Methods for Volterra Equations.* SIAM, 1985.
18. M. Lu, J. Wang, A. A. Ergin, and E. Michielssen. Fast evaluation of two–dimensional transient wave fields. *J. Comp. Phys.*, 158:161–185, 2000.
19. Ch. Lubich. On the multistep time discretization of linear initial–boundary value problems and their boundary integral equations. *Numerische Mathematik*, 67:365–389, 1994.
20. A. Pujols. Time–dependent integral method for Maxwell equations. In G. Cohen, L. Halpern, and P. Joly, editors, *Mathematical and Numerical Aspects of Wave Propagation Phenomena*, pages 118–126. SIAM, 1991.
21. S. M. Rao and D. R. Wilton. Transient scattering by conducting surfaces of arbitrary shape. *IEEE Trans. Ant. Prop.*, 39:56–61, 1991.
22. S. M. Rao, D. R. Wilton, and A. W. Glisson. Electromagnetic scattering by surfaces of arbitrary shape. *IEEE Trans. Ant. Prop.*, 30:409–418, 1982.
23. B. P. Rynne. Instabilities in time marching methods for scattering problems. *Electromagnetics*, 6:129–144, 1986.
24. B. P. Rynne. Time domain scattering from arbitrary surfaces using the electric field equation. *J. Electromag. Waves & Appl.*, 5:93–112, 1991.
25. B. P. Rynne. The well–posedness of the electric field integral equation for transient scattering from a perfectly conducting body. *Math. Meth. Appl. Sci.*, 22:619–631, 1999.
26. B. P. Rynne and P. D. Smith. Stability of time marching algorithms for the electric field equation. *J. Electromag. Waves & Appl.*, 4:1181–1205, 1990.

27. V. Thomée. Convergence estimates for semi-discrete Galerkin methods for initial-value problems. In A. Dold and B. Eckmann, editors, *Numerische, insbesondere approximationstheoretische Behandlung von Funktionalgleichungen (Lecture Notes in Mathematics, 333)*, pages 243–262. Springer-Verlag, 1973.

hp-Adaptive Finite Elements for Maxwell's Equations

Leszek Demkowicz

Texas Institute for Computational and Applied Mathematics, The University of Texas at Austin, SHC 304, 105 Dean Keaton Street, Austin, TX 78712

Summary. This is a progress report on our current work on *hp*-adaptive finite elements for Maxwell's equations. I recall the main definitions [2], and show how the recent progress on the *hp* interpolation error estimates [3] has led to a fully automatic *hp* adaptivity based on the idea of minimizing the *hp*-interpolation error for a reference solution corresponding to a globally *hp*-refined grid [5]. Critical to the implementation of these ideas is a our new data structure for *hp* discretizations [4], supporting anisotropic refinements, and the calculation of prolongation operator for multigrid operations.

1 Introduction

The goal of this research is to work out a *fully automatic hp*-adaptive strategy that would deliver meshes with a minimum number of degrees of freedom (d.o.f.) in the full range of error level, especially in the preasymptotic range.

The presentation will be set up in two dimensions only, as I do not have 3D examples to show yet, and there are still major gaps in understanding the underlying theory of interpolation in 3D.

Critical to the *hp* discretizations for Maxwell's equations is the so called de Rham diagram relating two exact sequences of spaces, on both continuous and discrete levels, and corresponding commuting interpolation operators,

$$\begin{array}{ccccccccc}
\mathbb{R} & \longrightarrow & H^{1+\epsilon} & \xrightarrow{\nabla} & \boldsymbol{H}^{\epsilon} \cap \boldsymbol{H}(\mathrm{curl}) & \xrightarrow{\nabla\times} & L^2 & \longrightarrow & \mathbf{0} \\
\big\downarrow id & & \big\downarrow \Pi & & \big\downarrow \Pi^{\mathrm{curl}} & & \big\downarrow P & & \\
\mathbb{R} & \longrightarrow & \mathcal{P}^{p+1}_{p_e+1} & \xrightarrow{\nabla} & \mathbf{P}^{p}_{p_e} & \xrightarrow{\nabla\times} & \mathcal{P}^{p-1} & \longrightarrow & \mathbf{0}\,.
\end{array} \qquad (1)$$

All functional spaces are defined on the equilateral, master triangular element[1]

$$T = \left\{ (x_1, x_2) \,:\, x_2 > 0,\; x_2 < \sqrt{3}(x_1 + \frac{1}{2}),\; x_2 < -\sqrt{3}(x_1 - \frac{1}{2}) \right\}\,.$$

The polynomial spaces present in the diagram are defined as follows.

[1] We shall restrict our presentation to triangular elements only

- $\mathcal{P}^p_{p_e}$ - space of polynomials of order p, defined on the triangle, whose traces to edges e reduce to (possibly lower) order $p_e, e = 1, 2, 3$.
- $\boldsymbol{P}^p_{p_e}$ - space of vector-valued polynomials of order p, defined on triangle T, with traces of their tangential components on edges e of (possibly lower) order p_e.
- $\mathcal{P}^p$ - space of polynomials of order p, defined on triangle T.

In particular, $\mathcal{P}^p_{-1}$ denotes the space of polynomials of order p, vanishing on the boundary of the triangle, and $\boldsymbol{P}^p_{-1}$ stands for the space of vector-valued polynomials of order p, with the trace of the tangential component on the boundary equal zero. The assumption that edge orders p_e should not exceed polynomial order p, $p_e \leq p, e = 1, 2, 3$, is realized in practice by implementing the *minimum rule* that sets an edge order p_e to the minimum of orders p corresponding to the adjacent elements.

H^1-conforming interpolation. Given a function $u \in H^{1+\epsilon}(T)$, we define the corresponding interpolant $\Pi u := u^p \in \mathcal{P}^p_{p_e}(T)$ as the sum of three components,

$$u^p = u_1 + \underbrace{\sum_{e=1}^{3} u^p_{2,e}}_{u^p_2} + u^p_3 \,. \tag{2}$$

Step 1: Linear interpolant. We construct $u_1 \in \mathcal{P}^1(T)$ by the standard, linear interpolation at the vertices,

$$u_1 \in \mathcal{P}^1(T), \quad u_1(v) = u(v), \quad \text{for every vertex } v \,.$$

Step 2: Edge interpolants. For each edge e, we project trace of difference $u - u_1$, onto space $\mathcal{P}^{p_e}_{-1}(e)$ of polynomials of order p_e, vanishing at the edge endpoints, using the $H^{\frac{1}{2}}_{00}(e)$-norm.

$$\begin{cases} u_{2,e} \in \mathcal{P}^{p_e}_{-1}(e) \\ \|u_{2,e} - (u - u_1)|_e\|_{H^{\frac{1}{2}}_{00}(e)} \to \min \end{cases}$$

Next, we extend $u_{2,e}$ to a polynomial $u^p_{2,e}$ from $\mathcal{P}^p_{p_e}(T)$, vanishing along the two remaining edges. The sum of edge interpolants $u^p_{2,e}$ will be denoted by u^p_2.

Step 3: Interior interpolant. We project difference $u - u_1 - u^p_2$ onto space $\mathcal{P}^p_{-1}(T)$ of polynomials of order p, vanishing on ∂T, in the H^1-seminorm.

$$\begin{cases} u^p_3 \in \mathcal{P}^p_{-1}(T) \\ |u^p_3 - (u - u_1 - u^p_2)|_{H^1(T)} \to \min \end{cases}$$

H(curl)*-conforming interpolation.* Given a function $\boldsymbol{E} \in \boldsymbol{H}^{\epsilon} \cap \boldsymbol{H}(\text{curl}, T)$, we construct interpolant $\Pi^{curl}\boldsymbol{E} := \boldsymbol{E}^p \in \boldsymbol{P}^p_{p_e}(T)$ again in three steps,

$$\boldsymbol{E}^p = \boldsymbol{E}_1 + \underbrace{\sum_{e=1}^{3} \boldsymbol{E}^p_{2,e}}_{\boldsymbol{E}^p_2} + \boldsymbol{E}^p_3 \,. \tag{3}$$

Step 1: Whitney's (lowest order) interpolant. For each edge e, let $\boldsymbol{\phi}^e \in \boldsymbol{P}^1(T)$ denote the vector-valued, linear polynomial such that:

$$\phi^e_t = \boldsymbol{n} \times \boldsymbol{\phi}^e = \begin{cases} 1 \text{ along edge } e \\ 0 \text{ along the remaining edges.} \end{cases}$$

Here $\boldsymbol{n}$ is the outward normal unit vector to ∂T, and $\phi_t = \boldsymbol{n} \times \boldsymbol{\phi} = (-n_2)\phi_1 + n_1\phi_2$ denotes the trace of the tangential component of vector-valued function $\boldsymbol{\phi}$ to boundary ∂T. The Whitney interpolant is then defined as

$$\boldsymbol{E}_1 = \sum_e \left(\int_e E_t \right) \boldsymbol{\phi}^e \,.$$

Step 2: Edge interpolants. It follows from the construction of the Whitney interpolant that the trace of tangential component $\boldsymbol{n} \times (\boldsymbol{E} - \boldsymbol{E}_1)$ has zero average over each edge e. Thus, we can introduce a scalar-valued function ψ, defined on boundary ∂T, such that

$$\frac{\partial \psi}{\partial s} = \boldsymbol{n} \times (\boldsymbol{E} - \boldsymbol{E}_1), \quad \psi = 0 \text{ at vertices.}$$

For each edge e then, we project restriction $\psi|_e$ in the $H^{\frac{1}{2}}_{00}$ norm onto polynomials $\mathcal{P}^{p_e+1}_{-1}(e)$,

$$\begin{cases} \psi_{2,e} \in \mathcal{P}^{p_e+1}_{-1}(e) \\ \|\psi_{2,e} - \psi|_e\|_{H^{\frac{1}{2}}_{00}(e)} \to \min \,. \end{cases}$$

We take then any polynomial extension $\psi^{p+1}_{2,e} \in \mathcal{P}^{p+1}_{p_e+1}(T)$ that vanishes along the two remaining edges, and define the edge interpolant by the gradient of the extension,

$$\boldsymbol{E}^p_{2,e} = \boldsymbol{\nabla}\psi^{p+1}_{2,e} \in \boldsymbol{P}^p_{p_e}(T) \,.$$

Step 3: Interior interpolant. We solve the constrained minimization problem,

$$\begin{cases} \boldsymbol{E}^p_3 \in \boldsymbol{P}^p_{-1}(T) \\ \|\text{curl}(\boldsymbol{E}^p_3 - (\boldsymbol{E} - \boldsymbol{E}_1 - \boldsymbol{E}^p_2))\| = \|\text{curl}(\boldsymbol{E}^p_3 - (\boldsymbol{E} - \boldsymbol{E}_1))\| \to \min \\ (\boldsymbol{E}^p_3 - (\boldsymbol{E} - \boldsymbol{E}_1 - \boldsymbol{E}^p_2), \boldsymbol{\nabla}\phi) = 0 \quad \forall \phi \in \mathcal{P}^{p+1}_{-1}(T) \,. \end{cases}$$

The third operator P in (1) is simply the L^2 projection. The following interpolation error estimates have been proved in [3].

Theorem 12. *There exist constants C, dependent upon ϵ, but independent of p, p_e such that*

$$\begin{aligned}
||u - \Pi u||_{H^1(T)} &\leq C \inf_{v \in \mathcal{P}^p_{p_e}} ||u - v||_{H^{1+\epsilon}(T)} \\
&\leq C p_{min}^{-(r-\epsilon)} ||u||_{H^{1+r}(T)} , \\
||\boldsymbol{E} - \Pi^{curl} \boldsymbol{E}||_{H(\text{curl},T)} &\leq C \inf_{\boldsymbol{F} \in \boldsymbol{P}^p_{p_e}} (||\boldsymbol{E} - \boldsymbol{F}||^2_{H^\epsilon} + ||\text{curl}(\boldsymbol{E} - \boldsymbol{F})||^2)^{\frac{1}{2}} \\
&\leq C p_{min}^{-(r-\epsilon)} (||\boldsymbol{E}||^2_{H^r} + ||\text{curl}\boldsymbol{E}||^2_{H^r})^{\frac{1}{2}} ,
\end{aligned}$$

for every $r > 1$ and $0 < \epsilon < r$. Here $p_{min} = \min_e p_e$.

Remark 5.

1. The edge interpolant can be defined directly using the projection in a $H^{-\frac{1}{2}}(\partial T)$ norm, see [3] for details.
2. The described interpolation procedures generalize the *hp*-interpolation proposed in [7, 2]. The original *hp*-interpolation uses stronger, $H^1_0(e)$ and $L^2(e)$ - norms along the edges, and require more regularity ($u \in H^{3/2+\epsilon}(T)$). Consequently, they do not yield optimal (up to ϵ) convergence rates.

By the standard scaling argument, the *p*-interpolation estimates lead to ϵ-optimal *hp* interpolation error estimates over a mesh element K of size h and order $p = \min_e p_e$.

$$\begin{aligned}
||u - \Pi u||_{H^1(K)} &\leq C \frac{h^s}{p^{(r-\epsilon)}} ||u||_{H^{1+r}(T)} , \\
||\boldsymbol{E} - \Pi^{curl} \boldsymbol{E}||_{H(\text{curl},T)} &\leq C \frac{h^s}{p^{(r-\epsilon)}} (||\boldsymbol{E}||^2_{H^r} + ||\text{curl}\boldsymbol{E}||^2_{H^r})^{\frac{1}{2}} ,
\end{aligned}$$

where $s = \min\{p, r\}$.

The interpolation error estimates and the commuting diagram property are crucial in proving the *discrete compactness property* which in turn allows for proving stability of FE approximation and paves a way for proving exponential rates of convergence.

In the past several months though, we focused on a more practical issue, and tried to verify whether the idea of the *hp* interpolation might serve as a basis for establishing an *hp*-adaptive algorithm. Ideally, we would like to satisfy the following conditions.

- The algorithm should be *fully automatic*. Once the boundary-value problem is set up, and an initial mesh generated, executing the code should not require any interaction with the user.
- The algorithm should deliver exponential convergence rates, for both regular and singular solutions, as predicted by the existing theory [10].

– The algorithm should produce meshes that are always competitive with meshes obtained using h-adaptivity only. Here, from the practical point of view, the strongest competitor are quadratic elements.

I hesitate to add that, as always, we want the algorithm to have solid theoretical foundations and be free of *fetch in*, problem dependent, parameters tuned by the user.

I believe that the algorithm presented in the next section satisfies all of the conditions above. Unfortunately, we have not implemented it yet for the Maxwell equations (even in 2D...) and I will present it only in context of Laplace equation only.

2 The *hp*-algorithm

Given a domain $\Omega \in \mathbb{R}^2$ with boundary Γ consisting of three disjoint (Dirichlet, Neumann, and Cauchy) parts $\Gamma_1, \Gamma_2, \Gamma_3$, we focus on a standard variational problem,

$$\begin{cases} u \in \hat{u}_1 + V \\ b(u,v) = l(v), \quad \forall v \in V \,, \end{cases}$$

where $\hat{u}_1$ is a lift of Dirichlet boundary condition data u_1 defined on Γ_1, V is the space of test functions,

$$V = \{v \in H^1(\Omega) \, : \, v = 0 \text{ on } \Gamma_1\} \,,$$

and bilinear and linear forms are defined in the standard way,

$$\begin{aligned} b(u,v) &= \int_\Omega \left\{ \sum_{i,j=1}^{2} a_{ij}(\boldsymbol{x}) \frac{\partial u}{\partial x_i} \frac{\partial v}{\partial x_j} + \sum_{i=1}^{2} b_i(\boldsymbol{x}) \frac{\partial u}{\partial x_i} v + c(\boldsymbol{x}) uv \right\} d\boldsymbol{x} + \\ &\qquad \int_{\Gamma_3} \beta(\boldsymbol{x}) uv \, ds \\ l(v) &= \int_\Omega f(\boldsymbol{x}) v \, d\boldsymbol{x} + \int_{\Gamma_2 \cup \Gamma_3} g(\boldsymbol{x}) v \, ds \,. \end{aligned}$$

Coefficients a_{ij}, b_j, c, β, and *load data* f, g satisfy the usual assumptions assuring well-posedness of the problem. We shall assume that domain Ω can be covered exactly with (possibly curvilinear, isoparametric) finite elements or, equivalently speaking, we shall neglect the error due to the approximation of the geometry.

Suppose now that we are given an existing FE *hp* mesh. We shall call it the *coarse mesh*. We begin by performing a global *hp*-refinement, i.e. we split every element into four element sons, and we raise the order of approximation by one, uniformly throughout the whole mesh. We shall call the resulted mesh - the *fine mesh*. We solve then the problem of interest on the fine mesh using a two- grid solver. We shall discuss shortly the solver and related, practical issues in the next section.

The main idea is now as follows. Given the fine mesh solution, we determine optimal refinement of the coarse grid, by minimizing the *hp*-interpolation error for the fine mesh solution interpolated on the next, optimal coarse mesh. The *hp*-interpolation is understood here in a generalized sense explained next.

Step 1: Linear (bilinear) interpolant. We interpolate fine mesh solution $u_{h/2,p+1}$ at coarse grid nodes, and subtract resulting linear (bilinear for quads) interpolant $u^1_{h,p}$ from the fine mesh solution.

Step 2: Optimal refinement of edges. We first identify which edges should possibly be refined. This is done by computing for each edge the corresponding *edge interpolant*, obtained by projecting difference of $u_{h/2,p+1} - u^1_{h,p}$ on the space of edge shape functions $V_{hp}(e)$, vanishing at the edge endpoints,

$$w^e_{hp} \in V_{hp}(e)$$

$$\|w - w^1_{hp} - w^e_{hp}\| \to \min \; .$$

Ideally, we should use here the $H^{\frac{1}{2}}_{00}(e)$-norm. Instead, in the practical implementation, we have replaced it with a weighted H^1_0 norm,

$$\|u\|^2_e = \int_e \left(\frac{\partial u}{\partial s}\right)^2 \left(\frac{\partial s}{\partial \xi}\right)^{-1} ds = \int_0^1 \left(\frac{\partial u}{\partial \xi}\right)^2 d\xi \; .$$

Here $\boldsymbol{x} = \boldsymbol{x}(\xi)$, $\xi \in (0,1)$ is the FE parametrization for the edge, and

$$\frac{\partial s}{\partial \xi} = \sqrt{\sum_{i=1}^{2} \left(\frac{\partial x_i}{\partial \xi}\right)^2} \; .$$

Note that this weighted H^1_0 norm scales with the length of the edge the same way as the $H^{\frac{1}{2}}_{00}$ norm.

The projection problem is equivalent to solving a system of equations,

$$\begin{cases} u^e_{hp} \in V_{hp}(e) \\ (u^e_{hp}, v_{hp})_e = ((u_{h/2,p+1} - u^1_{hp}), v_{hp})_e \quad \forall v_{hp} \in V_{hp}(e) \; , \end{cases}$$

where inner product $(\cdot,\cdot)_e$ corresponds to the norm discussed above.

Having solved the projection problems for *each edge* in the mesh, we compute the corresponding projection errors (squared),

$$\eta_e = \|u_{h/2,p+1} - u^1_{hp} - u^e_{hp}\|^2_e \; ,$$

determine the maximum error $\eta_{max} = \max_e \eta_e$, and identify all edges e for which

$$\eta_e \geq \frac{1}{3}\eta_{max} \, .$$

These are the edges marked for a possible refinement.

Next, for each element edge e, we compute the corresponding *edge interpolants* corresponding to a p-refined edge, and a sequence of *competitive* h-refinements. By the competitive h refinement we understand here breaking the edge, and assigning such degrees of polynomial shape functions for the sons that the total number of *degrees of freedom* (d.o.f.) for the h-refined element is the same as for the p-refined element. Thus, if the original order of approximation for the edge is p_e, then the new orders of approximation for the broken edge sons p_e^1, p_e^2, must satisfy the condition

$$p_e^1 + p_e^2 = p_e + 1 \, .$$

For example, for a linear edge element, the choice is between the quadratic element and two linear elements, for a quadratic edge, the choice is between the cubic edge and two refinements resulting in quadratic/linear of linear/quadratic approximations for the edge. In general, one has to solve $p_e + 1$ projection problems.

Once the optimal refinements for the preselected edges have been determined, we compute the corresponding projection (interpolation) errors and redefine the edge error indicators as the rates with which the interpolation errors decrease,

$$\eta_e = \|u_{h/2,p+1} - u_{hp}^1 - u_{hp}^e\|_e^2 - \|u_{h/2,p+1} - u_{hp}^1 - u_{opt}^e\|_e^2 \, .$$

Here u_{opt}^e denotes the edge interpolant corresponding to the optimal refinement of the edge, producing the smallest projection error. Using the new edge error indicators, we use the same criterion as before, to establish which edges should be refined. At this point, we know their desired optimal refinement and the corresponding interpolant (projection) u_{opt}^e.

Step 3: Breaking elements. Elements adjacent to edges marked for h-refinement are next h-refined. Triangular elements are just broken into four congruent element sons. In the case of quadrilaterals, we determine first an *isotropy flag* for each element by analyzing the difference

$$u_{h/2,p+1} - u_{hp} \, .$$

Roughly speaking, anisotropic refinements are selected only if the error function exhibits essentially one-dimensional behavior. This allows for an optimal refinement along boundary layers and edge singularities (3D), see [5] for details.

Mesh regularity rules (we enforce one-irregular meshes) imply some involuntary h refinements of edges. This may apply to edges marked for p-refinement or marked for no refinement at all. We have to revisit those edges and determine optimal orders of approximation for the resulted edge elements. Please see again [5] for details.

In summary, the decision on breaking elements is made exclusively by determining optimal refinements of edges and, for quads, analyzing the possible anisotropy of the error function. If none of element edges is selected for h-refinement, the element may only be p-refined. One can, of course, think of academic counterexamples showing the optimality of an h-refinement, even for an element with unrefined edges. However, we think that the assumption is very reasonable in practical situations. Leaving the possibility of h- refinements in the second stage of the algorithm, complicates it beyond control, especially in 3D.

Step 4: Determining optimal orders of approximation for refined elements. After the third step we know the topology of the mesh and optimal order of approximation for all element edges that overlap with edges of the coarse mesh. Given this information we set to determine optimal orders of approximation for all involved elements. The algorithm is again local - we consider one element of the coarse grid at the time. If none of the element edges has been changed, we do not change the approximation. If any of the element edges has been modified in any way, the element may have been left without any h-refinement, or it may have been h-refined in one (quads only) or two directions. In either of the cases, we need to determine new order of approximation for the element or its sons. We do it by monitoring the *interpolation error decrease rates*, defined as

$$\eta_K = \frac{|u_{h/2,p+1} - u^1_{hp} - u^2_{opt} - u^3_{hp}|^2_{H^1(K)} - |u_{h/2,p+1} - u^1_{hp} - u^2_{opt} - u^3_{opt}|^2_{H^1(K)}}{\Delta \text{nrdof}} .$$

Here

- $u^2_{opt} = \hat{u}^e_{opt}$ denotes a lift of edge interpolants corresponding to the optimal edge refinements determined in the previous step.
- u^3_{hp} denotes the projection of difference $u_{h/2,p+1} - u^1_{hp} - u^2_{opt}$ onto $\mathcal{P}^p_{-1}(K)$,

$$\begin{aligned} & u^3_{hp} \in \mathcal{P}^p_{-1}(K) \\ & |u_{h/2,p+1} - u^1_{hp} - u^2_{opt} - u^3_{hp}|_{H^1(K)} \to \min \; . \end{aligned}$$

This is again equivalent to solving a system of equations

$$\begin{cases} u^3_{hp} \in \mathcal{P}^p_{-1}(K) \\ \displaystyle\int_K \boldsymbol{\nabla} u^3_{hp} \boldsymbol{\nabla} v_{hp} \, d\boldsymbol{x} = \int_K \boldsymbol{\nabla}(u_{h/2,p+1} - u^1_{hp} - u^2_{opt}) \boldsymbol{\nabla} v_{hp} \, d\boldsymbol{x}, \\ \qquad\qquad\qquad\qquad\qquad\qquad\qquad\qquad \forall v_{hp} \in \mathcal{P}^p_{-1}(K) \, . \end{cases}$$

Projection u^3_{hp} serves for a reference *element interior interpolant* with respect which the error decrease rate is defined.

- Δnrdof denotes the increase in the number of the element interior d.o.f., compared with the coarse mesh.

Finally, u^3_{opt} is determined by solving again the coarse element interior projection problem with $\mathcal{P}^p_{-1}(K)$ replaced with *space of element bubble functions* $V_{hp}(K)$ defined as the span of all (new coarse grid) basis functions with support in $\bar{K}$. This is done recursively, starting with the order of approximation for the element(s) interior(s) determined by orders for the edges and the *minimum rule.* The projection problem is solved, and local contributions to the projection error analyzed with the order p increased accordingly. [2] We monitor interpolation error decrease rate η_K defined above, and stop the iteration when two conditions are satisfied:

- The error corresponding to the new mesh must not exceed the error corresponding to the existing mesh.
- The rate is greater or equal than a prespecified minimum rate,

$$\eta_K \geq \eta_{min} \, .$$

In global terms, this is an investment problem. We invest with refinements only in those elements that produce sufficiently good error decrease rates. The minimum 'investment rate' η_{min} is determined by identifying the element for which the edge optimization has resulted in the largest error decrease. We then solve the p-optimization problem for this problem, increasing order p all the way to the order of the fine grid, and determining the maximum rate for the element. The minimum rate is then set to 1/3 of the maximum rate for the element.

In the end of the fourth step, we have determined the new coarse mesh and we continue the process until a stopping criterion is met. For the stopping criterion we use simply the error between the coarse and fine grid solutions measured in the H^1_0 seminorm, relative to the H^1_0-seminorm of the fine grid solution,

$$\frac{|u_{h/2,p+1} - u_{hp}|_{H^1(\Omega)}}{|u_{h/2,p+1}|_{H^1(\Omega)}} \leq \text{admissible error} \, .$$

Example: L-shape domain problem. As an illustrating example, we present a solution of the classical L-shape domain problem. The Laplace equation was solved with Dirichlet boundary conditions imposed using the exact solution. The initial mesh consisted of four triangles and one quad, all of second order, and the mesh optimization procedure was continued until an error tolerance of 0.001 (.1 percent) was reached. The corresponding optimal *hp* mesh is shown in Fig. 1. Fig. 2 presents the corresponding convergence history, compared with the convergence history for just h-adaptivity using quadratic elements

[2] This is really a miniature p-adaptivity problem. Given an approximation on the boundary, determine the best p in the interior...

only. The scale on the horizontal axis corresponds to the optimal, exponential convergence rates predicted by the theory of *hp* discretizations [10], and confirmed by the algorithm.

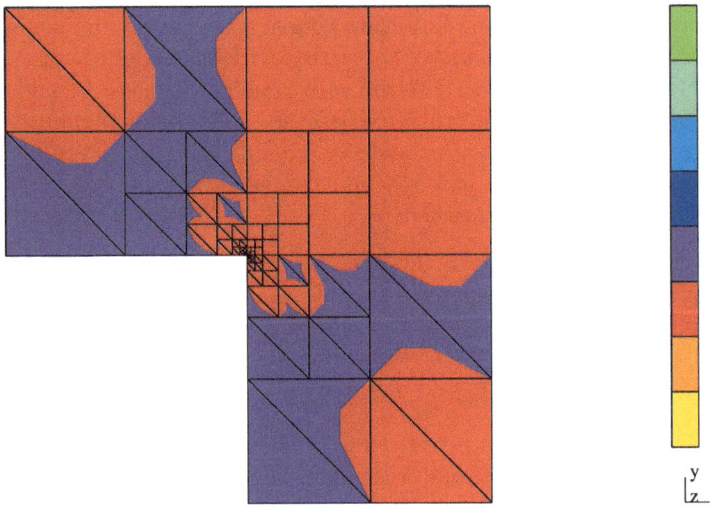

Fig. 1. L-shape domain example: final, optimal *hp* mesh

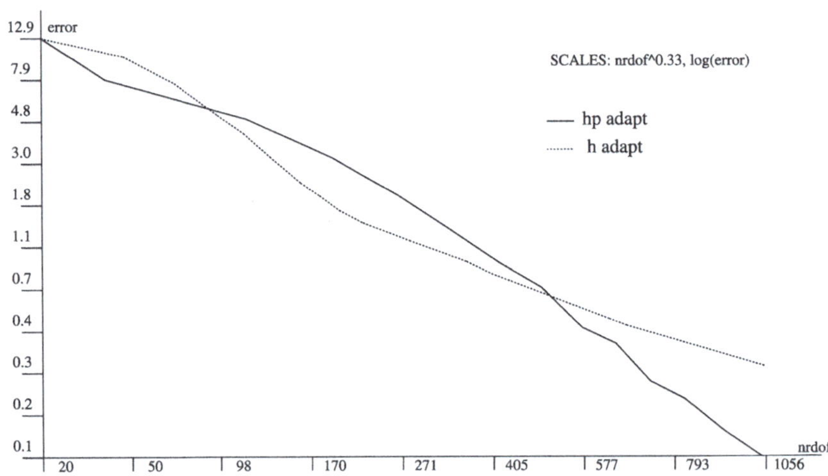

Fig. 2. L-shape domain example: Convergence history

Note that, in the preasymptotic range, the pure h-adaptive strategy can deliver slightly better results. This is related to the fact that our *hp* algorithm starts seeing the need for h-refinements only after a sufficient (spectral) information about the solution is available. In simple words, order p has to go up first ($p = 3$ minimum in this example) before the edges converging at the singular corner are broken. Asymptotically, however, *hp*-adaptivity always wins, as the fixed order discretization can only deliver algebraic rates of convergence. In problems with smoother solutions, the *hp* strategy wins earlier, see [5] for more examples.

Example: Radiation from a coil antenna into a dispersive medium. As a final illustration, we present a solution to the radiation problem for a single loop coil antenna wrapped around a metallic cylinder and radiating into a dispersive medium. The problem is axisymmetric, and it can be reduced to a single elliptic equation for transversal component E_ϕ of the electric field, in terms of cylindrical coordinates r, z. Due the exponential decay of the solution away from the antenna, both the metallic mandrel, and the soil domain can be truncated at a sufficiently large distance, and the equation is solved in a bounded domain shown, together with the corresponding solution, in Fig 3.

In Figures 4,5, we zoom on the antenna (compare the scale on the y-axis) and show the corresponding optimal *hp* mesh generated automatically by the *hp* algorithm. Note heavy h-refinements capturing the boundary layer in the mandrel (part of the mandrel is modeled as a medium with high conductivity, not just by a boundary condition), and a characteristic radiation pattern corresponding to a strong reflection from the mandrel.

3 Concluding Remarks

Multigrid solver and an integrated approach. I believe that a successful and competitive implementation of a fully automatic *hp* method must be based on a fully integrated approach where both the error estimation and mesh optimization are embedded into a multigrid solver, in a similar way as this has been done for h-adaptive methods [1, 6]. At present, we have only implemented a two-grid solver to solve the fine grid problem. The solver combines a block-Jacobi smoother on the *hp*-refined mesh, with a direct solve on the coarse mesh. Critical to a successful implementation of such a solver is the choice of patches defining blocks for the smoother and the definition of the prolongation operator. For the globally refined mesh, the corresponding finite element spaces are nested, and the definition of the prolongation operator is straightforward, although quite involved on the coding side. Implementation of a multi-grid solver on a sequence of *hp* meshes involves two technical issues.

Definition of the prolongation operator for non-nested hp meshes. The *hp* mesh optimization algorithm does not produce nested meshes. This can

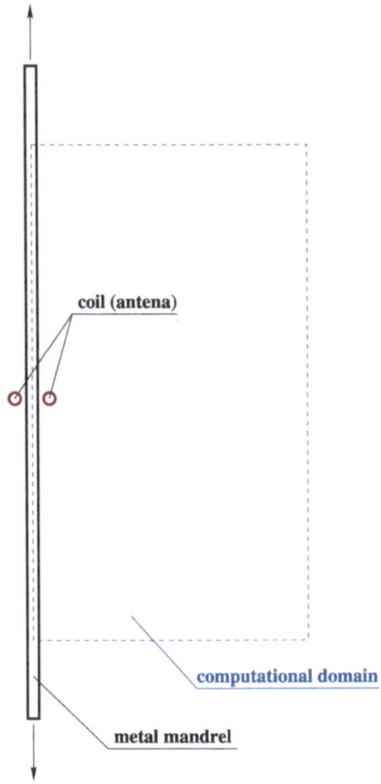

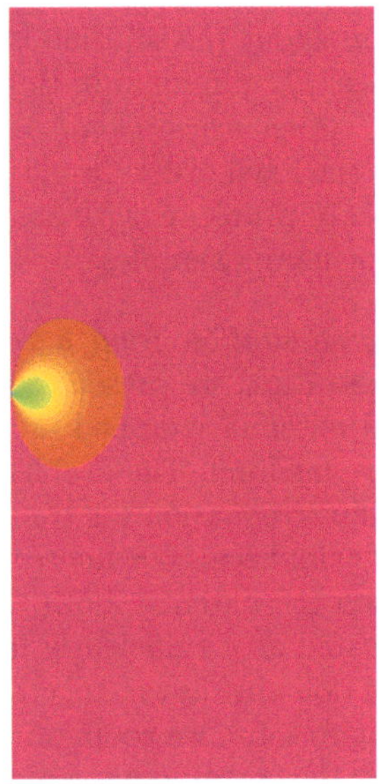

Fig. 3. Radiation from a coil antenna into a dispersive medium. Geometry of the domain and the solution

be easily seen in 1D case when, e.g. a quadratic element is refined into a linear and quadratic element sons. We hope to use again the idea of *hp*-interpolation to cope with this situation.

Selection of hp meshes for the solver. Some of the steps during the optimization process result in a large increase of number of d.o.f., some in only very few. It is not clear what criteria should be used when making the selection of intermediate meshes for the solver.

We are in process of implementing the solver and investigating the related technical questions, and we hope to report our findings in a forthcoming paper.

Is the solution on the fine grid necessary? In 3D, the global *hp* refinement increases the problem size by one order of magnitude. When the number of d.o.f. jumps from 200k to 2M, it is clear that one cannot survive without a two-grid solver at minimum. It is a natural question to ask whether one could replace the fine mesh solution with some other reference solution that would be suitable for the *hp* algorithm. In principle, one could use various postprocessed solutions. In our original work [8] we used Babuška's extraction

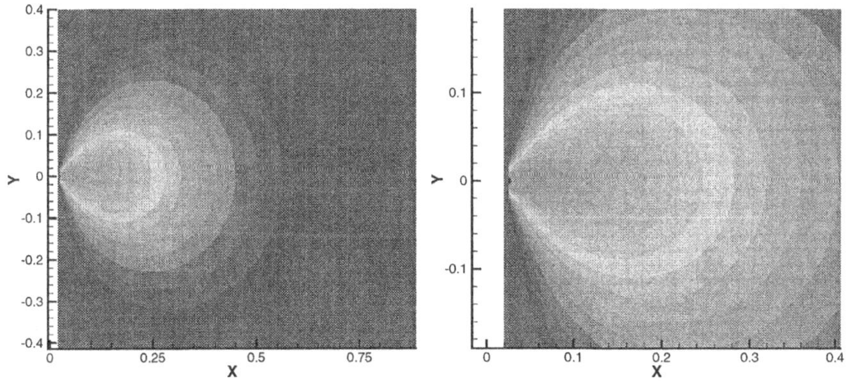

Fig. 4. Radiation from a coil antenna into a dispersive medium. Zooming on the antenna

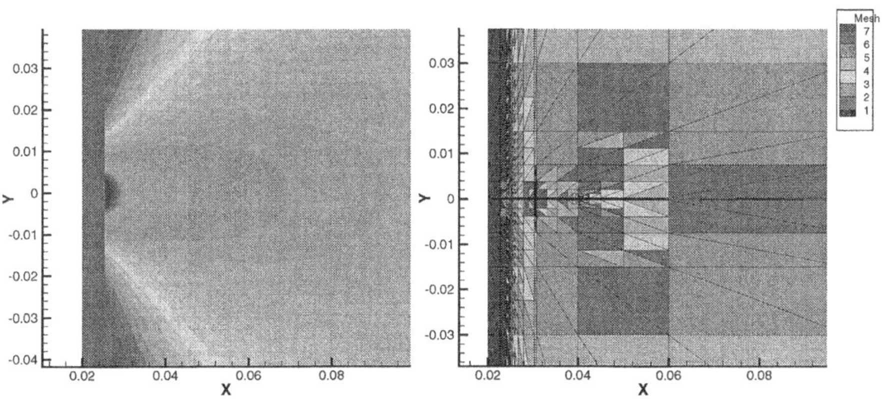

Fig. 5. Radiation from a coil antenna into a dispersive medium. Final zoom on the antenna and the corresponding optimal *hp* mesh

formulas, in boundary elements corresponding to equations of second type, standard postprocessing should yield the expected results. With the fine mesh concept, one could use just a few smoothing operations to pick up only higher spectral components of the solution, necessary for choosing between the h and p refinements. We are convinced, however, that from the efficiency point of view, it makes sense to converge fully to the fine mesh solution. This is because a majority of the CPU time is spent on calculating the stiffness and smoother matrices (+ prolongation matrix). Once these have been computed, the cost of the following matrix-vector multiplies is secondary. Besides, we can use the fine mesh solution as our final solution!

hp data structures. The work on 3D implementation of the presented algorithm, and necessary logical support for constructing the prolongation operators, have forced us to completely revise our previous data structures and rewrite our codes. Referring to [4] for details we mention two essential points of our new implementation:

- h-refinements are supported by 'growing' trees *not* for elements but rather for nodes (edges,faces, element interiors),
- nodal connectivities (including the constrained approximation info) are *not* stored but rather *reconstructed* from the nodal trees.

The necessity of the new concepts can be learned only from the struggle with a 3D implementation. In 2D, one can survive with alternative concepts.

Electromagnetics. 3D implementations. We are in process of implementing the algorithm in our 2D *hp* Maxwell code and for 3D elliptic problems. The principle is the same, first solve the problem on the fine grid, and then construct the next coarse grid by minimizing the *hp*-interpolation error for the fine grid solution. There are major theoretical differences between 2D and 3D situations. First of all, we have not generalized yet our 2D *hp*-interpolation results to the 3D setting. In 3D non- local interpolation must be used. First results in this direction have been recently reported in [9]. Lack of theoretical results does not prevent us from implementing a 3D version of the algorithm and we hope to report our 3D numerical results soon.

Acknowledgment

The work has been supported by Air Force under Contract F49620-98-1-0255. Results reported in this note have been obtained in collaboration with Ivo Babuška, Phillipe Devloo, Waldek Rachowicz, and with a help from my students: Andrzej Bajer and David Pardo.

References

1. R. Beck, P. Deuflhard, R. Hiptmair, R.H.W. Hoppe, and B. Wohlmuth, "Adaptive Multilevel Methods for Edge Element Discretizations of Maxwell's Equations", *Surveys on Mathematics for Industry*, **8**, 271-312, 1999.
2. L. Demkowicz, P. Monk, L. Vardapetyan, and W. Rachowicz. "De Rham Diagram for hp Finite Element Spaces" *Mathematics and Computers with Applications*, **39**, 7-8, 29-38, 2000.
3. L. Demkowicz and I. Babuška, "Optimal p Interpolation Error Estimates for Edge Finite Elements of Variable Order in 2D", *TICAM Report 01-11*, submitted to *SIAM Journal on Numerical Analysis*, 2001.
4. L. Demkowicz, D. Pardo, "The Ultimate Data Structure for Three Dimensional, Anisotropic *hp* Refinements', *TICAM Report*, in preparation.
5. L. Demkowicz, W. Rachowicz, and Ph. Devloo, "A Fully Automatic *hp* Adaptivity", *TICAM Report 01-28*, accepted to *Journal of Scientific Computing.*
6. G. Haase, M. Kuhn, U. Langer, "Parallel Multigrid 3D Maxwell Solvers", Johannes Kepler University Linz, SFB Report 99-23, 1999.

7. J.T. Oden, L. Demkowicz, W. Rachowicz and T.A. Westermann, "Toward a Universal *hp* Adaptive Finite Element Strategy, Part 2. A Posteriori Error Estimation", *Computer Methods in Applied Mechanics and Engineering*, **77**,113-180,1989.
8. W. Rachowicz, J.T. Oden, and L. Demkowicz, "Toward a Universal h-p Adaptive Finite Element Strategy, Part 3. Design of h-p Meshes," *Computer Methods in Applied Mechanics and Engineering*, **77**, 181-212, 1989.
9. J. Schoeberl, "Commuting Quasi-Interpolation Operators for Mixed Elements", Second European Conference on Computational Mechanics, Cracow, June 26-29, 2001.
10. Ch. Schwab, *p and hp-Finite Element Methods*, Clarendon Press, Oxford 1998.

Coupled Calculation of Eigenmodes

H.-W. Glock, K. Rothemund, and U. van Rienen

Institut für Allgemeine Elektrotechnik, Universität Rostock, Albert Einstein-Straße 2, D-18059 Rostock, Germany

Summary. In many technical applications the electromagnetic eigenmodes - frequency spectrum *and* field distributions - of rf-components are to be determined during the design process. There are numerous cases where the studied component is too complex to allow for a detailed enough simulation on usual servers. One way out of this situation is domain decomposition and parallelization of the field simulation. Yet, this demands for a parallelized solver. In our approach, we combine the use of commercial single processor-based software for the field simulation with a tool based on scattering parameter description. The studied component is decomposed in several sections. The scattering matrices of these sections are computed in time domain for instance with a FDTD field solver. A linear system is set up to compute the eigenfrequencies of the complete system and the field amplitudes at the internal ports common to a pair of sections. With the knowledge of these amplitudes the fields of the eigenmodes can be computed with help of a frequency domain field solver. This approach is denoted as Coupled S-Parameter Calculation (CSC). Some advantages of this procedure are the possibility of easy exploitation of symmetries in the studied components and the use of very different granularities in discretization of the single sections. This paper presents the method, its validation using a standard eigenmode solver and applications in the field of accelerator physics. Special attention is given to the eigenmodes of structures with slight deviations from rotational symmetry.

1 Introduction

The analysis of rf-components often is based on the knowledge of their eigenmodes, i.e. resonant electromagnetic fields. These are electromagnetic waves with harmonic time dependence. Pure – undamped – resonance is a special property of loss-free and therefore entirely closed structures. The discrete spectrum of resonance frequencies and corresponding spatial field distributions is found as solution of the boundary value problem of the Helmholtz equation:

$$\begin{aligned}(\nabla^2 + \underline{k}^2)\underline{\mathbf{E}}(\mathbf{r}) = 0\\ (\nabla^2 + \underline{k}^2)\underline{\mathbf{H}}(\mathbf{r}) = 0.\end{aligned} \tag{1}$$

with vanishing tangential electric fields at the resonator's boundaries. Analytic solutions exist only for very simple resonator geometries like the rectangular or cylindrical box.

For practically used rf-components discrete Helmholtz equations are solved, often called Curl-Curl-equation, with numerical methods like the Finite Integration Technique (FIT)[13], [4] or the Finite Element Method (FEM) [6]. With FIT the following homogeneous discrete Curl-Curl-equation results (see e.g. [14], [11]):

$$(\text{rot}\ \frac{1}{\underline{\mu}}\ \text{rot}\ - \omega^2 \underline{\varepsilon}')\,\underline{\mathbf{E}} = 0 \quad \Longleftrightarrow \quad (\tilde{C}\underline{D}_{\mu}^{-1}C - \omega^2 \underline{D}_{\varepsilon})\underline{e} = 0. \qquad (2)$$

In this paper we will refer to the solution of the homogenous Curl-Curl-equation in one single computational run for a complete rf-structure as 'direct' eigenmode computation. There are several commercial programs at hand for that purpose. In our studies we used the eigenvalue solver E of MAFIA [17] and the eigenvalue solver of MicroWaveStudio^{TM} [19], two CAE-tools based on FIT for the solution of the homogeneous Curl-Curl-equation.

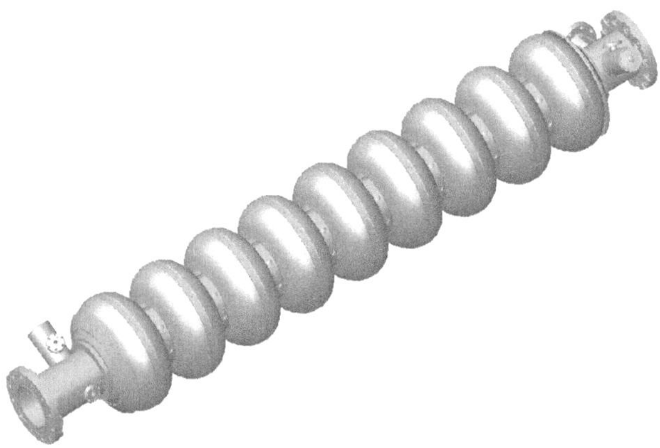

Fig. 1. CAD plot of the TESLA 9-cell cavity with attached input coupler and HOM-couplers. Plot: courtesy to DESY

It has to be remarked, that any technical device violates the criterion of vanishing losses. Nevertheless are the field properties of devices with surface resistances as small as those of commonly used metals very similar to the ideal loss-free case, which justifies the broad use of the concept of eigen-resonances.

Yet, some rf-structures of practical interest are too complex to allow for a detailed enough simulation on usual computing environments. Figures 1 and 2 show two examples from accelerator physics, here from the TESLA-project [2], [20]. The TESLA-cavity shown in Fig. 1 with one input coupler and two so-called HOM-couplers (HOM standing for Higher Order Modes, which are parasitic modes excited by passing electron bunches) is one example which both in a direct eigenmode computation and in an S-parameter

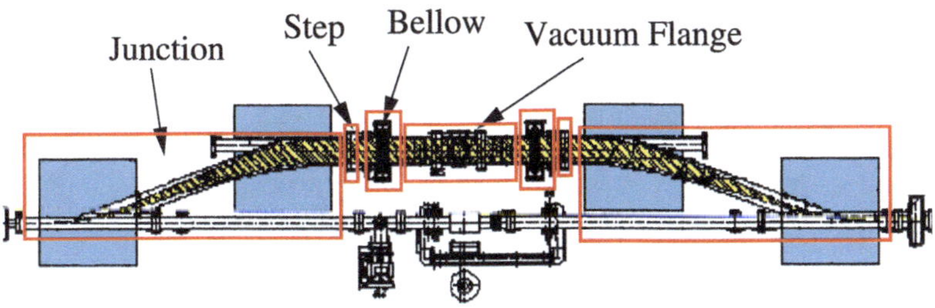

Fig. 2. CAD drawing of bunch compressor II of the TESLA Test Facility (TTF). The shaded areas are inside a dipole magnet, each. The lines shall indicate the different sections chosen for the CSC calculations. Drawing: courtesy to DESY

calculation of the entire structure would need a complete three-dimensional discretization. In an appropriate resolution this amounts to a number N of grid points of about $15 \cdot 10^6$. With our CSC approach we can make use of the fact that the cavity itself is cylindrically symmetric and thus allows for a two-dimensional discretization if it is taken as one section in CSC. Then the cavity needs only about $72 \cdot 10^3$ grid points, each of the coupler sections modelled in 3D approximately $0.5 \cdot 10^6$ grid points. Thus we solve for three problems which are two to three orders of magnitude smaller than the one in the direct computation. The CSC approach allows even to calculate the properties of chains of several accelerating cavities, reported in [9], [5].

Another example of a complex structure for which the rf-properties are of interest is the beam line of one of the so-called bunch compressors in the TESLA-FEL [2]. Parasitic modes might also be excited in such kind of structure and therefore the eigenmode computation is one among others in the design phase. The bunch compressor shown in Fig. 2 has an overall length of 4.5 m but also tiny substructures of 5 mm which have to be taken into account. It has certain symmetries which can be exploited. This results in 107k grid points needed for the junction, 296k mesh points for the bellow, 34k for the step and 72k grid points for discretizing the flange section. A minimal discretisation of the entire structure would take 1.5M points, though neglecting several geometrical details.

In the CSC approach, we combine the use of any commercial software for the field simulation with a tool based on scattering parameter description. This tool has been realized in form of a *Mathematica*™ [18] package. The principal idea is explained below and sketched in Fig. 3: usually we would choose some CAE-tool and first discretize the geometry of the studied structure, next use its eigenmode solver and finally display eigenfrequencies, visualize the fields and compute secondary quantities of interest. For those structures which would need too long computing times and/or too big storage capacities on a given computing environment the computation shall be partitioned into several smaller ones. With CSC the studied component is de-

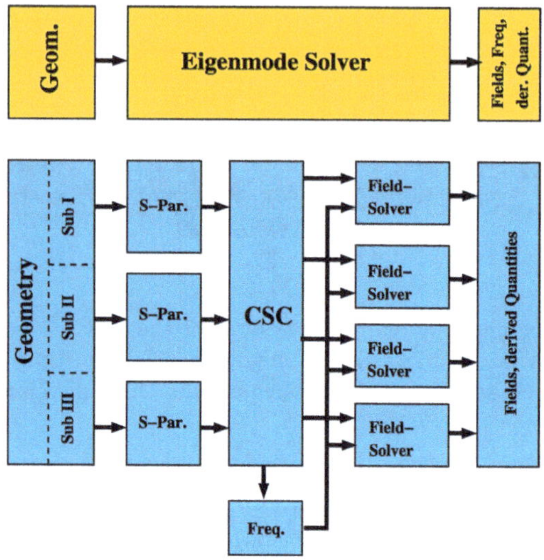

Fig. 3. Calculation of eigenmodes using CSC: The geometry is divided into several sections. Then their S-parameters are determined. CSC combines them and provides the eigenfrequencies and wave amplitudes at the section's ports to compute the field distributions of the eigenmodes in the sections

composed in several parts to which we will reference as 'sections'. Again we choose a CAE-tool (e.g. MAFIA [17], MicroWaveStudio^{TM} [19], HFSS [16]), now using it several times to discretize each of the sections first. In a second step its FDTD[1] solver [15], [13] is used to compute separately the broadband scattering parameters of each section. Those are exported to the kernel of the CSC procedure which in a third step computes the eigenfrequencies of the entire rf-structure from the S-parameters of the single sections. Besides the frequencies the wave amplitudes in all common ports of the sections are delivered. In a fourth step the wave amplitudes are used as excitation in a FDFD[2] solver [10] which finally also yields the field distributions of the eigenmodes. These are displayed in a fifth step and further secondary quantities may be computed in the post-processor of the chosen CAE-tool.

The partitioning can be chosen in such a way that geometrical properties like symmetries or repetitions of certain sections or groups of sections can be exploited. Also, very simple sections would allow for analytical solution.

[1] FDTD = Finite Difference Time Domain: the general time dependent Maxwell's equations, an inititial value problem, is solved by explicit time-stepping scheme.

[2] FDFD = Finite Difference Frequency Domain, i.e. excited time-harmonic problems in which the inhomogeneous Helmholtz resp. Curl-Curl-equation is solved.

2 Theory

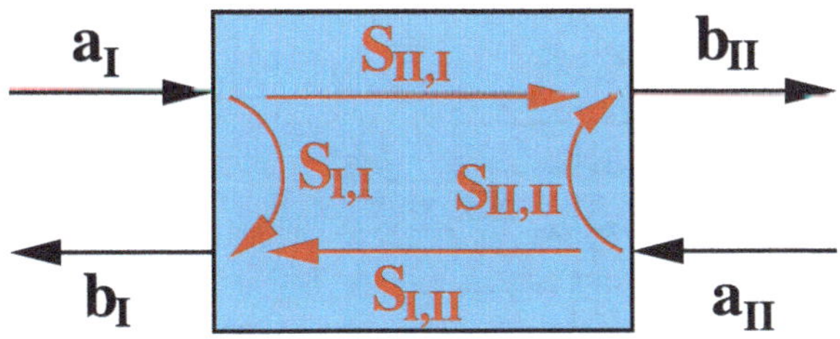

Fig. 4. Description of signal reflection and transmission between two ports: Amplitudes $\underline{\mathbf{a}}$ and $\underline{\mathbf{b}}$ of incident and scattered waves are related by the scattering parameters, collected in the so-called S-matrix

Wave reflection and transmission within the ports of any rf-system can be described by scattering-(S)-parameters. S-parameters are complex quantities hence both expressing amplitude and phase correlation of incident and scattered waves. Let us regard as most simple model any object with two single-moded ports as shown in Fig. 4. Then as an example the scattered - outgoing - wave b_I at port I is the superposition of the reflected signal at port I $S_{I,I}\, a_I$ and the transmission $S_{I,II}\, a_{II}$ from port II to port I. Doing this superposition a linear behaviour of the object is presumed. This linearity is essential for the entire theory presented here. Nevertheless it is a property common to a wide range of technically used devices. Setting up vectors $(a_I, a_{II}, \ldots)^T$ of all (complex) wave amplitudes of the incident waves and $(b_I, b_{II}, \ldots)^T$ of all scattered ones with a common port indexing one usually expresses the scattering properties of an object by its scattering matrix:

$$\begin{pmatrix} b_I \\ b_{II} \\ \vdots \end{pmatrix} = \begin{pmatrix} S_{I,I} & S_{I,II} & \cdots \\ S_{II,I} & S_{II,II} & \cdots \\ \vdots & \vdots & \ddots \end{pmatrix} \begin{pmatrix} a_I \\ a_{II} \\ \vdots \end{pmatrix} . \tag{3}$$

The S-matrix depends on the geometry *and* the actual frequency. For the majority of objects, that do not contain non-reciprocal materials like magnetized plasma or ferrites, the S-matrix is symmetric. If a loss-free object is considered, which preserves electromagnetic power, the S-matrix furthermore possesses unitarity (e.g. [3]).

Each port may carry more than a single mode, as it is commonly known from waveguide theory. The modes may not necessarily propagate. Every mode represents an independent channel and therefore requires its own row and column in the S-matrix.

Even more general is a n-port structure. It is represented by a $(n \times n)$-matrix $\mathbf{S}$ if all ports are single-moded. Again, the entry S_{ij} describes the

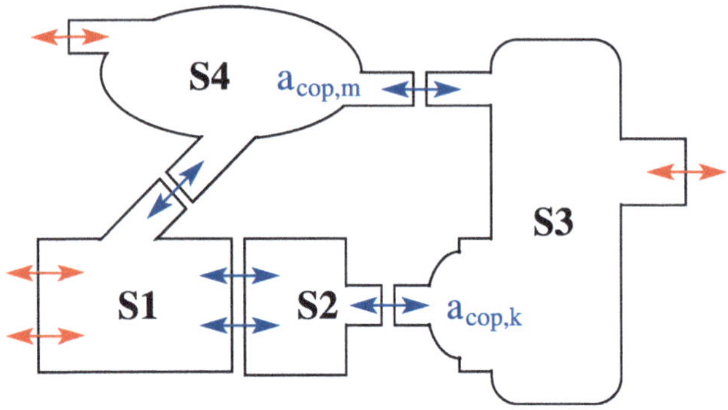

Fig. 5. RF-system with four sections, several external and internal ports

transmission of a signal from port j to port i. If there are ports with more than a single mode an appropriate port-mode-indexing is needed, and n is to understood as total number of modes summed up over all ports. Beside this, such objects are treated like any others. (In fact, this appears to be the usual situation.) Later on, we will discuss groups of sections combined to an entire structure. Then, the k-th section's S-matrix forms the relation

$$\mathbf{b}_k = \mathbf{S}_k(\omega)\ \mathbf{a}_k = \begin{pmatrix} S_{11}(\omega) & \cdots & S_{1n}(\omega) \\ \vdots & & \vdots \\ S_{n1}(\omega) & \cdots & S_{nn}(\omega) \end{pmatrix}_k \mathbf{a}_k . \tag{4}$$

2.1 Concatenation of S-matrices

Our goal is the determination of the overall S-matrix of some studied rf-system by determination of the S-matrices of sections. For a linear arrangement of sections we could use the well known concatenation of S-matrices described e.g. in [12]; for further details see [3].

Yet, for general arrangements of the sections (an example is given in Fig. 5) this method is not applicable, which is why we introduce another procedure, the Coupled S-Parameter Calculation (CSC). A comparison between the traditional way of handling S-matrices and CSC is given in [7] or [8].

2.2 CSC-procedure

The following procedure is based on a description that is also found in [1]. We arrange all N S-matrices $\mathbf{S}_k$ of the sections in one block diagonal matrix $\mathbf{S}$. This matrix couples vector $\mathbf{a}$ of all incident signals with vector $\mathbf{b}$ of all scattered signals:

$$\mathbf{b} = \mathbf{S}\,\mathbf{a} = \begin{pmatrix} \mathbf{S_1} & & \mathbf{0} \\ & \ddots & \\ \mathbf{0} & & \mathbf{S_N} \end{pmatrix} \begin{pmatrix} \mathbf{a_1} \\ \vdots \\ \mathbf{a_N} \end{pmatrix} = \begin{pmatrix} \mathbf{b_1} \\ \vdots \\ \mathbf{b_N} \end{pmatrix} \tag{5}$$

Next, vector $\mathbf{a}$ of incident signals is reordered such that all signals coming from outside the rf-system are collected in a vector $\mathbf{a}_{inc}$ whereas those signals that are incident in one of the sections, but outgoing from a neighbouring one are collected in the coupling vector $\mathbf{a}_{cop}$. A permutation matrix $\mathbf{P}$ links this reordered vector with $\mathbf{a}$:

$$\mathbf{a} = \mathbf{P} \begin{pmatrix} \mathbf{a}_{cop} \\ \mathbf{a}_{inc} \end{pmatrix} . \tag{6}$$

A second permutation matrix $\mathbf{F}$ is used to describe the feedback in the system, namely the fact that the outgoing signals of one port are incident at another port. All signals leaving the rf-system are kept identically.

Now we order the scattered signals in the same manner as the incoming signals in (6), too. This is achieved by applying the inverse permutation leading to

$$\begin{pmatrix} \mathbf{a}_{cop} \\ \mathbf{a}_{sct} \end{pmatrix} = \mathbf{P}^{-1}\,\mathbf{F}\,\mathbf{b}. \tag{7}$$

Combining (4), (6) and (7) results in the following matrix vector equation:

$$\begin{pmatrix} \mathbf{a}_{cop} \\ \mathbf{a}_{sct} \end{pmatrix} = \underbrace{\mathbf{P}^{-1}\,\mathbf{F}\,\mathbf{S}\,\mathbf{P}}_{\mathbf{G}} \begin{pmatrix} \mathbf{a}_{cop} \\ \mathbf{a}_{inc} \end{pmatrix} \tag{8}$$

where $\mathbf{G} = \mathbf{P}^{-1}\,\mathbf{F}\,\mathbf{S}\,\mathbf{P}$ describes the structure of the whole system.

The system matrix $\mathbf{G}$ can be split into four block matrices $\mathbf{G}_{ij}$ where the dimensions of the $\mathbf{G}_{ij}$ correspond to the dimensions of $\mathbf{a}_{cop}$, $\mathbf{a}_{inc}$ and $\mathbf{a}_{sct}$. Thus, (8) can be written as the following linear system:

$$\mathbf{a}_{cop} = \mathbf{G}_{11}\,\mathbf{a}_{cop} + \mathbf{G}_{12}\,\mathbf{a}_{inc}, \tag{9a}$$

$$\mathbf{a}_{sct} = \mathbf{G}_{21}\,\mathbf{a}_{cop} + \mathbf{G}_{22}\,\mathbf{a}_{inc}. \tag{9b}$$

Herein $\mathbf{a}_{inc}$ and $\mathbf{a}_{sct}$ represent the incident and scattered waves at the external ports. Hence the coupling between these signals is given by

$$\mathbf{a}_{sct} = \left(\mathbf{G}_{21}\,\left(\mathbf{1} - \mathbf{G}_{11}\right)^{-1}\,\mathbf{G}_{12} + \mathbf{G}_{22}\right)\mathbf{a}_{inc}. \tag{10}$$

The overall S-matrix, denoted by $\mathbf{S}^{(T)}$, can thus be written as

$$\mathbf{S}^{(T)} = \mathbf{G}_{21}\,\left(\mathbf{1} - \mathbf{G}_{11}\right)^{-1}\,\mathbf{G}_{12} + \mathbf{G}_{22}. \tag{11}$$

So far our considerations are applicable both for structures with open ports and resonators. The analysis of open structures, like cavities for particle acceleration with attached couplers, is described e.g. in [5].

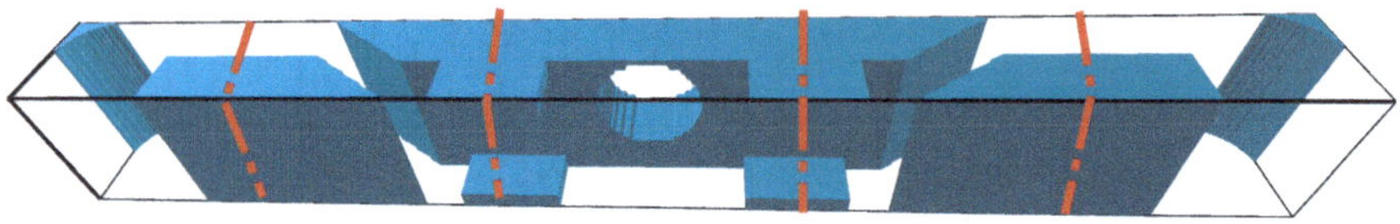

Fig. 6. Geometry of the test resonator used to verify the CSC-formulation. Vertical lines indicate separation planes

2.3 Determination of Eigenmodes

In the case of a resonator problem we have no open (external) ports. Thus $dim(\mathbf{a}_{sct}) = dim(\mathbf{a}_{inc}) = 0$ holds and only the coupling between the internal ports remains. Simultaneously the block matrices $\mathbf{G}_{12}$, $\mathbf{G}_{21}$ and $\mathbf{G}_{22}$ vanish and (9) reduces to

$$(\mathbf{1} - \mathbf{G}_{11}(\omega_0))\,\mathbf{a}_{cop} = \mathbf{0} \tag{12}$$

which has to be fulfilled. This is only valid for discrete frequencies ω_0 which are the resonant frequencies aimed for. In order to find ω_0 the eigenvalues of $(\mathbf{1} - \mathbf{G_{11}}(\omega))$ have to be determined repeatedly. Since the dimension of this matrix is equal to the number of internal signals, which is usually below 10^2, this can be done with standard solvers. A resonant frequency is found if at least one eigenvalue equals zero.

In that case the vector $\mathbf{a}_{cop}$ holds the amplitudes of the waveguide modes at the location of the internal ports. $\mathbf{a}_{cop}$ is found as basis of the eigenspace - usually one vector - of the eigenvalue zero which is the kernel of the matrix $(\mathbf{1} - \mathbf{G}_{11}(\omega_0))$.

Afterwards the eigenfields are computed in separate runs of a field solver for each section of the rf-system, using exciting waves with amplitudes given by $\mathbf{a}_{cop}$.

3 Validation

To validate the CSC technique a test model was set up, which is shown in Fig. 6. It was split into five sections. The S-parameters of each single section were calculated using the MAFIA time domain solver T3 in a frequency range of 1.2 GHz to 1.75 GHz. This frequency range was chosen since it keeps effort for comparative runs with a direct eigenmode solver within a reasonable range.

To monitor the in- and outgoing waves a field decomposition at the particular port planes is performed, using 2D-waveguide-modes that are determined from the port geometry except for an arbitrarily chosen factor. This is a standard procedure, that is prepared in the MAFIA time domain solver T3. In this procedure it has to be guaranteed that corresponding modes of neighbouring sections have the same orientation although they are calculated twice

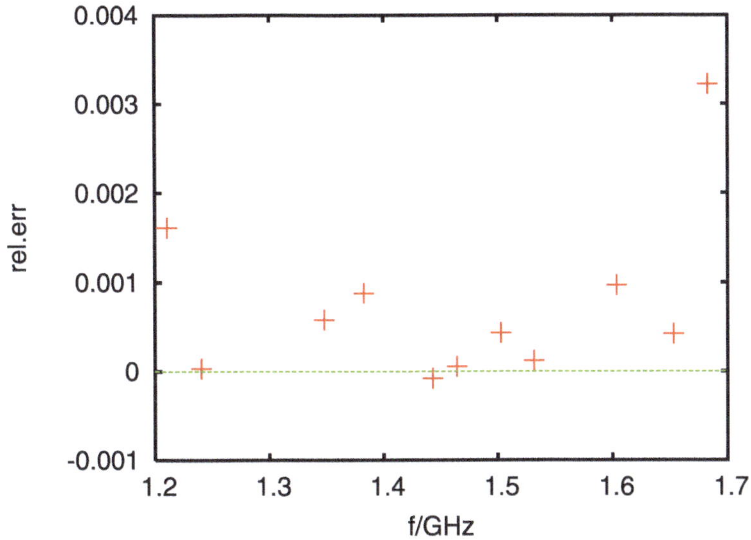

Fig. 7. Relative error of eigenfrequencies of the test geometry found by the CSC-technique and by MAFIA's E-module (double precision) in the frequency range of 1.2 GHz to 1.75 GHz

in different runs with probably different meshes. For that purpose a criterion was implemented to orientate all waveguide modes in a consistent manner. It applies either an integral $\int \mathbf{E}(x,y)\,dx\,dy$ on the waveguide mode fields, or, if this due to mode symmetry yields zero, an integral $\int x\,\mathbf{E}(x,y)\,dx\,dy$, which then is nonvanishing. Comparing signs of these integrals for waveguide modes of different runs allows to decide whether field orientation is identical or not. In the latter case odd-oriented fields are fliped. (In fact, it is not possible to find a "correct" mode orientation from the solution of the two-dimensional eigenproblem leading to the waveguide mode pattern. The value of the arbitrary factor is fixed by a power normalization, whereas the sign remains undetermined. Since any S-parameter is given with respect to *oriented* mode patterns of incident and scattered waves, both pattern *and* orientation are to be defined.)

A *Mathematica* [18] package is used to compute both the resonant frequencies and the amplitudes at each internal port. Then, the field distributions of the eigenmodes in all sections are calculated using MAFIA's frequency domain solver W3. The input power and phase of the incident waves are given by CSC and have to be specified for the particular ports of the section. During the field calculation again the same orientation of all 2D port modes is guaranteed like in the S-parameter calculation by the test procedure described above. This results in the resonant field pattern of the according resonator section.

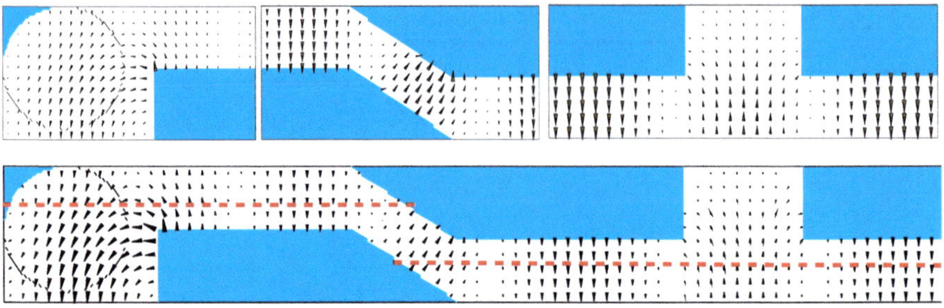

Fig. 8. The electric field of the $f = 1.44258$ GHz eigenmode computed by the CSC (upper plots) and directly with MAFIA's E-module (lower plot)

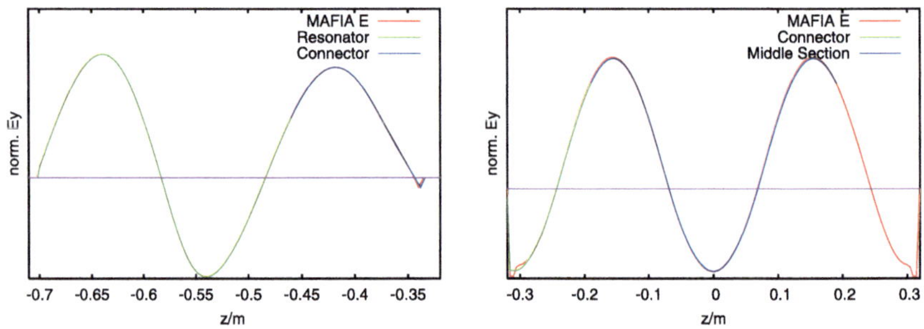

Fig. 9. E_y along some path (shown dashed in Fig. 8) in z-direction shown for the eigenmode with frequency $f = 1.44258$ GHz

Alternatively, the time domain solver T3 can be used to derive the eigenmode field patterns. In this case monochromatic waves of the given frequency are excited at every port with the appropriate amplitudes and phases. It must be secured that the calculation stabilized to steady state before the resonant fields are evaluated. This can be checked by testing whether all outgoing signals reached a constant amplitude. Nevertheless it may be difficult to decide whether the steady state is reached if resonant substructures with high quality factor are calculated, which makes this solver type less attractive for such objects.

Since this example was chosen for validation the eigenmodes of the complete structure were also computed with MAFIA's eigenmode solver E. As shown in Fig. 7 the frequencies found by the CSC technique match very well those calculated directly.

Figure 8 shows the field distribution of the two eigenmodes with frequency $f = 1.44258$ GHz calculated with CSC and the direct eigenmode solver MAFIA-E. Figure 9 compares the y-component of the electric field along some path crossing the port plane perpendicularly. As can be seen, the field distribution computed by MAFIA's E-module and CSC match extremely well.

4 Applications

4.1 Resonant Fields in Bunch Compressor

The Bunch Compressor II of the TESLA Test Facility was chosen as a real-life problem - in fact this structure gave first reason for the CSC development. This device is intended to compress electron bunches using a dispersive beam path arranged by four dipole magnets. A sketch of the geometry is shown in Fig. 2.

The straight part of the chicane is not modeled here because its lowest cut-off frequency of approximately 3.0 GHz lies well above the examined frequency range of 1.7...2.2 GHz. In this frequency range only one mode propagates and the junction sections act as frequency dependent closings of the structure, so, only their reflection parameter has to be calculated.

The sectioning is chosen in a way, that bellow and flange section are symmetric. Thus only one side of these devices needs to be excited in the T3-runs since their S-matrices reflect the geometrical symmetry in identical reflection parameters. This is not the case for the step section, that is geometrically asymmetric and therefore needs two runs to achieve full information.

CSC found 23 eigenfrequencies in the range of 1.7 GHz to 2.2 GHz. The wave amplitudes at the cutting planes of first eight modes are shown in Fig. 10.

Fig. 11 shows the field distribution of mode number eight, demonstrating the steady behaviour at the border of two neighbouring sections.

4.2 Eigenmode orientation in structures with weak deviations from rotational symmetry

The study of structures that experienced small deformations from ideal rotational symmetry is of practical interest with respect to the effect of tolerances during the manufacturing process of accelerating cavities. Furthermore it is an illustrative example for the capabilities of the CSC method.

The accelerating structure shown in Fig. 1 is simplified in order to need only a single deformation parameter, resulting in a symmetrical waist structure (Fig. 12). Both its outer diameters (100 mm radius) and its connecting cones are ideally circular, whereas the inner part is of a slightly elliptical shape (x-semiaxis: 55.7 mm, y: 54.5 mm). Chains of three to six identical sections are under consideration, each section rotated about a certain angle compared to its neighbour. The chain is closed with short planes at both ends. The determination of a single sections S-matrix, calculated with 86 k mesh

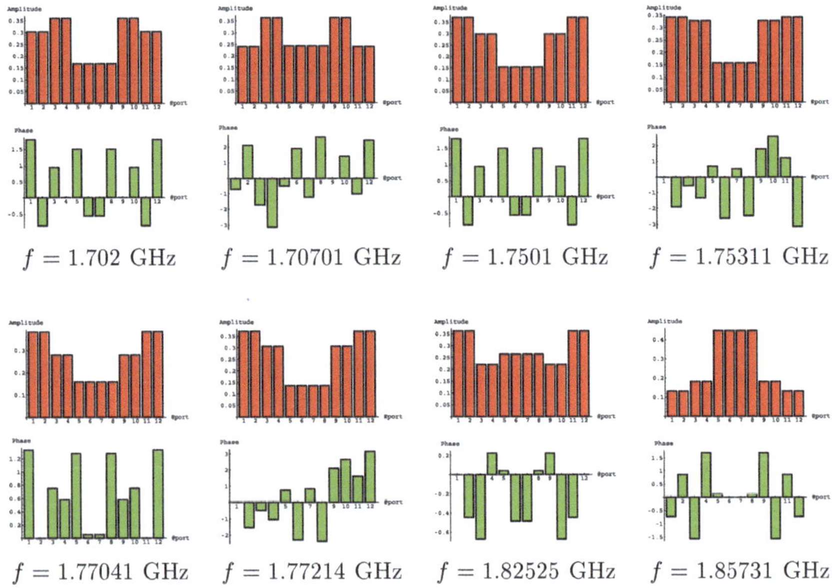

Fig. 10. Schematic illustration of the first eight resonant eigenmodes of the FEL bunch compressor beam line shown in Fig. 2. Representation uses amplitudes (dark) and phases (light grey bars) of the waves incident at the 12 internal ports of the entire resonator. Modes can be clearly divided in even shapes, symmetric to the mid plane, and odd ones with complementary phases adding up to π

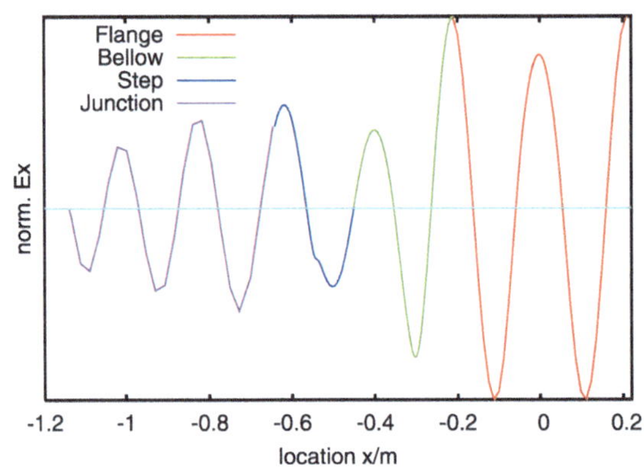

Fig. 11. E_y in the FEL bunch compressor beam line along a path in z-direction along the centre line of the middle sections, shown for the eigenmode number 8 with $f = 1.85731$ GHz

points in MicroWaveStudio^{TM}, takes approximately two hours on a 500MHz PIII. It is the only field solving calculation needed for this analysis.

Both polarizations of the TE11 circular waveguide mode are considered in the S-matrices. The monopol TM01 and higher waveguide modes are omitted, which is justified by their coupling to the TE11 modes, showing to be extremly small. Therefore only dipole-like eigenmodes of the chains arc cal culated. In a cavity of ideal rotational symmetry each of those modes is degenerated and appears in two polarizations. Due to the ellipticity both polarizations have slightly different eigenfrequencies, even without any rotation. Introducing a nonvanishing angular shift between the sections, which is assumed to be constant along the chain, will certainly cause a mixing of the polarisations. It is the aim of this analysis to study the behaviour of field polarization in skewed chains.

The CSC algorithm seems to be ideally suited for this purpose: All the sections are identical, so a field solving run is needed just once. Shorts and rotations are described analytically, the latter ones by virtual sections of vanishing length with a S-matrix (two modes, notation like Fig. 12, φ: angle of rotation):

$$\begin{pmatrix} b_{I,x} \\ b_{I,y} \\ b_{II,x} \\ b_{II,y} \end{pmatrix} = \begin{pmatrix} 0 & 0 & \cos\varphi & \sin\varphi \\ 0 & 0 & -\sin\varphi & \cos\varphi \\ \cos\varphi & -\sin\varphi & 0 & 0 \\ \sin\varphi & \cos\varphi & 0 & 0 \end{pmatrix} \begin{pmatrix} a_{I,x} \\ a_{I,y} \\ a_{II,x} \\ a_{II,y} \end{pmatrix} \tag{13}$$

Furthermore are the wave amplitudes that belong to the resonances and that are found directly as solutions of the eigenproblem (12) by themself sufficient for the determination of polarizational orientation: the field vector angle is found as $\arctan(Ay/Ax)$ (regarding the sign that is caused by temporal phase shifts of 0 or π). Here the difference of field angles at beginning and end of the skewed chain is analyzed. It is worth to mention that there appear only linear field polarizations at the cavity's ends; a circular or elliptical polarization cannot be observed.

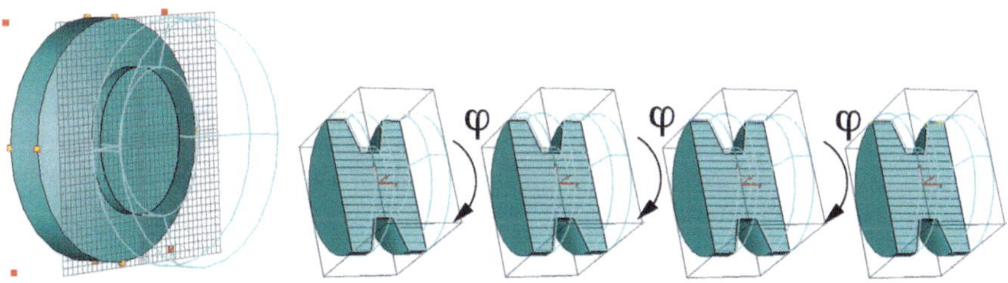

Fig. 12. Cavity with rotational cross section at the outer radius and weakly elliptically deformed waist (left figure); a chain of those cells, each twisted by the angle φ with respect to the previous one (right; no intermediate space in calculation)

The largest cavity under consideration forms a chain of six sections, six rotations (one additional at the end in order to reach the reference frame used at the entrance) and two shorts, so in total 14 objects, 12 with 4×4-matrices, the shorts with 2×2-matrices. Resonance search in a frequency range from 1.31 GHz to 1.69 GHz (a somewhat arbitrary choice after a pres-caning of "interesting" regions) with 760 frequency points for 46 different angles (0° to 90° in steps of 2°) takes approximately 2 hours on a 400 MHz G3 Macintosh, using the CSC-package with *Mathematica*™ 4.1. This should be compared with 46 3D field solving runs, each at least with 500k mesh points, assuming that the single section's discretization would be sufficient to map a slightly elliptical shape with odd angles onto the grid. To the best of author's knowledge those 3D calculations have been not undertaken yet.

Fig. 13 show the results of the calculation, which may be summarized as follows: The field vector experiences a rotation between the ends of a skewed cavity of up to some ten degrees. The rotation angle depends on the mode index. It increases with rising structure rotation angle between two sections and growing length of the chain. Often to observe is a "crossing" of modes that lead to a change of the frequency order of the mode pattern. Chains with odd numbers of cells cause zero field rotation both for 0° and 90° cell-to-cell rotation, whereas for a even numbered length a field rotation maximum may appear at 90° structure rotation, too. Pattern appearing in shorter chains are found similarily, but compressed to smaller ranges of structure rotation, in the results of longer resonators.

Dipole modes are responsible for the dominant contribution to the total deflecting forces on a particle beam in an accelerator. The application of the CSC procedure may give some further understanding for the field behaviour in special environments. Other constellations, different from the extremely regular one described above, could be considered with reasonable numerical effort, too.

5 Conclusion and Outlook

The presented CSC-technique allows the decomposition of long and/or complex rf-structures into smaller sections. Combination of their individual S-parameters yields the S-matrix of the complete structure (for structures with open ports) as well as the eigenfrequencies and the corresponding field distributions in the case of a resonant problem. CSC has been validated using a test example for comparison with some well established code for direct eigenmode calculation.

The main advantage of the CSC technique is the possibility to calculate the S-parameters of each section on different machines and/or in separate runs. It also allows to exploit possible symmetries or repetitions of particular sections or groups of sections; geometrically simple sections may be described analytically. Furthermore, CSC easily allows to specify frequency ranges of

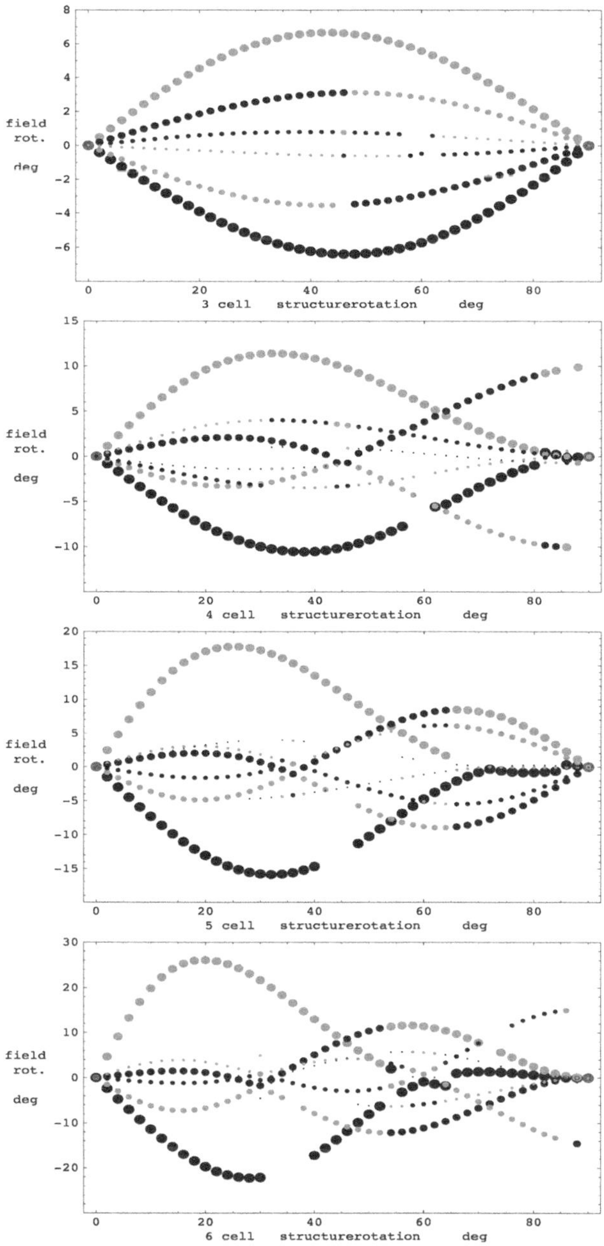

Fig. 13. Difference of field orientation angle between both cavity ends versus cell-to-cell twist angle. Thickest points indicate lowest mode index (different grey shadings for better distinction). Continuous curves are frequently composed by points from modes with different indices, due to changes in the order of frequency. Gaps in the curves are caused by points laying outside the calculation's frequency range

the eigenmodes searched for, which is not always possible in other eigenmode solvers.

If a direct calculation is possible, the overall effort of the CSC-procedure is of course essentially higher: One has to keep in mind that CSC was developed to provide the possibility of eigenmode computation for those structures which *cannot* be handled directly. Yet, even for less complex structures CSC might be the preferred choice if the design of sections shall be optimized: CSC allows to create an S-parameter data base for rf components. If a parameter study is carried for one section while all others are kept unchanged, CSC might save a tremendous amount of cpu time by making use of such a data base.

A typical example of a structure being far too large for usual calculation is the TTF bunch compressor. CSC was applied to calculate resonant fields. Since the eigenmodes cannot be validated with other codes on some computational environment available for us it is planned to validate them by measurements.

Special advantage is given if components have analytical descriptions. The example of rotated components illustrates that then even multi-dimensional parameter variations are in the range of the CSC capabilities.

References

1. Thomas-Alfred Abele. Über die streumatrix allgemein zusammengeschalteter mehrpole. *Mitteilung aus dem Institut für Hochfrequenztechnik der Technischen Hochschule Aachen*, pages 262–268, 1960. Über die Streumatrix allgemein zusammengeschalteter Mehrpole.
2. R. Brinkmann, K. Flöttmann, J. Rossbach, P. Schmüser, N. Walker, and H. Weise. Tesla — the superconducting electron-positron linear collider with an integrated x-ray laser laboratory - technical design report. Technical report, Deutsches Elektronen Synchrotron, March 2001.
3. Robert E. Collin. *Foundations for Microwave Engineering.* McGraw-Hill, 2. edition, 1992. sect. 4.9.
4. M. Dohlus, P. Thoma, and T. Weiland. Stability of finite difference time domain methods related to space and time discretisation. *IEEE-MTT*, submitted. Stability of Finite Difference Time Domain Methods related to Space and Time Discretisation.
5. H.-W. Glock, K. Rothemund, and U. van Rienen. CSC – a procedure for coupled s-parameter calculations. In *Proceedings of the Compumag 2001.* to be published.
6. W. Hackbusch. *Elliptic Differential Equations – Theory and Numerical Treatment*, volume 18 of *Springer Series in Computational Mathematics.* Springer, Berlin, 1993.
7. K. Rothemund, H.-W. Glock, M. Borecky, and U. van Rienen. Eigenmode calculation in long and complex rf-structures using the coupled s-parameter calculation technique. In *Proceedings of the ICAP 2000.* Eigenmode Calculation in Long and Complex RF-Structures Using the Coupled S-Parameter Calculation Technique.

8. K. Rothemund, H.-W. Glock, M. Borecky, and U. van Rienen. Eigenmode calculation in long and complex rf-structures using the coupled s-parameter calculation technique. Technical report. TESLA-Report 2000-33.
9. K. Rothemund, H.-W. Glock, and U. van Rienen. Calculations of HOMs in TESLA-cavities using the coupled s-parameter calculation method. In *Proceedings of the Particle Accelerator Conference 2001*. Calculations of HOMs in TESLA-Cavities Using the Coupled S-Parameter Calculation Method, to be published.
10. Peter Thoma. *Zur numerischen Lösung der Maxwell'schen Gleichungen im Zeitbereich.* PhD thesis, Technische Hochschule Darmstadt, 1997.
11. U. van Rienen. *Numerical Methods in Computational Electrodynamics - Linear Systems in Practical Applications*, volume 12 of *Lecture Notes in Computational Science and Engineering.* Springer, 2001.
12. Ursula van Rienen. Higher order mode analysis of tapered disc-loaded waveguides using the mode matching technique. *Particle Accelerators*, 41:173–201, 1993.
13. T. Weiland. Eine Methode zur Lösung der Maxwellschen Gleichungen für sechskomponentige Felder auf diskreter Basis. *AEÜ*, 31:116–120, 1977. Eine Methode zur Lösung der Maxwellschen Gleichungen für sechskomponentige Felder auf diskreter Basis.
14. T. Weiland. High precision eigenmode computation. *Part. Accel.*, 56:61, 1996. High precision eigenmode computation.
15. K.S. Yee. Numerical Solution of Initial Boundary Value Problems Involving Maxwell's Equations in Isotropic Media. *IEEE-AP*, 14:302–307, 1966.
16. *HP HFSS.* Agilent Technologies.
17. *MAFIA V4.0.* CST GmbH, Büdinger Straße 2a, D-64289 Darmstadt, Germany.
18. *Mathematica V4.1.* Wolfram Research, 100 Trade Center Drive, Champain, IL 61820-7237,USA.
19. *Micro Wave Studio V3.0.* CST GmbH, Büdinger Straße 2a, D-64289 Darmstadt, Germany.
20. `http://tesla.desy.de/`.

Boundary Element Methods for Eddy Current Computation

Ralf Hiptmair

Seminar für Angewandte Mathematik, ETH Zürich, CH-8092 Zürich
ralf@hiptmair.de

Summary. This paper studies numerical methods for eddy current problems in the case of homogeneous, isotropic, and linear materials. It provides a survey of approaches that entirely rely on boundary integral equations and their conforming Galerkin discretization. The pivotal role of potentials is discussed, as well as the topological issues raised by their use. Direct boundary integral equations and the so-called symmetric coupling of the integral equations corresponding to the conductor and the non-conducting regions is employed. It gives rise to coupled variational problems that are elliptic in suitable trace spaces. This implies quasi-optimal convergence of Galerkin boundary element schemes.

1 Introduction

The typical setting for an eddy current problem distinguishes between the bounded domain $\Omega_C \subset \mathbb{R}^3$ that is occupied by conductors and its complement $\Omega_e := \mathbb{R}^3 \setminus \overline{\Omega}_C$, the non-conducting air region. In most technical applications the conductors possess a piecewise smooth surface. More precisely, Ω_C is to be a curvilinear Lipschitz polyhedron in the sense of [26, Sect. 1]. Some excitation triggers electromagnetic fields, and the goal of a numerical simulation is to determine the behavior of these fields in space and time.

In this presentation the focus will exclusively be on boundary element methods based on a triangulation of $\Gamma := \partial\Omega_C$. Therefore we make the fundamental assumption that all material parameters are scalar constants both in Ω_C and Ω_e. In addition, we take advantage of the linearity of the materials and switch to the frequency domain, where all fields are supposed to display a sinusoidal temporal variation with a fixed angular frequency $\omega > 0$. This leads to equations for complex field amplitudes only depending on position in space.

We will also confine the possible excitations to two commonly used types:

1. The total current in a loop of the conductor is prescribed (lumped parameter excitation). Here, by loop we mean a connected component of Ω_C, whose first Betti number is equal to 1. Homeomorphic images of a torus are typical examples.
2. The complex amplitude of some compactly supported solenoidal "ghost current" $\mathbf{J}_s$ is prescribed in air region. This models a coil or antenna

creating a magnetic field, but not affected by the reaction fields. We rule out that $\mathbf{J}_s$ penetrates the surface of the conductor.

The behavior of electromagnetic fields is governed by Maxwell's equations. Yet, instead of using those, in special situations simplified *quasistatic models* supply sufficiently accurate approximations to the true fields (cf. [29, 3], [10, Ch. 8]). In particular, the eddy current model may be chosen, if the size of the conductor is small compared with the electromagnetic wavelength but large compared to the skin depth[1]. Formally, the eddy current model arises from Maxwell's equations by dropping the displacement current. In the frequency domain the eddy current model for scaled non-dimensional complex field amplitudes reads

$$\mathbf{curl}\,\mathbf{E} = -i\mu_r\mathbf{H} \quad \text{in } \mathbb{R}^3 \quad , \quad \mathbf{curl}\,\mathbf{H} = \begin{cases} \tau^2\mathbf{E} & \text{in } \Omega_C \,, \\ \mathbf{J}_s & \text{in } \Omega_e \,. \end{cases} \tag{1}$$

According to the assumptions made above the relative permeability μ_r is a positive constant on Ω_C and equal to 1 in the air region Ω_e. The normalized inverse skin depth τ is a positive constant, too[2]. The first equation is called Faraday's law, the second (reduced) Ampere's law. The equations have to be supplemented by the decay conditions

$$\mathbf{H}(\mathbf{x}) = O(|\mathbf{x}|^{-1}) \quad , \quad \mathbf{E}(\mathbf{x}) = O(|\mathbf{x}|^{-1}) \quad \text{uniformly for } |\mathbf{x}| \to \infty \,. \tag{2}$$

The eddy current model is a *magnetoquasistatic model* in the sense that the electric field energy is neglected. Necessarily, this breaks the symmetry between electric and magnetic fields that is a prominent feature of the full Maxwell equations.

Obviously, we cannot expect a solution for $\mathbf{E}$ to be unique, because it can be altered by any gradient supported in Ω_e and will still satisfy the equations. The solution for $\mathbf{H}$ will not be affected. This reflects the fact that in a magnetoquasistatic model $\mathbf{E}$ is relegated to the role of a fictitious quantity. Imposing the constraints

$$\operatorname{div}\mathbf{E} = 0 \quad \text{in } \Omega_e \quad \text{and} \quad \int_{\Gamma_k} \mathbf{E}\cdot\mathbf{n}\,dS = 0 \,, \tag{3}$$

where Γ_k, $k = 1,\dots,L$, are the connected components of Γ, will restore uniqueness of the solution for $\mathbf{E}$. Thus, one can single out a physically meaningful electric field in Ω_e. However, this is rather a gauging procedure, i.e. the selection of a representative from an equivalence class of meaningful fields

[1] This crude rule ignores the impact of the shape of Ω_C, which might rule out the application of the eddy current model, even if the rule of thumb might hold [10, Sect. 8.1.2]

[2] $\tau = \sqrt{\sigma\mu_0\omega}L$, σ conductivity, μ_0 permeability of empty space, L size of conductor

[40], than part of the generic eddy current model. When devising a numerical scheme, we should target **H** as principal variable.

The goal of this paper is a treatment of boundary element methods for (1) from a mathematical point of view. It aims to illustrate principles and to convey the fundamental differences of the two principal approaches, which are known as "**E**-based" and "**H**-based". For the sake of brevity proofs of most technical results will be skipped.

In engineering the use of boundary elements for the solution of eddy current problems has a long history [12, 11, 52, 56, 59, 46, 39, 38, 45, 58, 37, 57]. However, the mathematical analysis of resulting equations has remained cursory at best. In addition, all attempts to achieve a coupling between Ω_C and Ω_C were based on the ad-hoc non-symmetric coupling approach that so far has defied a rigorous analysis on non-smooth surfaces.

2 Spaces and Traces

All developments in this paper will be consistently set in a variational framework. The Hilbert spaces, on which the variational approach rests, have a very concrete physical meaning as spaces of fields with finite energy. Let $\Omega \subset \mathbb{R}^3$ be a generic domain, not necessarily bounded. The natural Hilbert space for magnetic fields with finite total energy on Ω is

$$\boldsymbol{H}(\mathbf{curl};\Omega) := \{\mathbf{V} \in \boldsymbol{L}^2(\Omega),\ \mathbf{curl}\,\mathbf{V} \in \boldsymbol{L}^2(\Omega)\}\ ,$$

equipped with the graph norm (cf. [32, Ch. 1]). In the context of the eddy current model the energy associated with the electric field is measured only by its **curl**. Of course, also the mean dissipated energy has to be finite, which entails square integrability over Ω_C, but in Ω_e the L^2-norm of the field need not be bounded. Therefore, weighted Beppo-Levi type spaces (cf. [31])

$$\boldsymbol{W}(\mathbf{curl},\Omega) := \{\frac{\mathbf{V}(\mathbf{x})}{\sqrt{1+|\mathbf{x}|^2}} \in \boldsymbol{L}^2(\Omega),\ \mathbf{curl}\,\mathbf{V} \in \boldsymbol{L}^2(\Omega)\}$$

are the proper choice for **E**. The property that their energy only depends on certain derivatives is characteristic for potentials. For them weighted spaces have to be used, for instance the standard Beppo Levi space (cf. [50, Sect. 2.5.4])

$$W^1(\Omega) := \{\frac{\Phi(\mathbf{x})}{\sqrt{1+|\mathbf{x}|^2}} \in \boldsymbol{L}^2(\Omega),\ \operatorname{grad}\Phi \in \boldsymbol{L}^2(\Omega)\}\ .$$

For each of the above spaces, the restrictions to Ω of smooth functions that are compactly supported in $\mathbb{R}^3$ form dense subsets.

Thanks to this density property we may wonder how to extend certain restrictions of smooth functions onto boundaries to continuous and surjective trace mappings. Now, assume that the boundary $\partial\Omega$ is compact and endowed with an exterior unit normal vectorfield $\mathbf{n} \in L^\infty(\partial\Omega)$. The pointwise restriction of functions in $C^\infty(\bar{\Omega})$ spawns the standard trace $\gamma : W^1(\Omega) \mapsto H^{\frac{1}{2}}(\partial\Omega)$. However, the relevant traces for electromagnetic fields are tangential traces of vectorfields. We can distinguish between the tangential components trace $\gamma_{\mathbf{t}}$ for $\mathbf{U} \in \boldsymbol{C}^\infty(\bar{\Omega})$ defined by $(\gamma_{\mathbf{t}}\mathbf{U})(\mathbf{x}) = \mathbf{n}(\mathbf{x}) \times (\mathbf{U}(\mathbf{x}) \times \mathbf{n}(\mathbf{x}))$ for almost all $\mathbf{x} \in \partial\Omega$, and the twisted tangential trace $(\gamma_\times \mathbf{U})(\mathbf{x}) := \mathbf{U}(\mathbf{x}) \times \mathbf{n}(\mathbf{x})$.

In eddy current computations we usually face piecewise smooth surfaces. This profoundly affects the smoothness of restrictions, in particular of tangential traces. Just keep in mind that even for smooth vectorfields their tangential traces will feature discontinuities at edges and corners of $\partial\Omega$. Therefore it takes sophisticated techniques to devise meaningful tangential trace operators on the function spaces. For domains with piecewise smooth boundary they were developed in [16, 17, 14]. These papers and, in particular [15], should be consulted as main references.

Before we tackle $\boldsymbol{W}(\mathbf{curl}, \Omega)$, we remind (see [16, Prop. 1.7]) that on piecewise smooth boundaries spaces $\boldsymbol{H}_{||}^{-\frac{1}{2}}(\partial\Omega)$ and $\boldsymbol{H}_{||}^{-\frac{1}{2}}(\partial\Omega)$ can be introduced so that the tangential traces become continuous and surjective operators $\gamma_{\mathbf{t}} : \boldsymbol{H}^1(\Omega) \mapsto \boldsymbol{H}_{||}^{\frac{1}{2}}(\partial\Omega)$, $\gamma_\times : \boldsymbol{H}^1(\Omega) \mapsto \boldsymbol{H}_{\perp}^{\frac{1}{2}}(\partial\Omega)$. Sloppily speaking, $\boldsymbol{H}_{||}^{-\frac{1}{2}}(\partial\Omega)$ contains the tangential surface vectorfields that are in $\boldsymbol{H}^{\frac{1}{2}}(\partial\Omega)$ for each smooth component of $\partial\Omega$ and feature a suitable "tangential continuity" across the edges. A corresponding "normal continuity" is satisfied by surface vectorfields in $\boldsymbol{H}_{\perp}^{-\frac{1}{2}}(\partial\Omega)$. The associated dual spaces will be denoted by $\boldsymbol{H}_{||}^{-\frac{1}{2}}(\partial\Omega)$ and $\boldsymbol{H}_{\perp}^{-\frac{1}{2}}(\partial\Omega)$, respectively.

Armed with these spaces and the density of smooth functions, the integration by parts formula

$$\int_{\Omega} \mathbf{curl}\, \mathbf{V} \cdot \mathbf{U} - \mathbf{V} \cdot \mathbf{curl}\, \mathbf{U} \, d\mathbf{x} = \int_{\partial\Omega} \gamma_\times \mathbf{U} \cdot \gamma_{\mathbf{t}} \mathbf{V} \, dS \tag{4}$$

is the key to establishing trace theorems for $\boldsymbol{W}(\mathbf{curl}, \Omega)$. Recall that the surface divergence operator div_Γ is the $\boldsymbol{L}^2(\partial\Omega)$-adjoint of the surface gradient grad_Γ. By rotating tangential surface vectorfields by $\frac{\pi}{2}$, we get the same relationship between curl_Γ and $\mathbf{curl}_\Gamma$. Using, first, $\mathbf{V} \in \boldsymbol{H}^1(\Omega)$, and, secondly, $\mathbf{V} \in \mathrm{grad}\, H^2(\Omega)$, we learn from (4) that

$$\gamma_{\mathbf{t}} : \boldsymbol{H}(\mathbf{curl}; \Omega) \mapsto \boldsymbol{H}_{\perp}^{-\frac{1}{2}}(\mathrm{curl}_\Gamma, \partial\Omega) := \{\mathbf{v} \in \boldsymbol{H}_{\perp}^{-\frac{1}{2}}(\partial\Omega), \mathbf{curl}_\Gamma \mathbf{v} \in H^{-\frac{1}{2}}(\partial\Omega)\},$$
$$\gamma_\times : \boldsymbol{H}(\mathbf{curl}; \Omega) \mapsto \boldsymbol{H}_{||}^{-\frac{1}{2}}(\mathrm{div}_\Gamma, \partial\Omega) := \{\boldsymbol{\lambda} \in \boldsymbol{H}_{||}^{-\frac{1}{2}}(\partial\Omega), \mathrm{div}_\Gamma \boldsymbol{\lambda} \in H^{-\frac{1}{2}}(\partial\Omega)\},$$

are continuous trace mappings. Moreover, according to Thm. 2.7 and Thm. 2.8 in [16], they are surjective, too. Thus, we have found the right tangential trace spaces for $\boldsymbol{H}(\mathbf{curl}; \Omega)$. By (4) the spaces $\boldsymbol{H}_{||}^{-\frac{1}{2}}(\mathrm{div}_\Gamma, \partial\Omega)$ and

$\boldsymbol{H}_{\perp}^{-\frac{1}{2}}(\mathrm{curl}_\Gamma, \partial\Omega)$ can be seen to be dual to each other (see [17, Sect. 4]). The sesqui-linear duality pairing will be denoted by $\langle\cdot,\cdot\rangle_{\boldsymbol{\tau}}$. Moreover, the rotation mapping $\mathsf{R}\mathbf{v} := \mathbf{v} \times \mathbf{n}$ can be extended to an isometry between the two spaces.

Integration by parts permits us to introduce several important weakly defined traces: The weak normal trace $\gamma_{\mathbf{n}}$ is defined for vectorfields $\mathbf{U} \in \boldsymbol{H}(\mathrm{div}; \Omega) := \{\mathbf{V} \in \boldsymbol{L}^2(\Omega), \mathrm{div}\,\mathbf{V} \in L^2(\Omega)\}$ by

$$\langle \gamma_{\mathbf{n}}\mathbf{U}, \gamma\Phi\rangle_{1/2,\partial\Omega} = \int_\Omega \mathrm{div}\,\mathbf{U}\,\overline{\Phi} + \mathbf{U}\cdot\mathrm{grad}\,\overline{\Phi}\,d\mathbf{x} \quad \forall\Phi \in H^1(\Omega)\,,$$

with $\langle\cdot,\cdot\rangle_{1/2,\partial\Omega}$ as duality pairing between $H^{-\frac{1}{2}}(\partial\Omega)$ and $H^{\frac{1}{2}}(\partial\Omega)$. The mapping $\gamma_{\mathbf{n}} : \boldsymbol{H}(\mathrm{div}; \Omega) \mapsto H^{-\frac{1}{2}}(\partial\Omega)$ is continuous and surjective, and an extension of the normal components trace $\gamma_{\mathbf{n}}\mathbf{U}(\mathbf{x}) := \mathbf{U}(\mathbf{x}) \cdot \mathbf{n}(\mathbf{x})$. Thus, the conormal trace $\partial_{\mathbf{n}} := \gamma_{\mathbf{n}} \circ \mathrm{grad}$ is continuous and surjective from $H(\Delta, \Omega) := \{\Phi \in W^1(\Omega), \Delta\Phi \in \boldsymbol{L}^2(\Omega)\}$ onto $H^{-\frac{1}{2}}(\partial\Omega)$. A **curl curl**-counterpart γ_N of this conormal trace for the Laplacian can be defined for

$$\mathbf{U} \in \boldsymbol{W}(\mathbf{curl}^2, \Omega) := \{\mathbf{V} \in \boldsymbol{W}(\mathbf{curl}, \Omega),\ \mathbf{curl}\,\mathbf{curl}\,\mathbf{V} \in \boldsymbol{L}^2(\Omega)\}$$

by demanding that for all $\mathbf{V} \in \boldsymbol{H}(\mathbf{curl}; \Omega)$

$$\langle \gamma_N \mathbf{U}, \gamma_{\mathbf{t}}\mathbf{V}\rangle_{\boldsymbol{\tau}} = \int_\Omega \mathbf{curl}\,\mathbf{U}\cdot\mathbf{curl}\,\overline{\mathbf{V}} - \mathbf{curl}\,\mathbf{curl}\,\mathbf{U}\cdot\overline{\mathbf{V}}\,d\mathbf{x}\,. \tag{5}$$

The trace γ_N furnishes a continuous and surjective mapping $\gamma_N : \boldsymbol{W}(\mathbf{curl}^2, \Omega) \mapsto \boldsymbol{H}_{||}^{-\frac{1}{2}}(\mathrm{div}_\Gamma, \partial\Omega)$ (cf. [35, Lemma 3.3]), which can be regarded as an extension of the restriction $(\gamma_N\mathbf{U})(\mathbf{x}) := \mathbf{curl}\,\mathbf{U}(\mathbf{x}) \times \mathbf{n}(\mathbf{x})$, $\mathbf{x} \in \partial\Omega$, for smooth $\mathbf{U}$.

Integration by parts also shows that a vectorfield in $\boldsymbol{C}^\infty(\overline{\Omega}_C) \cap \boldsymbol{C}^\infty(\overline{\Omega}_e)$ must feature tangential continuity in order to be contained in $\boldsymbol{W}(\mathbf{curl}, \mathbb{R}^3)$. Thus, both $\mathbf{E}$ and $\mathbf{H}$ can only belong to $\boldsymbol{W}(\mathbf{curl}, \mathbb{R}^3)$, if the following *transmission conditions* hold across $\Gamma := \partial\Omega_C$

$$[\gamma_{\mathbf{t}}\mathbf{E}]_\Gamma = 0 \quad \text{and} \quad [\gamma_\times\mathbf{H}]_\Gamma = 0\,. \tag{6}$$

Here, the "jump" $[\cdot]_\Gamma$ designates the difference of the values of a trace from Ω_e ("exterior") and from Ω_C ("interior"). We also stick to the convention that exterior traces will be labeled by a superscript $+$, whereas traces from Ω_C bear a superscript $-$.

3 Topological Prerequisites

Topological considerations come into play, when one wants to represent irrotational vectorfields on manifolds through gradients of scalar potentials. This is only possible, if the first cohomology group of the manifold is trivial [55, Ch. 6]. Otherwise, *cuts* have to be used to take care of irrotational vectorfields that are no gradients [10, Sect. 8.3.4].

Theorem 13. *For every domain $\Omega \subset \mathbb{R}^3$ with piecewise smooth boundary there exist piecewise smooth orientable embedded surfaces $\Sigma_1, \ldots, \Sigma_N \subset \bar{\Omega}$ (cuts), where N agrees with the first Betti number of Ω, such that*

- *the Σ_k, $k = 1, \ldots, N$, are mutually disjoint.*
- *the first cohomology group $H^1(\Omega', \mathbb{Z})$ of $\Omega' := \Omega \setminus (\Sigma_1 \cup \ldots \cup \Sigma_N)$ is trivial.*
- *Ω' is a generalized Lipschitz domain in the sense of [25], that is, when "seen from one side" its boundary $\partial\Omega'$ is Lipschitz continuous.*

Proof. The theorem is proved in [41]. □

In the sequel we are going to equip Ω_e with a set of cuts $\Sigma_1, \ldots, \Sigma_N$, according to Thm. 13. Each Σ_k has an orientation that translates into a crossing direction and thus we can distinguish between an "upper" surface Σ_k^+ and a "lower" surface Σ_k^-. Both surfaces are equipped with unit normal vectorfields $\mathbf{n}_k^+$, $\mathbf{n}_k^-$ pointing "away from Σ_k" into the interior of $\Omega' := \Omega_e \setminus (\Sigma_1 \cup \ldots \cup \Sigma_N)$. We fix $\mathbf{n}_{|\Sigma_k} := \mathbf{n}_k^+$ so that it agrees with the crossing direction.

The statement of Thm. 13 implies

$$\mathbf{V} \in \boldsymbol{H}(\mathbf{curl}; \Omega'),\ \mathbf{curl}\,\mathbf{V} = 0 \quad \Rightarrow \quad \exists \Phi \in H^1(\Omega') : \ \mathbf{V} = \operatorname{grad} \Phi \,.$$

It is even possible to characterize low dimensional spaces of vectorfields that fill the gap between $\mathrm{Ker}(\mathbf{curl}) \cap \boldsymbol{H}(\mathbf{curl}; \Omega_e)$ and $\operatorname{grad} H^1(\Omega_e)$. To that end, consider functions $\eta_k \in H^1(\Omega_e \setminus \Sigma_k)$, $k = 1, \ldots, N$, with $[\eta_k]_{\Sigma_k} = 1$. Here, $[\cdot]_S$ denotes the jump of some function across the externally oriented surface S, i.e. the difference of its value on the "+-side" and the "−-side".

Theorem 14. *Using the notations introduced above, we have the representation*

$$\mathrm{Ker}(\mathbf{curl}) \cap \boldsymbol{H}(\mathbf{curl}; \Omega_e) = \operatorname{grad} H^1(\Omega_e) + \mathrm{Span}\left\{\widetilde{\operatorname{grad}}\,\eta_1, \ldots, \widetilde{\operatorname{grad}}\,\eta_N\right\} ,$$

where $\widetilde{\operatorname{grad}}\,\eta_k \in \boldsymbol{L}^2(\Omega_e)$ is the gradient of η_k on $\Omega_e \setminus \Sigma_k$.

Proof. Compare Sect. 3 in [4]. □

¿From Thm. 14 we learn that

$$\mathrm{Ker}(\mathbf{curl}) \cap \boldsymbol{H}(\mathbf{curl}; \Omega_e) = \widetilde{\operatorname{grad}}\, H^1_\Sigma(\Omega_e) \,, \tag{7}$$

with $H^1_\Sigma(\Omega_e) := \{\varphi \in H^1(\Omega'), [\varphi]_{\Sigma_k} = \text{const.}, 1 \leq k \leq N\}$.

Thm. 13 may also be applied to Ω_C yielding N cutting surfaces $\widehat{\Sigma}_1, \ldots, \widehat{\Sigma}_N$, since the first Betti numbers of Ω_C and Ω_e agree. The boundaries $\sigma_1, \ldots, \sigma_N$,

$\widehat{\sigma}_1, \ldots, \widehat{\sigma}_N$ of Σ_k and $\widehat{\Sigma}_k$, $k = 1, \ldots, N$, respectively, represent a basis of the homology group $H_1(\Gamma, \mathbb{Z})$. In analogy to Thm. 14 we find that

$$\operatorname{Ker}(\operatorname{div}_\Gamma) \cap \boldsymbol{H}(\operatorname{div}; \Gamma) = \mathbf{curl}_\Gamma H^1(\Gamma) + \operatorname{Span}\left\{\mathbf{g}^1, \ldots, \mathbf{g}^N, \widehat{\mathbf{g}}^1, \ldots, \widehat{\mathbf{g}}^N\right\} , \tag{8}$$

where $\mathbf{g}^k$ is the vectorial surface rotation $\mathbf{curl}_\Gamma \varphi$ of some $\varphi \in H^1(\Gamma \setminus \sigma_k)$ that has a jump of constant height 1 across σ_i. The $\widehat{\mathbf{g}}^k$ are constructed analogously with respect to $\widehat{\sigma}_k$.

We remark that if Γ is equipped with some non-degenerate triangulation Γ_h (rendering it a cellular complex) the boundaries of interior and exterior cutting surfaces can be chosen such that they agree with edge cycles of Γ_h. Further, it is possible to pick piecewise smooth Lipschitz surfaces as related cuts . Such a choice of cuts will be a tacit assumption, whenever a triangulation of Γ has been fixed.

Remark 6. Please be aware that it is not the purpose of cuts to render Ω' simply connected, i.e., to ensure that it has a trivial first *homotopy* group. This is easily seen in the case of knotted geometries.

4 Formulations

Two basically different approaches to a variational formulation of (1) are conceivable. They can be distinguished by which equation is preserved in strong form and which is taken into account only in weak form [7]. This distinction parallels the primal and dual variational principles known from second order elliptic boundary value problems [13, Ch. 1].

The first approach involves Faraday's law in strong form. It is used to replace $\mathbf{H}$ in Ampere's law and the latter is multiplied by a test function in $\boldsymbol{H}(\mathbf{curl}; \mathbb{R}^3)$ and subjected to integration by parts according to (4). This results in the following variational problem (cf. [7, Sect. 3], [53], and [54]): Seek $\mathbf{E} \in \boldsymbol{W}(\mathbf{curl}, \mathbb{R}^3)$ such that for all $\mathbf{V} \in \boldsymbol{W}(\mathbf{curl}, \mathbb{R}^3)$

$$\left(\tfrac{1}{\mu_r}\,\mathbf{curl}\,\mathbf{E}, \mathbf{curl}\,\mathbf{V}\right)_{\boldsymbol{L}^2(\mathbb{R}^3)} + i\left(\tau^2 \mathbf{E}, \mathbf{V}\right)_{\boldsymbol{L}^2(\Omega_C)} = -i\left(\mathbf{J}_0, \mathbf{V}\right)_{\boldsymbol{L}^2(\Omega_e)} . \tag{9}$$

Theorem 15. *The variational problem (9) has a unique solution for* $\mathbf{H} := \frac{1}{i\mu_r}\,\mathbf{curl}\,\mathbf{E} \in \boldsymbol{H}(\mathbf{curl}; \mathbb{R}^3)$. *If it is posed on the constrained space*

$$\boldsymbol{\mathcal{W}} := \{\mathbf{V} \in \boldsymbol{W}(\mathbf{curl}, \mathbb{R}^3),\ \operatorname{div}\mathbf{V} = 0 \text{ in } \Omega_e,\ \int_{\Gamma_k} \gamma_{\mathbf{n}}\mathbf{V}\, dS = 0\, k = 1, \ldots, L\} ,$$

a unique solution $\mathbf{E} \in \boldsymbol{\mathcal{W}}$ *exists.*

Proof. The reader is referred to [3, Sect. 3] and [35, Sect. 2]. □

How can this so-called $\mathbf{E}$-based approach be reconciled with the asserted precedence of magnetic quantities? To understand this, recall that Faraday's law in strong form involves $\operatorname{div} \mu_r \mathbf{H} = 0$ everywhere. This makes it possible to introduce a magnetic vector potential $\mathbf{A} \in \boldsymbol{W}(\mathbf{curl}, \mathbb{R}^3)$ such that $\mathbf{curl}\,\mathbf{A} = \mu_r \mathbf{H}$ and to express $\mathbf{E}$ via a scalar potential $\Psi \in W^1(\mathbb{R}^3)$ as $\mathbf{E} = -\operatorname{grad} \Psi - i\mathbf{A}$. We have ample freedom to perform gauging and use it to set $\Psi = 0$. Thus, $\mathbf{E}$ turns out to be a scaled magnetic vector potential in disguise. I endorse this view as the proper reading of $\mathbf{E}$ in (9).

A crucial observation is that (9) is equivalent to a *transmission problem.* In order to state it for the case of an excitation by $\mathbf{J}_s$ a corresponding field is introduced by the Newton potential

$$\mathbf{E}_s(\mathbf{x}) = -i \int_{\mathbb{R}^3} \frac{\mathbf{J}_0(\mathbf{y})}{4\pi |\mathbf{x} - \mathbf{y}|} \, d\mathbf{y} \ .$$

Since $\operatorname{div} \mathbf{J}_s = 0$, it satisfies $\operatorname{div} \mathbf{E}_s = 0$ and $\mathbf{curl}\,\mathbf{curl}\,\mathbf{E}_s = -i\mathbf{J}_0$. The treatment of an excitation through a total loop current will be postponed until discretization is discussed in Sect. 5.3. Now, testing (9) with fields compactly supported in Ω_C or Ω_e, and taking into account (6), we get

$$\begin{aligned} \mathbf{curl}\,\mathbf{curl}\,\mathbf{E} + i\tau^2 \mu_r \mathbf{E} = 0 \quad \text{in } \Omega_C \ , \\ \mathbf{U} \in \boldsymbol{\mathcal{W}} \quad , \quad \mathbf{curl}\,\mathbf{curl}\,\mathbf{U} = 0 \quad \text{in } \Omega_e \ , \\ \gamma_{\mathbf{t}}^+ \mathbf{U} - \gamma_{\mathbf{t}}^- \mathbf{E} = -\gamma_{\mathbf{t}} \mathbf{E}_s \quad , \quad \gamma_N^+ \mathbf{U} - \tfrac{1}{\mu_r} \gamma_N^- \mathbf{E} = -\gamma_N \mathbf{E}_s \ . \end{aligned} \tag{10}$$

Here, $\mathbf{U} \in \boldsymbol{\mathcal{W}}$ is a *reaction field* that satisfies homogeneous equations in Ω_e. The electric field in Ω_e is recovered as $\mathbf{E} = \mathbf{U} + \mathbf{E}_s$. We emphasize that the gauge conditions implied by $\mathbf{U} \in \boldsymbol{\mathcal{W}}$ are not essential and may be relaxed.

The second option for a variational formulation is to keep Ampere's law strongly. Then, we have to use the trial space $\mathbf{H}_s + \boldsymbol{\mathcal{V}}$ with

$$\boldsymbol{\mathcal{V}} := \{\mathbf{V} \in \boldsymbol{H}(\mathbf{curl}; \mathbb{R}^3), \ \mathbf{curl}\,\mathbf{V} = 0 \text{ in } \Omega_e\}$$

for $\mathbf{H}$. There $\mathbf{H}_s \in \boldsymbol{H}(\mathbf{curl}; \mathbb{R}^3)$ is a source field such that $\operatorname{div} \mathbf{H}_s = 0$ and $\mathbf{curl}\,\mathbf{H}_s = \mathbf{J}_s$. It can be computed by means of the Biot-Savart law as

$$\mathbf{H}_s(\mathbf{x}) = \mathbf{curl}_{\mathbf{x}} \int_{\mathbb{R}^3} \frac{\mathbf{J}_s(\mathbf{y})}{4\pi |\mathbf{x} - \mathbf{y}|} \, d\mathbf{y} \ .$$

Now, testing the first equation of (1) with a compactly supported $\mathbf{V} \in \boldsymbol{\mathcal{V}}$, employing integration by parts on a ball with sufficiently large radius, and using the second equation inside Ω_C, we obtain: Seek $\mathbf{H} \in \boldsymbol{\mathcal{V}} + \mathbf{H}_s$ such that

$$\left(\tau^{-2}\,\mathbf{curl}\,\mathbf{H}, \mathbf{curl}\,\mathbf{V}\right)_{\boldsymbol{L}^2(\Omega_C)} + i\,(\mu_r \mathbf{H}, \mathbf{V})_{\boldsymbol{L}^2(\mathbb{R}^3)} = 0 \quad \forall \mathbf{V} \in \boldsymbol{\mathcal{V}} \ . \tag{11}$$

For a more detailed presentation of the considerations leading to (11) the reader is referred to [11], [7, Sect. 2], and [10, Ch. 8]. Existence and uniqueness of solutions of (11) immediately follow from the Lax-Milgram lemma.

Straight from (11) we infer $\mathrm{div}(\mu_r \mathbf{H}) = 0$ in all of $\mathbb{R}^3$. This involves the normal continuity of $\mu_r \mathbf{H}$ across Γ. We are led to the transmission problem

$$\begin{aligned} \mathbf{curl}\,\mathbf{curl}\,\mathbf{H} + i\tau^2\mu_r\mathbf{H} = 0 \quad \text{in } \Omega_C \;, \\ \mathbf{curl}\,\mathbf{H} = 0 \quad , \quad \mathrm{div}\,\mathbf{H} = 0 \quad \text{in } \Omega_e \;, \\ \gamma_{\mathbf{n}}^+\mathbf{H} - \mu_r\gamma_{\mathbf{n}}^-\mathbf{H} = -\gamma_{\mathbf{n}}\mathbf{H}_s \quad , \quad \gamma_{\mathbf{t}}^+\mathbf{H} - \gamma_{\mathbf{t}}^-\mathbf{H} = \gamma_{\mathbf{t}}\mathbf{H}_s \quad \text{on } \Gamma \;. \end{aligned} \tag{12}$$

However, if Ω_C has non-vanishing first Betti number, then there is no unique solution of (4) [51, 30]. To see this please notice that thanks to Thm. 14 the path integrals $f_k(\mathbf{H}) := \int_{\widehat{\sigma}_k} \mathbf{H} \cdot d\mathbf{s}$ supply continuous functionals on $\boldsymbol{\mathcal{V}}$. They do not vanish, because plugging in an extension to $\mathbb{R}^3$ of $\widetilde{\mathbf{grad}\,\eta_k}$ results in 1. Next, consider the variational problem (11) posed over $\boldsymbol{\mathcal{V}}$, but with f_k as non-homogeneous right hand side. A unique non-zero solution $\mathbf{H}_k \in \boldsymbol{\mathcal{V}}$ exists. From $f_k(\mathbf{grad}\,\Phi) = 0$ for all $\Phi \in W^1(\mathbb{R})$ we conclude that still $\mathrm{div}(\mu_r \mathbf{H}_k) = 0$. Hence, $\mathbf{H}_k$ satisfies all the transmission conditions of (4). Testing with smooth vectorfields that are compactly supported in Ω_C establishes the first equation of (4).

These considerations refute the equivalence of (4) and (11). The bottom line is that in general the **H**-based model does not allow a formulation as transmission problem, unless some extra coupling conditions that, however, fail to involve traces on Γ only, are taken into account. These additional conditions are formulated and investigated in [2] (see also [43]).

A third variational formulation, the so-called **A**-V-formulation [1, Sect.3.2], combines primal and dual variational principles, one kind applied in Ω_C the other in Ω_e. Faraday's law will be used in strong from in Ω_C. Conversely, Ampere's law is tested with $\mathbf{V} \in \boldsymbol{H}(\mathbf{curl}; \mathbb{R}^3)$, but integration by parts is performed on Ω_C only. Therefore, boundary terms have to be retained in the variational equation

$$\left(\frac{1}{\mu_r}\mathbf{curl}\,\mathbf{E}, \mathbf{curl}\,\mathbf{V}\right)_{\boldsymbol{L}^2(\Omega_C)} + i\tau^2\,(\mathbf{E}, \mathbf{V})_{\boldsymbol{L}^2(\Omega_C)} - \left\langle \frac{1}{\mu_r}\gamma_N^-\mathbf{E}, \gamma_{\mathbf{t}}^-\mathbf{V}\right\rangle_{\boldsymbol{\tau}} = 0$$

for $\mathbf{V} \in \boldsymbol{H}(\mathbf{curl}; \Omega_C)$. In Ω_e Ampere's law is incorporated strongly by zeroing in on $\mathbf{H} \in \mathbf{H}_s + \boldsymbol{\mathcal{V}}$. Faraday's law is tested with compactly supported irrotational fields only, and subsequently we integrate by parts. We end up with

$$\overline{\langle \gamma_\times^+\mathbf{V}, \gamma_{\mathbf{t}}^+\mathbf{E}\rangle_{\boldsymbol{\tau}}} + i\omega\,(\mathbf{H}, \mathbf{V})_{\boldsymbol{L}^2(\Omega_e)} = 0 \quad \forall \mathbf{V} \in \boldsymbol{\mathcal{V}} \;.$$

Both variational problems are linked through the transmission conditions, which enable us to replace $\frac{1}{\mu_r}\gamma_N^-\mathbf{E}$ by $i\gamma_\times^+\mathbf{H}$ in the boundary terms. This results in the variational problem [44]: Seek $\mathbf{E} \in \boldsymbol{H}(\mathbf{curl}; \Omega_C)$, $\mathbf{H} \in \mathbf{H}_s + \boldsymbol{\mathcal{V}}$ such that for all $\mathbf{W} \in \boldsymbol{H}(\mathbf{curl}; \Omega_C)$, $\mathbf{V} \in \boldsymbol{\mathcal{V}}$

$$\begin{aligned}\left(\tfrac{1}{\mu_r}\,\mathbf{curl}\,\mathbf{E},\mathbf{curl}\,\mathbf{W}\right)_{\boldsymbol{L}^2(\Omega_C)} + i\omega\,(\sigma\mathbf{E},\mathbf{W})_{\boldsymbol{L}^2(\Omega_C)} - i\,\langle\gamma_\times\mathbf{H},\gamma_\mathbf{t}\mathbf{W}\rangle_{\boldsymbol{\mathcal{T}}} &= 0\,,\\ i\,\langle\gamma_\times\mathbf{V},\gamma_\mathbf{t}\mathbf{E}\rangle_{\boldsymbol{\mathcal{T}}} \qquad + \omega\overline{(\mathbf{H},\mathbf{V})_{\boldsymbol{L}^2(\Omega_e)}} &= 0\,.\end{aligned} \tag{13}$$

Theorem 16. *The bilinear form associated with the variational problem (13) is* $\boldsymbol{H}(\mathbf{curl};\Omega_C)\times\boldsymbol{\mathcal{V}}$*-elliptic.*

Proof. Setting $\mathbf{W} := \mathbf{E}$, $\mathbf{V} := \mathbf{H}$, and adding both equations makes the "off-diagonal" terms cancel. □

Similarly as in the case of the **H**-based model, an equivalent transmission problem is also elusive for the **A**-V-formulation.

5 E-based Model

Now we discuss the steps leading to a symmetrically coupled boundary element formulation for the transmission problem (10).

5.1 Boundary Integral Operators

We follow the customary approach to the construction of boundary integral operators arising from differential operators [48]. A key notion is that of potentials, namely special mappings from boundary data to smooth functions off the boundary. Here, they are integral operators based on the Helmholtz kernel [48, Ch. 9]

$$G_\kappa(\mathbf{x},\mathbf{y}) := \frac{\exp(-\kappa|\mathbf{x}-\mathbf{y}|)}{4\pi|\mathbf{x}-\mathbf{y}|}\quad,\quad \mathbf{x}\neq\mathbf{y}\,,$$

for $\kappa\in\mathbb{C}$. In the sequel we will take for granted that $\mathrm{Re}\,\kappa\geq 0$. The kernel is used to define the scalar single layer potential

$$\Psi_V^\kappa(\varphi)(\mathbf{x}) = \int_\Gamma G_\kappa(\mathbf{x},\mathbf{y})\varphi(\mathbf{y})\,dS(\mathbf{y}) \quad \mathbf{x}\notin\Gamma\,,$$

the vectorial single layer potential

$$\boldsymbol{\Psi}_A^\kappa(\boldsymbol{\lambda})(\mathbf{x}) := \int_\Gamma G_0(\mathbf{x},\mathbf{y})\boldsymbol{\lambda}(\mathbf{y})\,dS(\mathbf{y}) \quad \mathbf{x}\notin\Gamma\,,$$

and the "Maxwell double layer potential"

$$\boldsymbol{\Psi}_M^\kappa(\mathbf{v}) := \mathbf{curl}\,\boldsymbol{\Psi}_A^\kappa(\mathsf{R}\mathbf{v})\,.$$

The potentials have the following mapping properties:

Theorem 17. *The operators*

$$\begin{aligned}
\Psi_V^\kappa &: H^{-\frac{1}{2}}(\Gamma) \mapsto W^1(\mathbb{R}^3) \cap H(\Delta, \Omega_C \cup \Omega_e) \, , \\
\boldsymbol{\Psi}_A^\kappa &: \boldsymbol{H}_{||}^{-\frac{1}{2}}(\mathrm{div}_\Gamma, \Gamma) \mapsto \boldsymbol{W}^1(\mathbb{R}^3) \cap \boldsymbol{W}(\mathbf{curl}^2, \Omega_C \cup \Omega_e) \, , \\
\boldsymbol{\Psi}_M^\kappa &: \boldsymbol{H}_{\perp}^{-\frac{1}{2}}(\mathrm{curl}_\Gamma, \Gamma) \mapsto \boldsymbol{W}(\mathbf{curl}^2, \Omega_C \cup \Omega_e) \, ,
\end{aligned}$$

are continuous.

Proof. For Ψ_V^κ see [24], for $\boldsymbol{\Psi}_A^\kappa$, $\boldsymbol{\Psi}_M^\kappa$ the proof can be looked up in [35, Sect. 5]. □

Important properties of the boundary integral operators are due to the following *jump relations.* They can be stated, because the previous theorem guarantees that the trace operator really make sense for the potentials.

Theorem 18. *The potentials satisfy the jump relations*

$$\begin{aligned}
[\gamma \Psi_V^\kappa]_\Gamma &= 0 \quad , & [\partial_{\mathbf{n}} \Psi_V^\kappa]_\Gamma &= -Id \, , \\
[\gamma_{\mathbf{t}} \boldsymbol{\Psi}_A^\kappa]_\Gamma &= 0 \quad , & [\gamma_N \boldsymbol{\Psi}_A^\kappa]_\Gamma &= -Id \, , \\
[\gamma_{\mathbf{t}} \boldsymbol{\Psi}_M^\kappa]_\Gamma &= -Id \quad , & [\gamma_N \boldsymbol{\Psi}_M^\kappa]_\Gamma &= 0 \, , \\
[\gamma_{\mathbf{n}} \boldsymbol{\Psi}_A^\kappa]_\Gamma &= 0 \quad , & [\gamma_{\mathbf{n}} \boldsymbol{\Psi}_M^\kappa]_\Gamma &= 0 \, .
\end{aligned}$$

Proof. The assertion for the single layer potential is proved in [24], for the vector valued potentials in [35, Sect. 5]. For smooth domains the results are contained in [51], [50, Thm. 5.5.1], and [22, Thm. 6.11]. □

Appealing to Thm. 17 we introduce boundary integral operators by taking different traces of potentials.

Theorem 19. *The boundary integral operators*

$$\begin{aligned}
\mathsf{V}^\kappa &:= \gamma \Psi_V^\kappa && : H^{-\frac{1}{2}}(\Gamma) \mapsto H^{\frac{1}{2}}(\Gamma) \, , \\
\mathsf{K}^\kappa &:= \tfrac{1}{2}(\gamma_{\mathbf{n}}^- + \gamma_{\mathbf{n}}^+) \, \mathrm{grad} \, \Psi_V^\kappa && : H^{-\frac{1}{2}}(\Gamma) \mapsto H^{-\frac{1}{2}}(\Gamma) \, , \\
\mathsf{A}^\kappa &:= \gamma_{\mathbf{t}} \boldsymbol{\Psi}_A^\kappa && : \boldsymbol{H}_{||}^{-\frac{1}{2}}(\Gamma) \mapsto \boldsymbol{H}_{||}^{\frac{1}{2}}(\Gamma) \, , \\
\mathsf{B}^\kappa &:= \tfrac{1}{2}(\gamma_N^- + \gamma_N^+) \boldsymbol{\Psi}_A^\kappa && : \boldsymbol{H}_{||}^{-\frac{1}{2}}(\mathrm{div}_\Gamma, \Gamma) \mapsto \boldsymbol{H}_{||}^{-\frac{1}{2}}(\mathrm{div}_\Gamma, \Gamma) \, , \\
\mathsf{C}^\kappa &:= \tfrac{1}{2}(\gamma_{\mathbf{t}}^- + \gamma_{\mathbf{t}}^+) \boldsymbol{\Psi}_M^\kappa && : \boldsymbol{H}_{\perp}^{-\frac{1}{2}}(\mathrm{curl}_\Gamma, \Gamma) \mapsto \boldsymbol{H}_{\perp}^{-\frac{1}{2}}(\mathrm{curl}_\Gamma, \Gamma) \, , \\
\mathsf{N}^\kappa &:= \gamma_N \boldsymbol{\Psi}_M^\kappa && : \boldsymbol{H}_{\perp}^{-\frac{1}{2}}(\mathrm{curl}_\Gamma, \Gamma) \mapsto \boldsymbol{H}_{||}^{-\frac{1}{2}}(\mathrm{div}_\Gamma, \Gamma) \, ,
\end{aligned}$$

are well defined and continuous.

Proof. The theorem immediately follows from Thm. 17 along with the continuity of the trace mappings. □

The superscript κ will be omitted from the operator symbols, if $\kappa = 0$.

Theorem 20. *The boundary integral operators* B^κ *and* C^κ *satisfy*

$$\langle \mathsf{B}^\kappa \boldsymbol{\mu}, \mathbf{v} \rangle_{\boldsymbol{\tau}} = -\langle \boldsymbol{\mu}, \mathsf{C}^\kappa \mathbf{v} \rangle_{\boldsymbol{\tau}} \quad \forall \boldsymbol{\mu} \in \boldsymbol{H}_{||}^{-\frac{1}{2}}(\mathrm{div}_\Gamma, \Gamma),\ \mathbf{v} \in \boldsymbol{H}_{\perp}^{-\frac{1}{2}}(\mathrm{curl}_\Gamma, \Gamma) .$$

Proof. The proof can be accomplished as in [35, Sect. 6] and that of Thm. 3.9 in [19]. □

Theorem 21. *For* $\mathrm{Re}\,\kappa \geq 0$ *the following coercivity properties hold*

$$\begin{aligned} |\langle \boldsymbol{\mu}, \mathsf{A}^\kappa \boldsymbol{\mu} \rangle_{\boldsymbol{\tau}}| &\geq c \left\|\boldsymbol{\mu}\right\|^2_{\boldsymbol{H}_{||}^{-\frac{1}{2}}(\Gamma)} \quad && \forall \boldsymbol{\mu} \in \boldsymbol{H}_{||}^{-\frac{1}{2}}(\mathrm{div}_\Gamma, \Gamma) , \\ |\langle \varphi, \mathsf{V}^\kappa \varphi \rangle_{1/2,\Gamma}| &\geq c \left\|\varphi\right\|^2_{H^{-\frac{1}{2}}(\Gamma)} \quad && \forall \varphi \in H^{-\frac{1}{2}}(\Gamma) . \end{aligned}$$

with positive constants $c > 0$ *depending on* κ *and* Γ.

Proof. The assertion is taken from [18, Thm. 4.2]. □

It is worth noting that (5) yields the identity (cf. [21, Formula (2.86)])

$$\langle \mathsf{N}^\kappa \mathbf{u}, \mathbf{v} \rangle_{\boldsymbol{\tau}} = \kappa^2 \langle \mathsf{A}^\kappa(\mathsf{R}\mathbf{u}), \mathsf{R}\mathbf{v} \rangle_{\boldsymbol{\tau}} + \langle \mathsf{V}^\kappa(\mathrm{curl}_\Gamma \mathbf{u}), \mathrm{curl}_\Gamma \mathbf{v} \rangle_{1/2,\Gamma} . \tag{14}$$

Potentials and their properties are of great interest, because they occur as the building blocks of representation formulas for solutions of both the interior and exterior problem in (10).

Theorem 22. *If* $\mathrm{Re}\,\kappa > 0$ *and a distribution* $\mathbf{E} \in \boldsymbol{H}(\mathbf{curl}; \Omega_C)$ *satisfies* $\mathbf{curl}\,\mathbf{curl}\,\mathbf{E} + \kappa^2 \mathbf{E} = 0$ *in* Ω_C, *then it has a representation*

$$\mathbf{E}(\mathbf{x}) = \boldsymbol{\Psi}_A^\kappa(\gamma_N^- \mathbf{E})(\mathbf{x}) + \boldsymbol{\Psi}_M^\kappa(\gamma_{\mathbf{t}}^- \mathbf{E})(\mathbf{x}) - \frac{1}{\kappa^2}\,\mathrm{grad}\,\Psi_V^\kappa(\mathrm{div}_\Gamma(\gamma_N^- \mathbf{E}))(\mathbf{x})$$

for all $\mathbf{x} \in \Omega_C$.

Proof. The assertion of the theorem is a variant of the Stratton-Chu representation formula (see [22, Thm 6.1], [47, Sect. 2], and [50, Sect. 5.5]). □

Theorem 23. *A distribution* $\mathbf{U} \in \boldsymbol{W}(\mathbf{curl}, \Omega_e)$ *satisfying* $\mathbf{curl}\,\mathbf{curl}\,\mathbf{U} = 0$ *and* $\mathrm{div}\,\mathbf{U} = 0$ *in* Ω_e *together with* $\mathbf{u}(\mathbf{x}) = O(\frac{1}{|\mathbf{x}|})$ *uniformly for* $|\mathbf{x}| \to \infty$ *possesses the representation*

$$\mathbf{U}(\mathbf{x}) = -\boldsymbol{\Psi}_A^0(\gamma_N^+ \mathbf{U}) - \boldsymbol{\Psi}_M^0(\gamma_{\mathbf{t}}^+ \mathbf{U}) - \mathrm{grad}\,\Psi_V^0(\gamma_{\mathbf{n}}^+ \mathbf{U}) , \quad \mathbf{x} \in \Omega_e .$$

Proof. Compare Sect. 5 in [35]. □

The presence of normal boundary data in the representation formula of Thm. 23 constitutes a stark difference of the two representation formulas. This can be blamed on the divergence constraint by which the double-**curl** operator had to be supplemented in Thm. 23. In a sense, we are dealing with a regularized problem in Ω_e, for which the underlying differential operator is essentially different from that in Ω_C.

5.2 Coupled Problem

Now, let $(\mathbf{E},\mathbf{U})$ stand for the solution of the transmission problem (10) in Ω_C and Ω_e, respectively. Suitable trace operators can be applied to the representation formulas and this procedure yields the *Calderon identities.* From Thm. 22 we get

$$\begin{aligned}\gamma_{\mathbf{t}}^{-}\mathbf{E} &= (\mathsf{A}^{\kappa} - \mathrm{grad}_{\Gamma}\circ\mathsf{V}^{\kappa}\circ\mathrm{div}_{\Gamma})(\gamma_N^{-}\mathbf{E}) + (\tfrac{1}{2}Id + \mathsf{C}^{\kappa})(\gamma_{\mathbf{t}}^{-}\mathbf{E})\,,\\ \gamma_N^{-}\mathbf{E} &= (\tfrac{1}{2}Id + \mathsf{B}^{\kappa})(\gamma_N^{-}\mathbf{E}) + \mathsf{N}^{\kappa}(\gamma_{\mathbf{t}}^{-}\mathbf{E})\,,\end{aligned} \tag{15}$$

where $\kappa = \frac{1}{2}\sqrt{2}(1+i)\tau\sqrt{\mu_r}$. By Thm. 23 we have

$$\begin{aligned}\gamma_{\mathbf{t}}^{+}\mathbf{U} &= -\mathsf{A}(\gamma_N^{+}\mathbf{U}) + (\tfrac{1}{2}Id - \mathsf{C})(\gamma_{\mathbf{t}}^{+}\mathbf{U}) - \mathrm{grad}_{\Gamma}\,\mathsf{V}(\gamma_{\mathbf{n}}^{+}\mathbf{U})\,,\\ \gamma_N^{+}\mathbf{U} &= (\tfrac{1}{2}Id - \mathsf{B})(\gamma_N^{+}\mathbf{U}) - \mathsf{N}(\gamma_{\mathbf{t}}^{+}\mathbf{U})\,,\\ \gamma_{\mathbf{n}}^{+}\mathbf{U} &= -\gamma_{\mathbf{n}}^{+}\boldsymbol{\Psi}_A^0(\gamma_N^{+}\mathbf{U}) - \gamma_{\mathbf{n}}^{+}\boldsymbol{\Psi}_M^0(\gamma_{\mathbf{t}}^{+}\mathbf{U}) + (\tfrac{1}{2}Id - \mathsf{K})(\gamma_{\mathbf{n}}^{+}\mathbf{U})\,.\end{aligned} \tag{16}$$

The boundary data for any solution of the interior/exterior $\mathbf{E}$-based eddy current equations will fulfill (15) and (16), respectively.

The gist of the symmetric coupling approach according to Costabel [23] is to use all of the equations of the Calderon identities in conjunction with the transmission conditions. However, here we have to grapple with a mismatch of interior and exterior boundary data due to the presence of $\gamma_{\mathbf{n}}^{+}\mathbf{U}$ in (16). A remedy is motivated by the observation

$$\mathbf{curl}\,\mathbf{curl}\,\mathbf{U} = 0 \quad \text{in } \Omega_e \quad \Rightarrow \quad \mathrm{div}_{\Gamma}(\gamma_N^{+}\mathbf{U}) = 0\,,$$

which is an immediate consequence of the identity $\mathrm{div}_{\Gamma}(\gamma_{\times}\mathbf{V}) = \gamma_{\mathbf{n}}\,\mathbf{curl}\,\mathbf{V}$ for boundary differential operators. We observe that $\gamma_N^{+}\mathbf{U}$ has to be sought in the space

$$\boldsymbol{H}_{||}^{-\frac{1}{2}}(\mathrm{div}_{\Gamma}0,\Gamma) := \{\boldsymbol{\mu}\in\boldsymbol{H}_{||}^{-\frac{1}{2}}(\mathrm{div}_{\Gamma},\Gamma),\ \mathrm{div}_{\Gamma}\boldsymbol{\mu} = 0\}\,.$$

By the transmission condition for γ_N and the fact that $\mathbf{curl}\,\mathbf{curl}\,\mathbf{E}_s = 0$ in a neighborhood of Γ, $\gamma_N^{-}\mathbf{E}$ has to be div_{Γ}-free, as well. Hence, we can restrict our attention to boundary data $\gamma_N^{-}\mathbf{E},\gamma_N^{+}\mathbf{U}$ in $\boldsymbol{H}_{||}^{-\frac{1}{2}}(\mathrm{div}_{\Gamma}0,\Gamma)$ throughout. Recalling the dualities, this is the proper test space for those equations of the Calderon identities that are set in $\boldsymbol{H}_{\perp}^{-\frac{1}{2}}(\mathrm{curl}_{\Gamma},\Gamma)$. Since div_{Γ} is the $\boldsymbol{L}^2(\Gamma)$-adjoint of grad_{Γ}, we find

$$\boldsymbol{\mu}\in\boldsymbol{H}_{||}^{-\frac{1}{2}}(\mathrm{div}_{\Gamma}0,\Gamma) \quad\Rightarrow\quad \left\langle \boldsymbol{\mu}, \underset{\Gamma}{\mathrm{grad}}\,\varphi\right\rangle_{\tau} = 0 \quad \forall\varphi\in H^{\frac{1}{2}}(\Gamma)\,.$$

This makes the undesirable terms disappear, when switching to a weak form of the top equations in the Calderon identities (15) and (16). For all $\boldsymbol{\mu}\in\boldsymbol{H}_{||}^{-\frac{1}{2}}(\mathrm{div}_{\Gamma}0,\Gamma)$ we obtain

$$\begin{aligned}\left\langle\boldsymbol{\mu},\gamma_{\mathbf{t}}^{-}\mathbf{E}\right\rangle_{\tau} &= \left\langle\boldsymbol{\mu},\mathsf{A}^{\kappa}(\gamma_N^{-}\mathbf{E})\right\rangle_{\tau} + \left\langle\boldsymbol{\mu},(\tfrac{1}{2}Id+\mathsf{C}^{\kappa})\gamma_{\mathbf{t}}^{-}\mathbf{E}\right\rangle_{\tau} ,\\ \left\langle\boldsymbol{\mu},\gamma_{\mathbf{t}}^{+}\mathbf{U}\right\rangle_{\tau} &= \left\langle\boldsymbol{\mu},-\mathsf{A}\gamma_N^{+}\mathbf{U}\right\rangle_{\tau} + \left\langle\boldsymbol{\mu},(\tfrac{1}{2}Id-\mathsf{C})\gamma_{\mathbf{t}}^{+}\mathbf{U}\right\rangle_{\tau} .\end{aligned}$$

¿From the transmission conditions we know $\gamma_{\mathbf{t}}^{+}\mathbf{U}-\gamma_{\mathbf{t}}^{-}\mathbf{E}=-\gamma_{\mathbf{t}}\mathbf{E}_s$. Thus, subtracting the above equations leads to

$$-\left\langle\boldsymbol{\mu},\mathsf{A}(\gamma_N^{+}\mathbf{U})+\mathsf{A}^{\kappa}(\gamma_N^{-}\mathbf{E})\right\rangle_{\tau} - \left\langle\boldsymbol{\mu},\mathsf{C}(\gamma_{\mathbf{t}}^{+}\mathbf{U})+\mathsf{C}^{\kappa}(\gamma_{\mathbf{t}}^{-}\mathbf{U})\right\rangle_{\tau} = -\left\langle\boldsymbol{\mu},\gamma_{\mathbf{t}}\mathbf{E}_s\right\rangle_{\tau}$$

for all $\boldsymbol{\mu}\in\boldsymbol{H}_{||}^{-\frac{1}{2}}(\mathrm{div}_{\Gamma}0,\Gamma)$. From the transmission condition $\gamma_N^{+}\mathbf{U}-\frac{1}{\mu_r}\gamma_N^{-}\mathbf{E}=-\gamma_N\mathbf{E}_s$ and the second equations of the Calderon identities we directly conclude

$$(\tfrac{1}{2}Id-\mathsf{B})\gamma_N^{+}\mathbf{U}-\mathsf{N}(\gamma_{\mathbf{t}}^{+}\mathbf{U})-\tfrac{1}{\mu_r}(\tfrac{1}{2}Id+\mathsf{B}^{\kappa})\gamma_N^{-}\mathbf{E}-\tfrac{1}{\mu_r}\mathsf{N}^{\kappa}(\gamma_{\mathbf{t}}^{-}\mathbf{E}) = -\gamma_N\mathbf{E}_s .$$

As final unknown quantities we introduce the tangential trace of the electric field $\mathbf{u}:=\gamma_{\mathbf{t}}^{-}\mathbf{E}\in\boldsymbol{H}_{\perp}^{-\frac{1}{2}}(\mathrm{curl}_{\Gamma},\Gamma)$ and the tangential trace of the magnetic field $\boldsymbol{\lambda}:=\frac{1}{\mu_r}\gamma_N^{-}\mathbf{E}\in\boldsymbol{H}_{||}^{-\frac{1}{2}}(\mathrm{div}_{\Gamma}0,\Gamma)$. The latter is also known as *equivalent surface current.* The transmission conditions enable us to express the exterior traces in these unknowns. We end up with the coupled variational problem: Seek $\mathbf{u}\in\boldsymbol{H}_{\perp}^{-\frac{1}{2}}(\mathrm{curl}_{\Gamma},\Gamma)$, $\boldsymbol{\lambda}\in\boldsymbol{H}_{||}^{-\frac{1}{2}}(\mathrm{div}_{\Gamma}0,\Gamma)$ such that

$$\begin{aligned}\left\langle(\mathsf{N}+\tfrac{1}{\mu_r}\mathsf{N}^{\kappa})\mathbf{u},\mathbf{v}\right\rangle_{\tau} &+ \left\langle(\mathsf{B}+\mathsf{B}^{\kappa})\boldsymbol{\lambda},\mathbf{v}\right\rangle_{\tau} = f(\mathbf{v}) ,\\ \left\langle\boldsymbol{\mu},(\mathsf{C}+\mathsf{C}^{\kappa})\mathbf{u}\right\rangle_{\tau} &+ \left\langle\boldsymbol{\mu},(\mathsf{A}+\mu_r\mathsf{A}^{\kappa})\boldsymbol{\lambda}\right\rangle_{\tau} = g(\boldsymbol{\mu})\end{aligned} \tag{17}$$

for all $\mathbf{v}\in\boldsymbol{H}_{\perp}^{-\frac{1}{2}}(\mathrm{curl}_{\Gamma},\Gamma)$, $\boldsymbol{\mu}\in\boldsymbol{H}_{||}^{-\frac{1}{2}}(\mathrm{div}_{\Gamma}0,\Gamma)$. The right hand side is given by

$$\begin{aligned}f(\mathbf{v}) &:= \left\langle(\tfrac{1}{2}Id+\mathsf{B})\gamma_N\mathbf{E}_s,\mathbf{v}\right\rangle_{\tau} + \left\langle\mathsf{N}(\gamma_{\mathbf{t}}\mathbf{E}_s),\mathbf{v}\right\rangle_{\tau} ,\\ g(\boldsymbol{\mu}) &:= \left\langle\boldsymbol{\mu},(\tfrac{1}{2}Id+\mathsf{C})\gamma_{\mathbf{t}}\mathbf{E}_s\right\rangle_{\tau} + \left\langle\boldsymbol{\mu},\mathsf{A}(\gamma_N\mathbf{E}_s)\right\rangle_{\tau} .\end{aligned}$$

Theorem 24. *The bilinear form* $\mathbf{d}$ *associated with the variational problem (17) is* $\boldsymbol{H}_{\perp}^{-\frac{1}{2}}(\mathrm{curl}_{\Gamma},\Gamma)\times\boldsymbol{H}_{||}^{-\frac{1}{2}}(\mathrm{div}_{\Gamma}0,\Gamma)$*-elliptic.*

Proof. As a simple consequence of the block skew-symmetric structure of the variational problem (cf. Thm 20) we find for $\mathbf{v}\in\boldsymbol{H}_{\perp}^{-\frac{1}{2}}(\mathrm{curl}_{\Gamma},\Gamma)$, $\boldsymbol{\mu}\in\boldsymbol{H}_{||}^{-\frac{1}{2}}(\mathrm{div}_{\Gamma}0,\Gamma)$ that

$$\mathbf{d}((\mathbf{v},\boldsymbol{\mu}),(\mathbf{v},\boldsymbol{\mu})) = \left\langle(\mathsf{N}+\tfrac{1}{\mu_r}\mathsf{N}^{\kappa})\mathbf{v},\mathbf{v}\right\rangle_{\tau} + \left\langle\boldsymbol{\mu},(\mathsf{A}+\mu_r\mathsf{A}^{\kappa})\boldsymbol{\mu}\right\rangle_{\tau} .$$

A close scrutiny of the integral operators based on (5) reveals that for all $\boldsymbol{\eta}\in\boldsymbol{H}_{||}^{-\frac{1}{2}}(\mathrm{div}_{\Gamma},\Gamma)$, $\varphi\in H^{-\frac{1}{2}}(\Gamma)$ holds $\mathrm{Re}\left\langle\boldsymbol{\eta},\mathsf{A}^{\kappa}\boldsymbol{\eta}\right\rangle_{\tau}\geq 0$, $\mathrm{Re}\left\langle\varphi,\mathsf{V}^{\kappa}\varphi\right\rangle_{1/2,\Gamma}\geq 0$ and $\mathrm{Im}\left\langle\boldsymbol{\eta},\mathsf{A}^{\kappa}\boldsymbol{\eta}\right\rangle_{\tau}\geq 0$, $\mathrm{Im}\left\langle\varphi,\mathsf{V}^{\kappa}\varphi\right\rangle_{1/2,\Gamma}\geq 0$. This permits us to conclude ellipticity of the whole bilinear form from separate estimates for the individual terms. These separate estimates are provided by Thm. 21 in conjunction with (14). □

Corollary 2. *The variational problem (17) has a unique solution* $(\mathbf{u}, \boldsymbol{\lambda}) \in \boldsymbol{H}_{\perp}^{-\frac{1}{2}}(\mathrm{curl}_{\Gamma}, \Gamma) \times \boldsymbol{H}_{||}^{-\frac{1}{2}}(\mathrm{div}_{\Gamma} 0, \Gamma)$.

By the derivation of the boundary integral equations we can be certain that traces $\gamma_{\mathbf{t}}^{-}\mathbf{E}$ and $\gamma_{\times}^{-}\mathbf{H}$ will always give rise to solutions of (17). Their uniqueness then confirms that we get the traces of solutions of the **E**-based eddy current model (9). These traces are fixed regardless of the gauging of **E** employed in Ω_e.

5.3 Galerkin Discretization

By Thm. 24 and Cea's lemma [20, Thm. 2.4.1] a conforming Galerkin boundary element discretization of the elliptic variational problem (17) will immediately produce approximate solutions that are quasi-optimal in the norms of the underlying Hilbert spaces $\boldsymbol{H}_{\perp}^{-\frac{1}{2}}(\mathrm{curl}_{\Gamma}, \Gamma)$ and $\boldsymbol{H}_{||}^{-\frac{1}{2}}(\mathrm{div}_{\Gamma}, \Gamma)$.

The choice of boundary elements should be guided by the insight that we aim to approximate traces. More precisely, $\mathbf{u}$ is the tangential trace of the electric field, $\boldsymbol{\lambda}$ the twisted tangential trace of the magnetic field. The most adequate mathematical description of these two physical quantities is supplied by 1-forms [28, 6, 5]. Correspondingly, they should be approximated by *discrete 1-forms.* Those are provided by $\boldsymbol{H}(\mathbf{curl}; \Omega)$-conforming finite elements (see e.g. [49]), whose lowest order specimens are known as edge elements [8, 9]. Taking tangential traces $\gamma_{\mathbf{t}}$ and $\gamma_{\times}$ of edge element functions will spawn conforming boundary element subspaces of $\boldsymbol{H}_{\perp}^{-\frac{1}{2}}(\mathrm{curl}_{\Gamma}, \Gamma)$ and $\boldsymbol{H}_{||}^{-\frac{1}{2}}(\mathrm{div}_{\Gamma}, \Gamma)$, respectively. Given a triangulation Γ_h of Γ, those will be denoted by $\boldsymbol{\mathcal{E}}_h(\Gamma_h) \subset \boldsymbol{H}_{\perp}^{-\frac{1}{2}}(\mathrm{curl}_{\Gamma}, \Gamma)$ and $\boldsymbol{\mathcal{F}}_h(\Gamma_h) \subset \boldsymbol{H}_{||}^{-\frac{1}{2}}(\mathrm{div}_{\Gamma}, \Gamma)$. Please note that $\boldsymbol{\mathcal{F}}_h(\Gamma_h)$ turns out to be a space of 2D $\boldsymbol{H}(\mathrm{div}; \Omega)$-conforming finite elements on Γ_h (cf. [13, Ch. 3]). We also point out that R takes any $\boldsymbol{\mathcal{F}}_h(\Gamma_h)$ to a possible $\boldsymbol{\mathcal{E}}_h(\Gamma_h)$ and vice versa. Yet, the reader should be aware that absolutely no relationship between $\boldsymbol{\mathcal{F}}_h(\Gamma_h)$ and $\boldsymbol{\mathcal{E}}_h(\Gamma_h)$ is stipulated.

However, ellipticity of (17) only holds provided that $\mathrm{div}_{\Gamma}\boldsymbol{\lambda} = 0$. One way to enforce this is through piecewise polynomial Lagrangian multipliers, as it is commonly done in the context of mixed finite element schemes. The (cheaper) alternative is the use of a potential representation according to (8). This is made possible by the existence of *discrete scalar potentials* in a suitable $H^1(\Gamma)$-conforming boundary element space $\mathcal{S}_h(\Gamma_h)$ [34]. In other words, $\boldsymbol{\mathcal{F}}_h^0(\Gamma_h) := \boldsymbol{\mathcal{F}}_h(\Gamma_h) \cap \boldsymbol{H}_{||}^{-\frac{1}{2}}(\mathrm{div}_{\Gamma} 0, \Gamma)$ is available as

$$\boldsymbol{\mathcal{F}}_h^0(\Gamma_h) = \mathbf{curl}_{\Gamma}\mathcal{S}_h(\Gamma_h) + \mathrm{Span}\left\{\mathbf{g}_h^1, \dots, \mathbf{g}_h^{2N}\right\} , \tag{18}$$

where for $j = 1, \dots, N$ we set $\mathbf{g}_h^j = \mathbf{curl}_{\Gamma}\varphi_h$ with Γ_h-piecewise linear $\varphi_h \in C^0(\Gamma \setminus \sigma_j)$ such that $[\varphi]_{\sigma_j} = 1$. The construction of $\mathbf{g}_h^j$ for $j = N+1, \dots, 2N$ is based on $\widehat{\sigma}_{j-N}$ (Throughout, we assume that $\sigma_j, \widehat{\sigma}_j$ are composed of edges of Γ_h). The final discrete variational problem reads: Seek $\mathbf{u}_h \in \boldsymbol{\mathcal{E}}_h(\Gamma_h)$, $\varphi_h \in \mathcal{S}_h(\Gamma_h)/\mathbb{R}$, $(\alpha_1, \dots, \alpha_{2N})^T \in \mathbb{C}^{2N}$ such that

$$
\begin{aligned}
\left\langle \widetilde{\mathsf{N}}\mathbf{u}_h, \mathbf{v}_h \right\rangle_{\mathcal{T}} + \left\langle \widetilde{\mathsf{B}}\mathbf{curl}_\Gamma \varphi_h, \mathbf{v}_h \right\rangle_{\mathcal{T}} + \sum_{k=1}^{2N} \alpha_k \left\langle \widetilde{\mathsf{B}}\mathbf{g}_h^k, \mathbf{v}_h \right\rangle_{\mathcal{T}} &= \\
&= f(\mathbf{v}_h) , \\
-\left\langle \widetilde{\mathsf{B}}\,\mathbf{curl}\,\psi_h, \mathbf{u}_h \right\rangle_{\mathcal{T}} + \left\langle \mathbf{curl}_\Gamma \psi_h, \widetilde{\mathsf{A}}\mathbf{curl}_\Gamma \varphi_h \right\rangle_{\mathcal{T}} + \sum_{k=1}^{2N} \alpha_k \left\langle \mathbf{curl}_\Gamma \psi_h, \widetilde{\mathsf{A}}\mathbf{g}_k \right\rangle_{\mathcal{T}} &= \\
&= g(\mathbf{curl}_\Gamma \psi_h) , \\
-\left\langle \widetilde{\mathsf{B}}\mathbf{g}_h^j, \mathbf{u}_h \right\rangle_{\mathcal{T}} + \left\langle \mathbf{g}_h^j, \widetilde{\mathsf{A}}\mathbf{curl}_\Gamma \varphi_h \right\rangle_{\mathcal{T}} + \sum_{k=1}^{2N} \alpha_k \left\langle \mathbf{g}_h^j, \widetilde{\mathsf{A}}\mathbf{g}_k \right\rangle_{\mathcal{T}} &= \\
&= g(\mathbf{g}_h^j) ,
\end{aligned}
$$

for all $\mathbf{v}_h \in \boldsymbol{\mathcal{E}}_h(\Gamma_h)$, $\psi_h \in \mathcal{S}_h(\Gamma_h)/\mathbb{R}$, $j = 1, \dots, 2N$. We abbreviated $\widetilde{\mathsf{A}} := \mathsf{A} + \mu_r \mathsf{A}^\kappa$, $\widetilde{\mathsf{B}} = \mathsf{B} + \mathsf{B}^\kappa$, $\widetilde{\mathsf{N}} := \mathsf{N} + \frac{1}{\mu_r}\mathsf{N}^\kappa$.

It remains to take into account lumped parameter excitation. In this case $\mathbf{E}_s = 0$, which leads to $f = 0$ and $g = 0$. Moreover, currents are confined to the conductors. Therefore, by Ampere's law, $\int_{\sigma_j} \mathbf{H} \cdot d\mathbf{s} = 0$, $j = 1, \dots, N$, since σ_j is the boundary of an oriented surface in Ω_e. Recalling the interpretation of $\boldsymbol{\lambda}$ this amounts to $\int_{\sigma_j} \mathbf{n} \times \boldsymbol{\lambda} \cdot d\mathbf{s} = 0$, $j = 1, \dots, L$. In terms of the representation (18) we conclude that the surface vectorfields $\mathbf{g}_h^{N+1}, \dots, \mathbf{g}_h^{2N}$ will not contribute to $\boldsymbol{\lambda}_h$. We can simply drop them in the discrete variational problem.

Thanks to our definition of a loop, exactly one surface edge cycle σ_l belongs to each loop of the conductor. If the current in a loop is fixed, Ampere's law determines the contribution of the associated $\mathbf{g}_h^l$ to $\boldsymbol{\lambda}_h$. Hence, α_l is known a priori and the related expressions can be moved to the right hand side of the discrete variational problem.

Remark 7. Integer combinations of surface edge cycles σ_k can be constructed with a computational effort proportional to the square of the number of edges of Γ_h [36].

6 H-based Model

For want of a transmission problem, the derivation of symmetrically coupled boundary integral equations starts from the variational problem (11).

6.1 Boundary Reduction

In order to be able to perform a reduction to the boundary through integration by parts we have to resort to scalar potentials. Therefore we use (7) to replace $\boldsymbol{\mathcal{V}}$ by

$$\mathcal{V}[\mathbf{H}_s] = \{(\mathbf{V},\Phi) \in \boldsymbol{H}(\mathbf{curl};\Omega_C)\times H^1_\Sigma(\Omega_e), \gamma_\mathbf{t}^-\mathbf{V} - \gamma_\mathbf{t}^+\widetilde{\mathrm{grad}}\,\Phi = \gamma_\mathbf{t}\mathbf{H}_s \text{ on } \Gamma\}.$$

Thus, (11) is converted into: Seek $(\mathbf{H},\Psi) \in \mathcal{V}[\mathbf{H}_s]$ such that

$$\begin{aligned}\left(\tau^{-2}\,\mathbf{curl}\,\mathbf{H}, \mathbf{curl}\,\mathbf{V}\right)_{\boldsymbol{L}^2(\Omega_C)} + i\mu_r\,(\mathbf{H},\mathbf{V})_{\boldsymbol{L}^2(\Omega_C)} +& \\ + \left(\mathbf{H}_s + \widetilde{\mathrm{grad}}\Psi, \widetilde{\mathrm{grad}}\Phi\right)_{\boldsymbol{L}^2(\Omega_e)} &= 0\,, \quad (19)\end{aligned}$$

for all $(\mathbf{V},\Phi) \in \mathcal{V}[0]$. Testing with functions compactly supported either in Ω_C or Ω_e shows that for $k = 1,\ldots,N$

$$\mathbf{curl}\,\mathbf{curl}\,\mathbf{H} + i\tau^2\mu_r\mathbf{H} = 0 \quad \text{in } \Omega_C\,, \quad (20)$$

$$-\Delta\Psi = 0 \text{ in } \Omega' \quad , \quad [\partial_\mathbf{n}\,\mathrm{grad}\,\Psi]_{\Sigma_k} = 0,\ [\gamma\Psi]_{\Sigma_k} = \text{const.}\,. \quad (21)$$

Integration by parts can be carried out on both Ω_C and Ω'. Thus, (19) becomes

$$\left\langle \tau^{-2}\gamma_N^-\mathbf{H}, \gamma_\mathbf{t}^-\mathbf{V}\right\rangle_{\boldsymbol{\tau}} - \langle\partial'_\mathbf{n}\Psi, \gamma\Phi\rangle_{1/2,\partial\Omega'} = \langle\gamma_\mathbf{n}\mathbf{H}_s, \gamma'\Phi\rangle_{1/2,\partial\Omega'}\,. \quad (22)$$

Here, γ' and $\partial'_\mathbf{n}$ are the standard trace and conormal derivative onto $\partial\Omega'$. The definition of $\partial'_\mathbf{n}$ relies on the interior unit normal vectorfield on $\partial\Omega'$.

Remark 8. Splitting the duality pairing $\langle\gamma_\mathbf{n}\mathbf{H}_s, \gamma'\Phi\rangle_{1/2,\partial\Omega'}$ into contributions of Γ and of the cuts cannot be done immediately, because the individual integrals are no continuous functionals on the space $H^{\frac{1}{2}}(\partial\Omega')$. This procedure must be postponed until after discretization.

6.2 Coupled Problem

For both (20) and (21) we need a realization of the Dirichlet-to-Neumann operator by boundary integral operators. For (21) we can rely on the exterior Calderon projector for the Laplacian on Ω', which gives the identities

$$\begin{aligned}\gamma'\Psi &= (\tfrac{1}{2}Id + \mathsf{K}')(\gamma'\Psi) - \mathsf{V}'(\partial'_\mathbf{n}\Psi)\,,\\ \partial'_\mathbf{n}\Psi &= \mathsf{D}'(\gamma'\Psi) + (\tfrac{1}{2}Id - (\mathsf{K}')^*)(\partial'_\mathbf{n}\Psi)\,.\end{aligned} \quad (23)$$

K' is the double layer potential integral operator for Δ, $(\mathsf{K}')^*$ its $\boldsymbol{L}^2(\partial\Omega)$-adjoint, and D' stands for the (negative semidefinite) hypersingular operator (cf. [48, Ch. 7]). We stress that these boundary integral operators are defined on $\partial\Omega'$ and based on a unit normal vectorfield pointing into the interior of Ω'.

What is not reflected in the statement of the Calderon identities is the special nature of the traces $\gamma'\Psi$ and $\partial'_\mathbf{n}\Psi$ entailed by the transmission conditions of (21). They imply that

$$\gamma'\Psi \in H^{\frac{1}{2}}_\Sigma(\partial\Omega') := \{v \in H^{\frac{1}{2}}(\partial\Omega'),\ [v]_{\Sigma_j} = \text{const.},\ j = 1,\ldots,N\}\,,$$

$$\partial'_\mathbf{n}\Psi \in H^{-\frac{1}{2}}_\Sigma(\partial\Omega') := \{\phi \in H^{-\frac{1}{2}}(\partial\Omega'),\ \phi^+ + \phi^- = 0 \text{ on } \Sigma_j,\ j = 1,\ldots,N\}\,.$$

For the interior problem (20) we can reuse the Calderon identities (15) with $\kappa = \frac{1}{\sqrt{2}}(1+i)\tau\sqrt{\mu_r}$ and $\mathbf{H}$ instead of $\mathbf{E}$:

$$\begin{aligned} \gamma_{\mathbf{t}}^{-}\mathbf{H} &= (\mathsf{A}^{\kappa} - \operatorname{grad}_{\Gamma} \circ \mathsf{V}^{\kappa} \circ \operatorname{div}_{\Gamma})(\gamma_N^{-}\mathbf{H}) + (\tfrac{1}{2}Id + \mathsf{C}^{\kappa})(\gamma_{\mathbf{t}}^{-}\mathbf{H}) \;, \\ \gamma_N^{-}\mathbf{H} &= (\tfrac{1}{2}Id + \mathsf{B}^{\kappa})(\gamma_N^{-}\mathbf{H}) + \mathsf{N}^{\kappa}(\gamma_{\mathbf{t}}^{-}\mathbf{H}) \;. \end{aligned} \tag{24}$$

Now we can merge (22), (23), and (24), making use of $\gamma_{\mathbf{t}}^{-}\mathbf{V} = \operatorname{grad}_{\Gamma} \gamma^{+}\Phi$ and $\gamma_{\mathbf{t}}^{-}\mathbf{H} = \operatorname{grad}_{\Gamma} \gamma^{+}\Psi + \gamma_{\mathbf{t}}\mathbf{H}_s$. This results in: Seek $u \in H_{\Sigma}^{\frac{1}{2}}(\partial\Omega')/\mathbb{R}$, $\psi \in H_{\Sigma}^{-\frac{1}{2}}(\partial\Omega')$, $\boldsymbol{\eta} \in \boldsymbol{H}_{||}^{-\frac{1}{2}}(\operatorname{div}_{\Gamma}, \Gamma)$ such that

$$\begin{aligned} n'(u,v) \;+\; b(\boldsymbol{\eta}, v) \;-\; k'(\psi, v) &= f(v) \;, \\ -b(\boldsymbol{\mu}, u) + a(\boldsymbol{\eta}, \boldsymbol{\mu}) \qquad\qquad &= g(\boldsymbol{\mu}) \;, \\ k'(\phi, u) \qquad\qquad + d'(\psi, \phi) &= 0 \;. \end{aligned} \tag{25}$$

for all $v \in H_{\Sigma}^{\frac{1}{2}}(\partial\Omega')/\mathbb{R}$, $\phi \in H_{\Sigma}^{-\frac{1}{2}}(\partial\Omega')$, $\boldsymbol{\mu} \in \boldsymbol{H}_{||}^{-\frac{1}{2}}(\operatorname{div}_{\Gamma}, \Gamma)$, where

$$\begin{aligned} n'(u,v) &:= \tfrac{1}{\tau^2}\left\langle \mathsf{N}^{\kappa}(\operatorname*{grad}_{\Gamma} u), \operatorname*{grad}_{\Gamma} v \right\rangle_{\tau} - \langle \mathsf{D}'u, v\rangle_{1/2,\partial\Omega'} \;, \\ b(\boldsymbol{\eta}, v) &:= \tfrac{1}{\tau^2}\left\langle (\tfrac{1}{2}Id + \mathsf{B}^{\kappa})\boldsymbol{\eta}, \operatorname*{grad}_{\Gamma} v \right\rangle_{\tau} \;, \\ k'(\psi, v) &:= \langle \psi, (\tfrac{1}{2}Id - \mathsf{K}')v\rangle_{1/2,\partial\Omega'} \;, \\ a(\boldsymbol{\eta}, \boldsymbol{\mu}) &:= \tfrac{1}{\tau^2}(\langle \boldsymbol{\mu}, \mathsf{A}^{\kappa}\boldsymbol{\eta}\rangle_{\tau} + \langle \operatorname{div}_{\Gamma}\boldsymbol{\mu}, \mathsf{V}^{\kappa}\operatorname{div}_{\Gamma}\boldsymbol{\eta}\rangle_{1/2,\Gamma}) \;, \\ d'(\psi, \phi) &:= \langle \phi, \mathsf{V}'\psi\rangle_{1/2,\partial\Omega'} \;, \\ f(v) &:= \langle \gamma_{\mathbf{n}}'\mathbf{H}_s, v\rangle_{1/2,\partial\Omega'} - \left\langle \mathsf{N}^{\kappa}(\gamma_{\mathbf{t}}\mathbf{H}_s), \operatorname*{grad}_{\Gamma} v \right\rangle_{\tau} \;, \\ g(\boldsymbol{\mu}) &:= \tau^{-2}\left\langle \boldsymbol{\mu}, (\tfrac{1}{2}Id - \mathsf{C}^{\kappa})\gamma_{\mathbf{t}}\mathbf{H}_s \right\rangle_{\tau} \;. \end{aligned}$$

The surface gradient of the u-component of the solution of (25) provides the tangential trace of $\mathbf{H}$, whereas $\psi := \partial_{\mathbf{n}}'\Psi$ can be viewed as the magnetic flux through $\partial\Omega'$. The meaning of $\boldsymbol{\eta} := \gamma_N^{-}\mathbf{H}$ is that of a τ^2-scaled twisted tangential trace of the electric field.

Theorem 25. *The bilinear form associated with the variational problem (25) is* $H_{\Sigma}^{\frac{1}{2}}(\partial\Omega')/\mathbb{R} \times H_{\Sigma}^{-\frac{1}{2}}(\partial\Omega') \times \boldsymbol{H}_{||}^{-\frac{1}{2}}(\operatorname{div}_{\Gamma}, \Gamma)$*-elliptic.*

Proof. As in the proof of Thm. 24 we can exploit the block skew symmetric structure, because the bilinear forms on the diagonal are elliptic on their respective spaces. □

Remark 9. Actually, the coupled variational problem for the $\mathbf{H}$-based model fails the condition that only equations on Γ may be involved, because some integral operators rely on cutting surfaces, too. This is an enormous practical obstacle to the use of the **H**-based model, because the construction of cutting surfaces requires a triangulation of some part of the air region and turns out to be prohibitively expensive [42, 33].

One might wonder why this drawback is inevitable with the **H**-based model but not encountered in the case of the **E**-based model. We owe this to the second nature of **E** as a vector potential. For this reason we do not have to introduce another potential to carry out boundary reduction. On top of that a vector potential always exists and is not tied to any topological constraints.

6.3 Galerkin Discretization

A assume that a combined triangulation Γ'_h of Γ and the cuts Σ_k, $k = 1, \ldots, N$, is supplied. As before, we write Γ_h for its restriction to Γ. Thanks to Thm. 25 a conforming Galerkin discretization will yield quasi-optimal approximations of solutions u, ψ, and $\boldsymbol{\eta}$ of (25).

In particular, the space $\boldsymbol{\mathcal{F}}_h(\Gamma_h) \subset \boldsymbol{H}_{||}^{-\frac{1}{2}}(\mathrm{div}_\Gamma, \Gamma)$ can be reused as trial space for $\boldsymbol{\eta}$. To approximate u and ψ we can employ the usual conforming boundary element spaces for $H^{\frac{1}{2}}(\partial\Omega')$ and $H^{-\frac{1}{2}}(\partial\Omega')$. Let $\mathcal{S}_h(\Gamma'_h)$ and $\mathcal{Q}_h(\Gamma'_h)$ stand for these.

A common trait of the boundary element spaces is that they offer far more regularity than required by mere conformity. For instance, all boundary element functions will belong to $L^\infty(\partial\Omega')$. Then the constraints inherent in the spaces $H^{\frac{1}{2}}_\Sigma(\partial\Omega')$ and $H^{-\frac{1}{2}}_\Sigma(\partial\Omega')$ permit us to restrict the operators V', K', and, D' to Γ: Straightforward manipulations using the integral operator representations of V', D', and K' show that for $u, v \in H^{\frac{1}{2}}_\Sigma(\partial\Omega') \cap L^\infty(\partial\Omega')$ and $\phi, \psi \in H^{-\frac{1}{2}}_\Sigma(\partial\Omega') \cap L^\infty(\partial\Omega')$

$$
\begin{aligned}
\langle\psi, \mathsf{V}'\phi\rangle_{1/2,\partial\Omega'} &= \langle\psi, \mathsf{V}\phi\rangle_{1/2,\Gamma} \quad , \quad \langle\mathsf{D}'u, v\rangle_{1/2,\partial\Omega'} = \langle\mathsf{D}u, v\rangle_{1/2,\Gamma} \ , \\
\langle\phi, \mathsf{K}'v\rangle_{1/2,\partial\Omega'} &= \langle\phi, \mathsf{K}v\rangle_{1/2,\Gamma} + \sum_{k=1}^{N} [v]_{\Sigma_k} \int_\Gamma \int_{\Sigma_k} \frac{\partial G_0(\mathbf{x}, \mathbf{y})}{\partial \mathbf{n}(\mathbf{y})} \phi(\mathbf{x}) \, dS(\mathbf{y}) dS(\mathbf{x}) \ , \\
\langle\phi, v\rangle_{1/2,\partial\Omega'} &= \langle\phi, v\rangle_{1/2,\Gamma} + \sum_{k=1}^{N} [v]_{\Sigma_k} \cdot \int_{\Sigma_k} \phi(\mathbf{x}) \, dS(\mathbf{x}) \ .
\end{aligned}
\tag{26}
$$

We observe that the cuts will enter the discrete variational problem only through some global integral quantities that are not sensitive to the choice of boundary elements on the cuts. Sloppily speaking, this permits us to cover each cut by only a single surface element. More precisely, we may choose

$$
\begin{aligned}
\mathcal{S}_h(\Gamma'_h) &= \mathcal{S}_h(\Gamma_h) + \mathrm{Span}\left\{c_h^1, \ldots, c_h^N\right\} \ , \\
\mathcal{Q}_h(\Gamma'_h) &= \mathcal{Q}_h(\Gamma_h) + \mathrm{Span}\left\{\chi_1, \ldots, \chi_N\right\} \ .
\end{aligned}
$$

Here, c_h^k is a Γ_h'-piecewise linear function $\in C^0(\partial\Omega') \cap H_\Sigma^{\frac{1}{2}}(\partial\Omega')$, whose restriction to Γ has a jump of height 1 across the edge cycle σ_k and is continuous across any other σ_j, $j \neq k$. The function $\chi_k \in L^\infty(\partial\Omega') \cap H_\Sigma^{-\frac{1}{2}}(\partial\Omega')$ assumes the values $+1$ and -1 on Σ_k^+ and Σ_k^-, respectively, and vanishes on $\partial\Omega' \setminus \Sigma_k$.

Using the identities (26), the discrete variational problem can be rephrased as: Seek $\tilde{u}_h \in \mathcal{S}_h(\Gamma_h)$, $\tilde{\psi}_h \in \mathcal{Q}_h(\Gamma_h)$, $\boldsymbol{\eta}_h \in \boldsymbol{\mathcal{F}}_h(\Gamma_h)$, $\alpha_1, \dots, \alpha_N \in \mathbb{C}$, $\beta_1, \dots, \beta_N \in \mathbb{C}$ such that

$$
\begin{aligned}
&n(\tilde{u}_h, v_h) + b(\boldsymbol{\eta}_h, v_h) - k(\tilde{\psi}_h, v_h) + \sum_k \alpha_k n(c_k, v_h) &&= f(v_h)\,,\\
&-b(\boldsymbol{\mu}_h, \tilde{u}_h) + a(\boldsymbol{\eta}_h, \boldsymbol{\mu}_h) - \sum_k \alpha_k b(\boldsymbol{\mu}_h, c_k) &&= g(\boldsymbol{\mu}_h)\,,\\
&k(\phi_h, \tilde{u}_h) + d(\psi_h, \phi_h) + \sum_k \alpha_k k'(\phi_h, c_k) &&= 0\,,\\
&n(\tilde{u}_h, c_j) + b(\boldsymbol{\eta}_h, c_j) - k'(\tilde{\psi}_h, c_j) + \sum_k \alpha_k n(c_k, c_j) - {}\\
&\qquad \sum_k \beta_k k'(\chi_k, c_j) &&= f(c_j)\,,\\
&\qquad \sum_k \alpha_k k'(\chi_l, c_k) &&= 0\,.
\end{aligned}
$$

for all $v_h \in \mathcal{S}_h(\Gamma_h)$, $\boldsymbol{\mu}_h \in \boldsymbol{\mathcal{F}}_h(\Gamma_h)$, $\phi_h \in \mathcal{Q}_h(\Gamma_h)$, $j = 1, \dots, N$, $l = 1, \dots, N$. Here we set, using [27, Thm. 7, Ch. XI],

$$
\begin{aligned}
n(u, v) &:= \tfrac{1}{\tau^2}\left(\mathsf{N}^\kappa(\operatorname{grad}_\Gamma u), \operatorname{grad}_\Gamma v\right)_{\boldsymbol{L}^2(\Gamma)} - \left(\mathsf{A}\widetilde{\mathbf{curl}_\Gamma u}, \widetilde{\mathbf{curl}_\Gamma v}\right)_{\boldsymbol{L}^2(\Gamma)}\,,\\
d(\psi, \phi) &:= (\phi, \mathsf{V}\psi)_{L^2(\Gamma)}\,, \quad k(\psi, v) := \left(\psi, (\tfrac{1}{2}Id - \mathsf{K})v\right)_{L^2(\Gamma)}\,,
\end{aligned}
$$

for bilinear forms induced by integral operators on Γ alone. The discrete solution can be obtained as $u_h = \tilde{u}_h + \sum_k \alpha_k c_k$ and $\psi_h = \tilde{\psi}_h + \sum_k \beta_k \chi_k$. A closer study of the boundary integral operators shows that the cuts only come into play through integrals of the form

$$
\int_\Gamma \int_{\Sigma_k} \frac{\partial G_0(\mathbf{x}, \mathbf{y})}{\partial \mathbf{n}(\mathbf{y})} \phi_h(\mathbf{x})\, dS(\mathbf{y}, \mathbf{x}) \quad , \quad \int_{\Sigma_k} \mathbf{H}_s \cdot \mathbf{n}\, dS\,,
$$

for $\phi_h \in \mathcal{Q}_h(\Gamma_h)$. Obviously, by Gauß' divergence theorem, Σ_k can be replaced by any other surface homologous in $H_2(\Omega_e, \mathbb{Z})$ without changing the values of the integrals. Paradoxically, information about the concrete geometry of the Σ_k seems to be indispensable for the evaluation of the integrals.

The case of lumped parameter excitation is treated in a similar fashion as in the case of the **E**-based model. First, note that α_k measures the jump of the magnetic scalar potential across Σ_k. According to Ampere's law the height of this jump agrees with the total current in the loop of the conductor corresponding to Σ_k. Hence, a prescribed total current in a loop of the conductor can be taken into account by fixing the value of α_k for the related cut.

Remark 10. The values of the β_k agree with the total magnetic flux through the cut Σ_k. By Faraday's law it is proportional to the electromotive force along σ_k. Hence, if the voltage around a loop of the conductor is to be imposed, we can do so by fixing the value of the associated β_k. The possibility to take into account lumped parameter voltage excitation is only available with the **H**-based model.

References

1. R. Albanese and G. Rubinacci, *Fomulation of the eddy-current problem*, IEE Proc. A, 137 (1990), pp. 16–22.
2. A. Alonso-Rodriguez, P. Fernandes, and A. Valli, *Weak and strong formulations for the time-harmonic eddy-current problem in general domains*, Report UTM 603, Dipartimento di Matematica, Universita degli Studi di Trento, Trento, Italy, September 2001.
3. H. Ammari, A. Buffa, and J.-C. Nédélec, *A justification of eddy currents model for the Maxwell equations*, SIAM J. Appl. Math., 60 (2000), pp. 1805–1823.
4. C. Amrouche, C. Bernardi, M. Dauge, and V. Girault, *Vector potentials in three-dimensional nonsmooth domains*, Math. Meth. Appl. Sci., 21 (1998), pp. 823–864.
5. D. Baldomir, *Differential forms and electromagnetism in 3-dimensional Euclidean space* $\mathbb{R}^3$, IEE Proc. A, 133 (1986), pp. 139–143.
6. D. Baldomir and P. Hammond, *Geometry of Electromagnetic Systems*, Clarendon Press, Oxford, 1996.
7. A. Bossavit, *Two dual formulations of the 3D eddy-currents problem*, COMPEL, 4 (1985), pp. 103–116.
8. ——, *Whitney forms: A class of finite elements for three-dimensional computations in electromagnetism*, IEE Proc. A, 135 (1988), pp. 493–500.
9. ——, *A new viewpoint on mixed elements*, Meccanica, 27 (1992), pp. 3–11.
10. ——, *Computational Electromagnetism. Variational Formulation, Complementarity, Edge Elements*, vol. 2 of Electromagnetism Series, Academic Press, San Diego, CA, 1998.
11. A. Bossavit and J. Vérité, *A mixed FEM-BIEM method to solve 3D eddy-current problems*, IEEE Trans. MAG, 18 (1982), pp. 431–435.
12. A. Bossavit and J. Verité, *The "Trifou" code: Solving the three-dimensional eddy-currents problem by using h as a state variable*, IEEE Trans. Mag., 19 (1983), pp. 2465–2470.
13. F. Brezzi and M. Fortin, *Mixed and hybrid finite element methods*, Springer, 1991.
14. A. Buffa, *Hodge decompositions on the boundary of a polyhedron: The multi-connected case*, Math. Mod. Meth. Appl. Sci., 11 (2001), pp. 1491–1504.
15. ——, *Traces theorems for functional spaces related to Maxwell equations: An overwiew*, this volume, pp. ??–??.
16. A. Buffa and P. Ciarlet, *On traces for functional spaces related to Maxwell's equations. Part I: An integration by parts formula in Lipschitz polyhedra.*, Math. Meth. Appl. Sci., 24 (2001), pp. 9–30.

17. ——, *On traces for functional spaces related to Maxwell's equations. Part II: Hodge decompositions on the boundary of Lipschitz polyhedra and applications*, Math. Meth. Appl. Sci., 24 (2001), pp. 31–48.
18. A. Buffa, M. Costabel, and C. Schwab, *Boundary element methods for Maxwell's equations on non-smooth domains*, Report 2001-01, Seminar für Angewandte Mathematik, ETH Zürich, Zürich Switzerland, 2001. To appear in Numer. Math.
19. A. Buffa, R. Hiptmair, T. von Petersdorff, and C. Schwab, *Boundary element methods for Maxwell equations on Lipschitz domains*, Numer. Math., (2002). To appear.
20. P. Ciarlet, *The Finite Element Method for Elliptic Problems*, vol. 4 of Studies in Mathematics and its Applications, North-Holland, Amsterdam, 1978.
21. D. Colton and R. Kress, *Integral equation methods in scattering theory*, Pure and Applied Mathematics, John Wiley & Sons, 1983.
22. ——, *Inverse Acoustic and Electromagnetic Scattering Theory*, vol. 93 of Applied Mathematical Sciences, Springer-Verlag, Heidelberg, 2nd ed., 1998.
23. M. Costabel, *Symmetric methods for the coupling of finite elements and boundary elements*, in Boundary Elements IX, C. Brebbia, W. Wendland, and G. Kuhn, eds., Springer-Verlag, Berlin, 1987, pp. 411–420.
24. ——, *Boundary integral operators on Lipschitz domains: Elementary results*, SIAM J. Math. Anal., 19 (1988), pp. 613–626.
25. M. Costabel and M. Dauge, *Singularities of Maxwell's equations on polyhedral domains*, in Analysis, Numerics and Applications of Differential and Integral Equations, M. Bach, ed., vol. 379 of Longman Pitman Res. Notes Math. Ser., Addison Wesley, Harlow, 1998, pp. 69–76.
26. ——, *Maxwell and Lamé eigenvalues on polyhedra*, Math. Methods Appl. Sci., 22 (1999), pp. 243–258.
27. R. Dautray and J.-L. Lions, *Mathematical Analysis and Numerical Methods for Science and Technology*, vol. 4, Springer, Berlin, 1990.
28. G. Deschamps, *Electromagnetics and differential forms*, Proc. IEEE, 69 (1981), pp. 676–695.
29. H. Dirks, *Quasi–stationary fields for microelectronic applications*, Electrical Engineering, 79 (1996), pp. 145–155.
30. P. Fernandes and G. Gilardi, *Magnetostatic and electrostatic problems in inhomogeneous anisotropic media with irregular boundary and mixed boundary conditions*, Math. Models Meth. Appl. Sci., M^3AS, 7 (1997), pp. 957–991.
31. V. Girault, *Curl–conforming finite element methods for Navier–Stokes equations with non–standard boundary conditions in $\mathbb{R}^3$*, vol. 1431 of Lecture Notes in Mathematics, Springer–Verlag, Berlin, 1989, pp. 201–218.
32. V. Girault and P. Raviart, *Finite element methods for Navier–Stokes equations*, Springer, Berlin, 1986.
33. P. Gross, *Efficient finite element-based algorithms for topological aspects of 3-dimensional magnetoquasistatic problems*, PhD thesis, College of Engineering, Boston University, Boston, USA, 1998.
34. R. Hiptmair, *Canonical construction of finite elements*, Math. Comp., 68 (1999), pp. 1325–1346.
35. ——, *Symmetric coupling for eddy current problems*, Tech. Rep. 148, Sonderforschungsbereich 382, Universität Tübingen, Tübingen, Germany, March 2000. To appear in SIAM J. Numer. Anal.

36. R. Hiptmair and J. Ostrowski, *Generators of $H_1(\Gamma_h, \mathbb{Z})$ for triangulated surfaces: Construction and classification*, Report 160, SFB 382, Universität Tübingen, Tübingen, Germany, 2001. To appear in SIAM J. Computing.
37. C. Huber, W. Rieger, M. Haas, and W. Rucker, *A boundary element formulation using higher order curvilinear edge elements*, IEEE Trans. Mag., 34 (1998), pp. 2441–2444.
38. K. Ishibashi, *Eddy current analysis by BEM utilizing edge boundary conditions*, IEEE Trans. Mag., 32 (1996), pp. 832–835.
39. ———, *Eddy current analysis by integral equation method utilizing loop electric and surface magnetic currents as unknowns*, IEEE Trans. Mag., 34 (1998), pp. 2585–2588.
40. L. Kettunen, K. Forsman, and A. Bossavit, *Gauging in Whitney spaces*, IEEE Trans. Magnetics, 35 (1999), pp. 1466–1469.
41. P. Kotiuga, *On making cuts for magnetic scalar potentials in multiply connected regions*, J. Appl. Phys., 61 (1987), pp. 3916–3918.
42. ———, *An algorithm to make cuts for magnetic scalar potentials in tetrahedral meshes based on the finite element method*, IEEE Trans. Magnetics, 25 (1989), pp. 4129–4131.
43. ———, *Topological considerations in coupling magnetic scalar potentials to stream functions describing surface currents*, IEEE Trans. Magnetics, 25 (1989), pp. 2925–2927.
44. M. Kuhn and O. Steinbach, *FEM-BEM coupling for 3d exterior magnetic field problems*, Math. Meth. Appl. Sci., (2002). To appear.
45. S. Kurz, J. Fetzer, G. Lehner, and W. Rucker, *A novel formulation for 3D eddy current problems with moving bodies using a Lagrangian description and BEM-FEM coupling*, IEEE Trans. Mag., 34 (1998), pp. 3068–3073.
46. I. Mayergoyz, *3D eddy current problems and the boundary integral equation method*, in Computational electromagnetics, Z. Cendes, ed., Elsevier, Amsterdam, 1986, pp. 163–171.
47. R. McCamy and E. Stephan, *Solution procedures for three-dimensional eddy-current problems*, J. Math. Anal. Appl., 101 (1984), pp. 348–379.
48. W. McLean, *Strongly Elliptic Systems and Boundary Integral Equations*, Cambridge University Press, Cambridge, UK, 2000.
49. J. Nédélec, *Mixed finite elements in $\mathbb{R}^3$*, Numer. Math., 35 (1980), pp. 315–341.
50. J.-C. Nédélec, *Acoustic and Electromagnetic Equations: Integral Representations for Harmonic Problems*, vol. 44 of Applied Mathematical Sciences, Springer-Verlag, Berlin, 2001.
51. M. Reissel, *On a transmission boundary-value problem for the time-harmonic Maxwell equations without displacement currents*, SIAM J. Math. Anal., 24 (1993), pp. 1440–1457.
52. Z. Ren, F. Bouillault, A. Razek, A. Bossavit, and J. Vérité, *A new hybrid model using electric field formulation for 3D eddy-current problems*, IEEE Trans. Mag., 36 (1990), p. 473.
53. Z. Ren, F. Bouillault, A. Razek, and J. Verité, *Comparison of different boundary integral formulations when coupled with finite elements in three dimensions*, IEE Proc. A, 135 (1988), pp. 501–505.
54. Z. Ren and A. Razek, *New techniques for solving three-dimensional multiply connected eddy-current problems*, IEE Proc. A, 137 (1990), pp. 135–140.

55. A. SCHWARZ, *Topology for Physicists*, vol. 308 of Grundlehren der mathematischen Wissenschaften, Springer-Verlag, Berlin, 1994.
56. J. SHEN, *Computational electromagnetics using boundary elements*, vol. 24 of Topics in Engineering, Computational Mechanics Publ., Southampton, Boston, 1995.
57. O. STERZ AND C. SCHWAB, *A scalar BEM for time harmonic eddy current problems with impedance boundary conditions*, in Scientific Computing in Electrical Engineering, U. van Rienen, M. Günther, and D. Hecht, eds., vol. 18 of Lecture Notes in Computational Science and Engineering, Springer, Berlin, Germany, 2001, pp. 129–136.
58. J. YUAN, X. MA, AND X. CUI, *Three-dimensional eddy current calculation by an adaptive three-component boundary element algorithm*, IEEE Trans. Magnetics, 33 (1997), pp. 1275–1278.
59. D. ZHENG, *Three-dimensional eddy current analysis by the boundary element method*, IEEE Trans. Magnetics, 33 (1997), pp. 1354–1357.

A Simple Proof of Convergence for an Edge Element Discretization of Maxwell's Equations

Peter Monk

Department of Mathematical Sciences, University of Delaware, Newark, DE 19716, USA

Summary. The time harmonic Maxwell's equations for a lossless medium are neither elliptic or definite. Hence the analysis of numerical schemes for these equations presents some unusual difficulties. In this paper we give a simple proof, based on the use of duality, for the convergence of edge finite element methods applied to the cavity problem for Maxwell's equations. The cavity is assumed to be a general Lipschitz polyhedron, and the mesh is assumed to be regular but not quasi-uniform.

1 Introduction

In this paper we are going to give a simple proof of convergence of edge finite element approximations to the cavity problem for Maxwell's equations. We start by describing this boundary value problem. Let Ω be a bounded Lipschitz smooth polyhedron in $\mathbb{R}^3$ with boundary $\Gamma = \partial\Omega$ and unit outward normal $\boldsymbol{\nu}$. We suppose that the boundary Γ consists of a single connected component and that Ω is connected and simply connected. In fact these topological assumptions are not necessary, but we introduce them to shorten the proofs. The results here can be modified to allow more boundary components and non simply connected domains (see for example [2], [7]).

We wish to approximate the electric field $\boldsymbol{E} = \boldsymbol{E}(\boldsymbol{x})$ that satisfies Maxwell's equations

$$\nabla \times (\nabla \times \boldsymbol{E}) - k^2\boldsymbol{E} = \boldsymbol{F} \text{ in } \Omega, \tag{1a}$$

$$\boldsymbol{\nu} \times \boldsymbol{E} = 0 \text{ on } \Gamma. \tag{1b}$$

Here $\boldsymbol{F}$ is a given function related to the imposed current sources and the parameter k is the wave-number assumed to be real and positive. Equation (1b) specifies a standard perfect conducting boundary condition on the boundary of Ω.

The problem can be posed variationally using the space

$$H_0(\text{curl};\Omega) = \left\{\boldsymbol{u} \in \left(L^2(\Omega)\right)^3 \mid \nabla \times \boldsymbol{u} \in \left(L^2(\Omega)\right)^3,\ \boldsymbol{\nu} \times \boldsymbol{u} = 0 \text{ on } \Gamma\right\}.$$

In particular suppose $\boldsymbol{F} \in H_0(\text{curl};\Omega)'$ were $H_0(\text{curl};\Omega)'$ is the dual space of $H_0(\text{curl};\Omega)$ with respect to the $\left(L^2(\Omega)\right)^3$ inner product. Then following the usual Galerkin strategy we arrive at the problem of finding $\boldsymbol{E} \in H_0(\text{curl};\Omega)$ such that

$$\int_\Omega \nabla \times \boldsymbol{E} \cdot \nabla \times \boldsymbol{\phi} - k^2 \boldsymbol{E} \cdot \boldsymbol{\phi} \, dV = \int_\Omega \boldsymbol{F} \cdot \boldsymbol{\phi} \, dV, \quad \forall \boldsymbol{\phi} \in H_0(\text{curl};\Omega). \quad (2)$$

Because k is real, we can assume that any solution of this problem is real, so all spaces and functions in this paper are real.

Using the Helmholtz decomposition [19] which states that

$$H_0(\text{curl};\Omega) = \tilde{X} \oplus \nabla H_0^1(\Omega) \quad (3)$$

where

$$\tilde{X} = \left\{ \boldsymbol{u} \in H_0(\text{curl};\Omega) \mid \int_\Omega \boldsymbol{u} \cdot \nabla p \, dV = 0, \quad \forall p \in H_0^1(\Omega) \right\}$$

problem (2) can be reduced to a problem on $\tilde{X}$. More precisely, let $\boldsymbol{E} = \boldsymbol{u} + \nabla p$ with $\boldsymbol{u} \in \tilde{X}$ and $p \in H_0^1(\Omega)$. Choosing $\boldsymbol{\phi} = \nabla \xi$ in (2) shows that p satisfies

$$-k^2 \int_\Omega \nabla p \cdot \nabla \xi \, dV = \int_\Omega \boldsymbol{F} \cdot \nabla \xi \, dV$$

for all $\xi \in H_0^1(\Omega)$. This uniquely determines p as the solution of a Poisson problem. Once p is determined, we choose $\boldsymbol{\phi} \in \tilde{X}$ in (2) and find that $\boldsymbol{u} \in \tilde{X}$ satisfies

$$\int_\Omega \nabla \times \boldsymbol{u} \cdot \nabla \times \boldsymbol{\phi} - k^2 \boldsymbol{u} \cdot \boldsymbol{\phi} \, dV = \int_\Omega \boldsymbol{F} \cdot \boldsymbol{\phi} \, dV, \quad \forall \boldsymbol{\phi} \in \tilde{X}. \quad (4)$$

The compact embedding of $\tilde{X}$ into $(L^2(\Omega))^3$ [28, 10] and the Fredholm alternative can then be used to show that for any $\boldsymbol{F} \in H_0(\text{curl};\Omega)'$, problem (2) has a unique solution $\boldsymbol{E} \in H_0(\text{curl};\Omega)$ depending continuously on the data $\boldsymbol{F}$ provided k is not an interior Maxwell eigenvalue for Ω. For the remainder of the paper we assume $k > 0$ is not such an eigenvalue.

The problem of approximating $\boldsymbol{E}$ by finite elements then reduces to constructing a finite element subspace $X_h \subset H_0(\text{curl};\Omega)$ and computing $\boldsymbol{E}_h \in X_h$ such that

$$\int_\Omega \nabla \times \boldsymbol{E}_h \cdot \nabla \times \boldsymbol{\phi}_h - k^2 \boldsymbol{E}_h \cdot \boldsymbol{\phi}_h \, dV = \int_\Omega \boldsymbol{F} \cdot \boldsymbol{\phi}_h \, dV, \quad \forall \boldsymbol{\phi}_h \in X_h. \quad (5)$$

The obvious choice of using vector continuous piecewise linear elements is dangerous since, if the domain has re-entrant corners, it is possible to compute finite element solutions that converge to a field that is not a solution of Maxwell's equations [11]. For this simple model problem modifications to the bilinear form to restore convergence are given in [12, 13], but further modifications are needed to handle, for example, discontinuous coefficients.

We prefer to construct X_h using the edge finite elements of Nédélec [23]. These avoid the problem of spurious solutions at the cost of increased complexity. Furthermore these elements can be applied to problems involving discontinuous coefficients (modeling different media) without modification.

Our goal in this paper is to derive estimates for $\boldsymbol{E}-\boldsymbol{E}_h$ in the $H_0(\text{curl};\Omega)$ norm given by

$$||\boldsymbol{E}-\boldsymbol{E}_h||_{curl} = \sqrt{||\boldsymbol{E}-\boldsymbol{E}_h||^2 + ||\nabla\times(\boldsymbol{E}-\boldsymbol{E}_h)||^2},$$

where $||\cdot||$ is the standard $\left(L^2(\Omega)\right)^3$ norm. Otherwise, for a Hilbert spaces $\mathcal{X}$, we denote the norm by $||\cdot||_{\mathcal{X}}$.

There have been three previous results in this direction. In [22], I proved convergence using the ideas of Schatz [25] concerning the compact perturbation of coercive bilinear forms. Due to limitations on the understanding of edge elements and the regularity theory for Maxwell's equations at that time, I had to assume that Ω was convex, and the mesh was quasi-uniform.

In [14], Demkowicz and I applied the theory of collectively compact operators to prove convergence on general Lipschitz polyhedra. We assumed quasi-uniformity of the mesh to provide a certain inverse inequality (which is actually not necessary). Moreover, using the results of [7] our proof extends to include rather general spatially dependent coefficients in the equations (for example piecewise constant coefficients).

Perhaps the most general result to date is due to Boffi and Gastaldi [6]. They use the general convergence theory of Rappaz [17], together with their estimates of Maxwell eigenvalue convergence, to prove convergence on general regular meshes. This follows the observation by Demkowicz and Vardapetyan that convergence of the source problem considered here is implied by convergence of discrete Maxwell eigenvalues [16].

Both the work of Boffi and Gastaldi and my own with Demkowicz can be criticized for being too complicated. The goal here is to give a simple proof of convergence not relying on any abstract operator theory.

The main tool we shall use is the improved understanding of edge element interpolation theory and regularity results provided by [2]. We also use results from [3] modified appropriately for a Lipschitz polyhedral domain. Our paper is motivated by the work of Gopalakrishnan and Pasciak [20] who use similar estimates in their analysis of Schwarz methods for Maxwell's equations.

The layout of the paper is as follows. We start by summarizing some of the properties of edge elements. We then derive a weak Garding inequality for the error. After analyzing discrete divergence free vector fields, we use duality theory to prove the desired estimate. The proof is an improved, and simplified, version of the one in [22], but the techniques and approach are very much from [20].

2 Finite Elements and Interpolation

Let τ_h, $h > 0$, be a regular family of tetrahedral finite element meshes on Ω [8]. We shall now briefly summarize the construction of the edge and face finite elements of Nédélec [23], and some of their relevant properties.

Let P_l, $l > 0$ integer, denote the set of polynomials of total degree at most l in x_1, x_2 and x_3. We shall also use the spaces $P_l(e)$ and $P_l(f)$ of polynomials in arc length on an edge e, or surface coordinates on a face f, more precisely $P_l(f) = \{q \mid q = p|_f, p \in P_l\}$. In addition let $\tilde{P}_l$ denote the set of homogeneous polynomials of degree exactly l in x_1, x_2 and x_3. We then define

$$S_l = \left\{ \boldsymbol{p} \in (\tilde{P}_l)^3 \mid \boldsymbol{x} \cdot \boldsymbol{p} = 0 \right\}$$

and

$$R_l = (P_{l-1})^3 \oplus S_l .$$

With these definitions, the finite element space we shall use is the standard edge element space [23] given by

$$X_h = \{ \boldsymbol{u}_h \in H_0(\text{curl}; \Omega) \mid \boldsymbol{u}_h|_K \in R_l, \ \forall K \in \tau_h \} .$$

This space has the following unisolvent set of degrees of freedom, defined for sufficiently smooth vector functions $\boldsymbol{u}$ on a tetrahedron K (we give a precise statement of the smoothness requirements later). In particular let

$$M_e = \left\{ \int_e \boldsymbol{u} \cdot \boldsymbol{\tau} q \, ds, \forall q \in P_{l-1}(e) \ \forall \text{ edges } e \text{ of } K \right\}$$

where $\boldsymbol{\tau}$ is a unit tangent to e. Let $(P_{l-2}(f))^2$ denote the set of vectors tangential to f, each component being in the restriction of P_{l-2} to f. Then we define

$$M_f = \left\{ \int_f \boldsymbol{u} \cdot \boldsymbol{q} \, dA, \ \forall \boldsymbol{q} \in (P_{l-2}(f))^2, \ \forall \text{ faces } f \text{ of } K \right\},$$

$$M_K = \left\{ \int_K \boldsymbol{u} \cdot \boldsymbol{q} \, dV, \quad \forall \boldsymbol{q} \in (P_{l-3}(K))^3 \right\}.$$

Then the degrees of freedom on K are

$$\Sigma_K = M_e \cup M_f \cup M_K .$$

For sufficiently smooth vector fields $\boldsymbol{u}$, these degrees of freedom define an interpolant $r_h \boldsymbol{u}$ element by element. In particular, from [2] we know that this interpolant is well defined provided that there is a $\delta > 0$ and integer $q > 2$ such that for each tetrahedron $K \in \tau_h$

$$\boldsymbol{u} \in (H^{1/2+\delta}(K))^3, \text{ and } \nabla \times \boldsymbol{u} \in (L^q(K))^3.$$

Let

$$H^s(\text{curl};\Omega) = \left\{\boldsymbol{u} \in (H^s(\Omega))^3 \mid \nabla \times \boldsymbol{u} \in (H^s(\Omega))^3\right\}.$$

Using scaling arguments the following estimate is proved in [1].

Theorem 26. *If τ_h, $h > 0$, is a regular family of meshes on Ω and if $\boldsymbol{u} \in H^s(\text{curl};\Omega)$, $\frac{1}{2} < s \leq l$, then there is a constant C depending on s but not on h or $\boldsymbol{u}$ such that*

$$\|\boldsymbol{u} - r_h\boldsymbol{u}\|_{curl} \leq Ch^s(\|\boldsymbol{u}\|_{H^s(\Omega)} + \|\nabla \times \boldsymbol{u}\|_{H^s(\Omega)}).$$

We can also define the $H_0(\text{curl};\Omega)$ orthogonal projection P_h: $H_0(\text{curl};\Omega) \to X_h$, such that if $\boldsymbol{u} \in H_0(\text{curl};\Omega)$ then $P_h\boldsymbol{u} \in X_h$ satisfies

$$\int_\Omega \nabla \times (\boldsymbol{u} - P_h\boldsymbol{u}) \cdot \nabla \times \boldsymbol{\phi}_h + (\boldsymbol{u} - P_h\boldsymbol{u}) \cdot \boldsymbol{\phi}_h \, dV = 0, \quad \forall \boldsymbol{\phi}_h \in X_h. \tag{6}$$

This projection satisfies the optimal error estimate

$$\|\boldsymbol{u} - P_h\boldsymbol{u}\|_{curl} = \inf_{\boldsymbol{v}_h \in X_h} \|\boldsymbol{u} - \boldsymbol{v}_h\|_{curl}.$$

If $\boldsymbol{u} \in H^s(\text{curl};\Omega)$, $s > 1/2$, Theorem 26 can then be used to provide order estimates for the right hand side of the above equality.

Now let

$$S_h = \left\{p_h \in H_0^1(\Omega) \mid p_h|_K \in P_l \quad \forall K \in \tau_h\right\}.$$

This is just the standard space of continuous, piecewise degree $l-1$ finite elements. We have ([23])

$$\nabla S_h \subset X_h.$$

This provides a large subspace of test functions in X_h. Using this space, we say a function $\boldsymbol{u} \in \left(L^2(\Omega)\right)^3$ is discrete divergence free if

$$\int_\Omega \boldsymbol{u} \cdot \nabla \xi_h \, dV = 0 \quad \forall \xi_h \in S_h.$$

We then have the following discrete Helmholtz decomposition analogous to (3)

$$X_h = \tilde{X}_h \oplus \nabla S_h$$

where $\tilde{X}_h$ is the space of discrete divergence free finite elements.

Using the test function $\boldsymbol{\phi}_h = \nabla \xi_h$, $\xi_h \in S_h$ in (6) shows that $\boldsymbol{u} - P_h\boldsymbol{u}$ is discrete divergence free since

$$\int_{\Omega} (\boldsymbol{u} - P_h \boldsymbol{u}) \cdot \nabla \xi_h \, dV = 0, \qquad \forall \xi_h \in S_h. \tag{7}$$

We shall also need some properties of a subspace of divergence conforming finite element functions in the space

$$H_0(\text{div}; \Omega) = \left\{ \boldsymbol{u} \in (L^2(\Omega))^3 \mid \nabla \cdot \boldsymbol{u} \in L^2(\Omega), \quad \boldsymbol{\nu} \cdot \boldsymbol{u} = 0 \text{ on } \Gamma \right\}.$$

These finite elements are also found in Nédélec [23]. Let

$$D_l = (P_{l-1})^3 \oplus \tilde{P}_{l-1} \boldsymbol{x}$$

and define $Y_h \subset H_0(\text{div}\,; \Omega)$ by

$$Y_h = \{ \boldsymbol{u}_h \in H_0(\text{div}; \Omega) \mid \boldsymbol{u}_h|_K \in D_l \quad \forall K \in \tau_h \}.$$

Although we will not need error estimates for this space, we shall need some properties of the interpolant. The degrees of freedom for this space are defined element by element as follows. For a face f in the mesh with normal $\boldsymbol{\nu}_f$, let

$$N_f = \left\{ \int_f \boldsymbol{u} \cdot \boldsymbol{\nu}_f q \, dA, \quad \forall q \in P_{l-1}(f) \text{ for each face } f \text{ of } K \right\},$$

and let

$$N_K = \left\{ \int_K \boldsymbol{u} \cdot \boldsymbol{q} \, dV, \quad \forall \boldsymbol{q} \in (P_{l-2})^3 \right\}.$$

The degrees of freedom on an element K are $\Sigma_K = N_K \cup N_f$. These degrees of freedom define an interpolation operator w_h element by element. This operator is well defined for example on functions $\boldsymbol{u} \in (H^{1/2+\delta}(\Omega))^3$, for some $\delta > 0$.

The only property of w_h we shall use is the "commuting diagram property" that if $\boldsymbol{u}$ is such that both the interpolants $r_h \boldsymbol{u}$ and $w_h \nabla \times \boldsymbol{u}$ are well defined then

$$\nabla \times r_h \boldsymbol{u} = w_h \nabla \times \boldsymbol{u}. \tag{8}$$

This commuting property is part of the discrete deRham diagram whose importance has been pointed out particularly by Boffi and co-workers [5, 6].

Now suppose that $\boldsymbol{u} \in (H^{1/2+\delta}(\Omega))^3$ is such that $\nabla \times \boldsymbol{u} \in Y_h$. Since functions in Y_h are piecewise polynmials of fixed degree, it follows that $\nabla \times \boldsymbol{u} \in (L^q(\Omega))^3$, for any $q \geq 2$ and hence the interpolant $r_h \boldsymbol{u}$ is well defined. Using a scaling argument along the lines of [1] and the equivalence of norms for a piecewise polynomial on the reference element as in [3] we have the following result.

Lemma 2. *Let τ_h be a regular mesh, and suppose $\boldsymbol{u} \in (H^{1/2+\delta}(\Omega))^3$ is such that $\nabla \times \boldsymbol{u} \in Y_h$. Then there exists a constant C independent of h and $\boldsymbol{u}$ such that*

$$\|\boldsymbol{u} - r_h \boldsymbol{u}\| \leq C \left(h^{1/2+\delta} \|\boldsymbol{u}\|_{H^{1/2+\delta}(\Omega)} + h \|\nabla \times \boldsymbol{u}\| \right).$$

One further remark is needed regarding the discrete divergence free space $\tilde{X}_h$. First, we note that

$$\nabla \times X_h \subset Y_h$$

(clearly $\nabla \times X_h \subset H_0(\mathrm{div};\Omega)$ and the piecewise polynomials in $\nabla \times X_h$ are vector functions of degree $l-1$, so in Y_h). Thus, as in [3], we can regard the curl as a bounded operator from X_h into Y_h. In X_h the null-space of the curl operator is denoted $N(\mathrm{curl})$. Let $\boldsymbol{u}_h \in N(\mathrm{curl})$. Since the domain Ω is simply connected and the boundary Γ is connected, the fact that $\nabla \times \boldsymbol{u}_h = 0$ in Ω implies $\boldsymbol{u}_h = \nabla p$ for some $p \in H_0^1(\Omega)$. In addition since $\boldsymbol{u}_h \in X_h$ then $p \in S_h$. Hence $N(\mathrm{curl}) = \nabla S_h$.

The discrete divergence free space $\tilde{X}_h$ is thus given by $\tilde{X}_h = N(\mathrm{curl})^{\perp}$ where $N(\mathrm{curl})^{\perp}$ is the orthogonal complement of $N(\mathrm{curl}) \subset X_h$ in the $(L^2(\Omega))^3$ inner product. Now, following [3], let $\nabla_h \times$ denote the discrete adjoint operator for the curl by which we mean that for each $\boldsymbol{w}_h \in Y_h$, $\nabla_h \times \boldsymbol{w}_h \in X_h$ is the unique function such that

$$(\nabla_h \times \boldsymbol{w}_h, \boldsymbol{\psi}_h) = (\boldsymbol{w}_h, \nabla \times \boldsymbol{\psi}_h), \quad \forall \boldsymbol{\psi}_h \in X_h.$$

By a standard theorem from functional analysis (see Theorem 4.6 of [9]) we know that

$$N(\mathrm{curl})^{\perp} = \nabla_h \times (\nabla \times X_h)$$

so that we have the following result.

Lemma 3. *For each $\boldsymbol{v}_h \in \tilde{X}_h$ there is a function $\boldsymbol{w}_h \in \nabla \times X_h \subset Y_h$ such that $\boldsymbol{v}_h = \nabla_h \times \boldsymbol{w}_h$ or*

$$\int_{\Omega} \boldsymbol{v}_h \cdot \boldsymbol{\phi}_h \, dV = \int_{\Omega} \boldsymbol{w}_h \cdot \nabla \times \boldsymbol{\phi}_h \, dV \quad \forall \boldsymbol{\phi}_h \in X_h.$$

This lemma is from [3] where it is pointed out that an alternative way to write the discrete Helmholtz decomposition is as follows. Any function $\boldsymbol{v}_h \in X_h$ may be written

$$\boldsymbol{v}_h = \nabla_h \times \boldsymbol{w}_h + \nabla p_h$$

for some $\boldsymbol{w}_h \in \nabla \times X_h \subset Y_h$ and $p_h \in S_h$. This makes the discrete Helmholtz decomposition look a little more like the continuous one.

3 Error Analysis

This section is devoted to proving our main Theorem 27. For convenience we use the notation

$$a(\boldsymbol{u}, \boldsymbol{v}) = \int_{\Omega} \nabla \times \boldsymbol{u} \cdot \nabla \times \boldsymbol{v} - k^2 \boldsymbol{u} \cdot \boldsymbol{v} \, dV$$

and

$$(\boldsymbol{u}, \boldsymbol{v}) = \int_{\Omega} \boldsymbol{u} \cdot \boldsymbol{v} \, dV.$$

At this stage we do not know that $\boldsymbol{E}_h$ exists, but if it does exist we define $\boldsymbol{e}_h = \boldsymbol{E} - \boldsymbol{E}_h$. Then by subtracting (5) from (2) we obtain the Galerkin error equation

$$a(\boldsymbol{e}_h, \boldsymbol{\psi}_h) = 0, \quad \forall \boldsymbol{\psi}_h \in X_h. \tag{9}$$

In particular, choosing $\boldsymbol{\psi}_h = \nabla \xi_h$ for $\xi_h \in S_h$ shows that $\boldsymbol{e}_h$ is discrete divergence free.

In [22] the problem of estimating $||\boldsymbol{E} - \boldsymbol{E}_h||_{curl}$ was approached via a classical Garding inequality. Our first lemma is a weaker form of the Garding inequality as used in [20].

Lemma 4. *There is a constant C independent of h, $\boldsymbol{E}$ and $\boldsymbol{E}_h$ such that*

$$||\boldsymbol{e}_h||_{curl} \leq ||\boldsymbol{E} - P_h \boldsymbol{E}||_{curl} + C \sup_{\boldsymbol{v}_h \in X_h} \frac{|(\boldsymbol{e}_h, \boldsymbol{v}_h)|}{||\boldsymbol{v}_h||_{curl}}. \tag{10}$$

Proof. Using a very slight modification of the proof of Lemma 4.4 of [20] we see that by the definition of the curl norm, and the definition of $a(\cdot, \cdot)$ we have

$$\begin{aligned} ||\boldsymbol{e}_h||^2_{curl} &= a(\boldsymbol{e}_h, \boldsymbol{e}_h) + (1 + k^2)(\boldsymbol{e}_h, \boldsymbol{e}_h) \\ &= a(\boldsymbol{e}_h, \boldsymbol{E} - P_h \boldsymbol{E}) + a(\boldsymbol{e}_h, P_h \boldsymbol{E} - \boldsymbol{E}_h) + (1 + k^2)(\boldsymbol{e}_h, \boldsymbol{e}_h) \end{aligned}$$

Now using the Galerkin condition (9), the definition of the curl norm, and the definition of $|| \cdot ||_{curl}$ we have

$$\begin{aligned} ||\boldsymbol{e}_h||^2_{curl} &= a(\boldsymbol{e}_h, \boldsymbol{E} - P_h \boldsymbol{E}) + (1 + k^2)(\boldsymbol{e}_h, \boldsymbol{e}_h) \\ &= (\nabla \times \boldsymbol{e}_h, \nabla \times (\boldsymbol{E} - P_h \boldsymbol{E})) + (\boldsymbol{e}_h, (\boldsymbol{E} - P_h \boldsymbol{E})) + \\ &\qquad (1 + k^2) \{(\boldsymbol{e}_h, \boldsymbol{e}_h) - (\boldsymbol{e}_h, (\boldsymbol{E} - P_h \boldsymbol{E}))\} \\ &= (\nabla \times \boldsymbol{e}_h, \nabla \times (\boldsymbol{E} - P_h \boldsymbol{E})) + (\boldsymbol{e}_h, (\boldsymbol{E} - P_h \boldsymbol{E})) + \\ &\qquad (1 + k^2)(\boldsymbol{e}_h, P_h \boldsymbol{E} - \boldsymbol{E}_h). \end{aligned}$$

Hence using the Cauchy-Schwarz inequality, and the boundedness of the projection $P_h : H(\text{curl}; \Omega) \to X_h$

$$\begin{aligned} ||\boldsymbol{e}_h||^2_{curl} &\leq ||\boldsymbol{E} - P_h \boldsymbol{E}||_{curl} \, ||\boldsymbol{e}_h||_{curl} + (1 + k^2) \sup_{\boldsymbol{v}_h \in X_h} \frac{|(\boldsymbol{e}_h, \boldsymbol{v}_h)|}{||\boldsymbol{v}_h||_{curl}} ||P_h \boldsymbol{E} - \boldsymbol{E}_h||_{curl} \\ &= ||\boldsymbol{E} - P_h \boldsymbol{E}||_{curl} \, ||\boldsymbol{e}_h||_{curl} + (1 + k^2) \sup_{\boldsymbol{v}_h \in X_h} \frac{|(\boldsymbol{e}_h, \boldsymbol{v}_h)|}{||\boldsymbol{v}_h||_{curl}} ||P_h \boldsymbol{e}_h||_{curl} \\ &\leq ||\boldsymbol{E} - P_h \boldsymbol{E}||_{curl} \, ||\boldsymbol{e}_h||_{curl} + (1 + k^2) \sup_{\boldsymbol{v}_h \in X_h} \frac{|(\boldsymbol{e}_h, \boldsymbol{v}_h)|}{||\boldsymbol{v}_h||_{curl}} ||\boldsymbol{e}_h||_{curl}. \end{aligned}$$

This proves the desired estimate with $C = 1 + k^2$.

Our error estimate will be finished if we can estimate the supremum on the right hand side of (10). This is done in in Lemma 6.

Before we prove this lemma we need to investigate discrete divergence free functions in more detail. For such functions we can construct a nearby exactly divergence free function. This construction was used for example by Girault [18] and myself [22] with an ad-hoc analysis. Very similar results also appear in [4]. However the clearest analysis is from Arnold et al. [3].

For a given discrete divergence free function $\boldsymbol{v}_h \in \tilde{X}_h$, let us define $\boldsymbol{v}^h \in H_0(\mathrm{curl};\Omega)$ by

$$\nabla \times \boldsymbol{v}^h = \nabla \times \boldsymbol{v}_h \text{ in } \Omega, \tag{11a}$$

$$\nabla \cdot \boldsymbol{v}^h = 0 \text{ in } \Omega. \tag{11b}$$

In [3] it is suggested to view $\boldsymbol{v}^h$ as part of the solution of the mixed problem of finding $\boldsymbol{v}^h \in H_0(\mathrm{curl};\Omega)$ and $\boldsymbol{w}^h \in \nabla \times H_0(\mathrm{curl};\Omega)$ such that

$$(\boldsymbol{v}^h, \boldsymbol{\phi}) + (\nabla \times \boldsymbol{\phi}, \boldsymbol{w}^h) = 0, \quad \forall \boldsymbol{\phi} \in H_0(\mathrm{curl};\Omega), \tag{12a}$$

$$(\nabla \times \boldsymbol{v}^h, \boldsymbol{\xi}) = (\nabla \times \boldsymbol{v}_h, \boldsymbol{\xi}), \quad \forall \boldsymbol{\xi} \in \nabla \times H_0(\mathrm{curl};\Omega). \tag{12b}$$

Both the coercivity condition and Babuška-Brezzi condition for mixed methods are obviously satisfied and so $(\boldsymbol{v}^h, \boldsymbol{w}^h)$ exists.

We have the following lemma:

Lemma 5. *Let $\boldsymbol{v}_h \in \tilde{X}_h$. Suppose $\boldsymbol{v}^h \in H_0(\mathrm{curl};\Omega)$ satisfies (12) then there are constants C and $\delta > 0$ independent of h and $\boldsymbol{v}_h$ and $\boldsymbol{v}^h$ such that*

$$\|\boldsymbol{v}^h - \boldsymbol{v}_h\| \le C h^{1/2+\delta} \|\nabla \times \boldsymbol{v}_h\|.$$

Proof. The proof follows [3] checking that their result, proved for convex domains, holds here. From [2] there is an exponent $\delta > 0$ such that $\boldsymbol{v}^h \in \left(H^{1/2+\delta}(\Omega)\right)^3$, and since $\nabla \times \boldsymbol{v}^h = \nabla \times \boldsymbol{v}_h$, we see that $\nabla \times \boldsymbol{v}^h \in (L^q(\Omega))^3$ for all $q > 2$. Hence using Lemma 2, $r_h \boldsymbol{v}^h$ is well defined. But then, using the commuting diagram property of edge elements

$$\nabla \times r_h \boldsymbol{v}^h = w_h \nabla \times \boldsymbol{v}^h = w_h \nabla \times \boldsymbol{v}_h = \nabla \times \boldsymbol{v}_h. \tag{13}$$

Since $\boldsymbol{v}_h$ is discrete divergence free, by Lemma 3 there is a function $\boldsymbol{w}_h \in \nabla \times X_h$ such that

$$(\boldsymbol{v}_h, \boldsymbol{\phi}_h) + (\nabla \times \boldsymbol{\phi}_h, \boldsymbol{w}_h) = 0, \quad \forall \boldsymbol{\phi}_h \in X_h, \tag{14a}$$

$$(\nabla \times \boldsymbol{v}_h, \xi_h) = (\nabla \times \boldsymbol{v}_h, \xi_h), \quad \forall \boldsymbol{\xi}_h \in \nabla \times X_h. \tag{14b}$$

Thus $(\boldsymbol{v}_h, \boldsymbol{w}_h)$ is nothing else than the mixed finite element approximation to $(\boldsymbol{v}^h, \boldsymbol{w}^h)$ defined by (12). Now selecting $\boldsymbol{\phi} = r_h \boldsymbol{v}^h - \boldsymbol{v}_h$ in (12a) and $\boldsymbol{\phi}_h = r_h \boldsymbol{v}^h - \boldsymbol{v}_h$ in (14a) and using the fact that $\nabla \times \boldsymbol{\phi}_h = 0$ (see (13)) we have

$$(\boldsymbol{v}^h - \boldsymbol{v}_h, r_h\boldsymbol{v}^h - \boldsymbol{v}_h) = 0.$$

Thus

$$(\boldsymbol{v}^h - \boldsymbol{v}_h, \boldsymbol{v}^h - \boldsymbol{v}_h) = (\boldsymbol{v}^h - \boldsymbol{v}_h, \boldsymbol{v}^h - r_h\boldsymbol{v}^h) + (\boldsymbol{v}^h - \boldsymbol{v}_h, r_h\boldsymbol{v}^h - \boldsymbol{v}_h).$$

Hence $\|\boldsymbol{v}^h - \boldsymbol{v}_h\| \le \|\boldsymbol{v}^h - r_h\boldsymbol{v}^h\|$ and using Corollary 2 we have

$$\|\boldsymbol{v}^h - \boldsymbol{v}_h\| \le C\left(h^{1/2+\delta}\|\boldsymbol{v}^h\|_{H^{1/2+\delta}(\Omega)} + h\|\nabla\times\boldsymbol{v}_h\|\right).$$

The a priori estimate $\|\boldsymbol{v}^h\|_{H^{1/2+\delta}(\Omega)} \le C\|\nabla\times\boldsymbol{v}_h\|$ completes the proof.

Now we can estimate the troublesome term in (10).

Lemma 6. *For all h small enough there exists constants C and δ with $0 < \delta \le 1/2$ such that*

$$\sup_{\boldsymbol{v}_h\in X_h}\frac{|(\boldsymbol{e}_h, \boldsymbol{v}_h)|}{\|\boldsymbol{v}_h\|_{curl}} \le Ch^{\delta+1/2}\|\boldsymbol{e}_h\|_{curl}.$$

Proof. This lemma is proved by a duality argument similar to the one in the proof of Lemma 4.3 of [20] and in [22]. Indeed the use of the continuous Helmholtz decomposition as a means of estimating the discrete error goes back to the pioneering work of Kikuchi [21]. Using the continuous Helmholtz decomposition there is a divergence free function $\boldsymbol{e}_0^h \in H_0(\text{curl};\Omega)$ and a scalar $p^h \in H_0^1(\Omega)$ such that

$$\boldsymbol{e}_h = \boldsymbol{e}_0^h + \nabla p^h.$$

Here $p^h \in H_0^1(\Omega)$ satisfies

$$(\nabla p^h, \nabla\xi) = (\boldsymbol{e}_h, \nabla\xi), \quad \forall\xi \in H_0^1(\Omega).$$

Thus, by choosing $\xi = p^h$, we see that $\|\nabla p^h\| \le \|\boldsymbol{e}_h\|$.

Using the discrete Helmholtz decomposition we also can write

$$\boldsymbol{v}_h = \boldsymbol{v}_{0,h} + \nabla\xi_h$$

for some $\boldsymbol{v}_{0,h} \in \tilde{X}_h$ and $\xi_h \in S_h$. Since we have already shown that $\boldsymbol{e}_h$ is discrete divergence free, we have

$$(\boldsymbol{e}_h, \boldsymbol{v}_h) = (\boldsymbol{e}_h, \boldsymbol{v}_{0,h}) = (\boldsymbol{e}_0^h, \boldsymbol{v}_{0,h}) + (\nabla p^h, \boldsymbol{v}_{0,h}). \tag{15}$$

The first term on the right hand side is estimated by

$$|(\boldsymbol{e}_0^h, \boldsymbol{v}_{0,h})| \le \|\boldsymbol{e}_0^h\|\,\|\boldsymbol{v}_{0,h}\| \le C\|\boldsymbol{e}_0^h\|\,\|\boldsymbol{v}_h\| \tag{16}$$

where we have made use of the fact that $\|\nabla\xi_h\| \le \|\boldsymbol{v}_h\|$. Thus we can estimate this term by estimating $\|\boldsymbol{e}_0^h\|$ which we do next.

We define the adjoint variable $z \in H_0(\text{curl};\Omega)$ such that

$$a(\boldsymbol{\phi}, \boldsymbol{z}) = (\boldsymbol{e}_0^h, \boldsymbol{\phi}), \quad \forall \boldsymbol{\phi} \in H_0(\text{curl};\Omega). \tag{17}$$

Clearly $\boldsymbol{z}$ is the weak solution in $H_0(\text{curl};\Omega)$ of

$$\nabla \times \nabla \times \boldsymbol{z} - k^2 \boldsymbol{z} = \boldsymbol{e}_0^h.$$

and the assumption that k is not an interior Maxwell eigenvalue implies that $\boldsymbol{z}$ is well defined and there is a constant C such that $||\boldsymbol{z}||_{curl} \leq C||\boldsymbol{e}_0^h||$.

Since $\boldsymbol{e}_0^h$ is divergence free, it follows that $\boldsymbol{z}$ is also divergence free (to see this take $\boldsymbol{\phi} = \nabla\xi$ for $\xi \in H_0^1(\Omega)$ in equation (17)). Thus we have

$$\nabla \times \boldsymbol{z} \in (L^2(\Omega))^3, \quad \nabla \cdot \boldsymbol{z} = 0 \text{ in } \Omega \text{ and } \boldsymbol{\nu} \times \boldsymbol{z} = 0 \text{ on } \Gamma.$$

By Proposition 3.7 of [2] we have $\boldsymbol{z} \in \left(H^{1/2+\delta}(\Omega)\right)^3$ for some δ with $0 < \delta \leq 1/2$ together with the norm bound

$$||\boldsymbol{z}||_{H^{1/2+\delta}(\Omega)} \leq C||\boldsymbol{e}_0^h||.$$

In addition we see that $\nabla \times \boldsymbol{z} \in (L^2(\Omega))^3$ is the weak solution of

$$\begin{aligned} \nabla \times (\nabla \times \boldsymbol{z}) &= k^2 \boldsymbol{z} + \boldsymbol{e}_0^h \in (L^2(\Omega))^3, \\ \nabla \cdot (\nabla \times \boldsymbol{z}) &= 0 \text{ in } \Omega, \\ \boldsymbol{\nu} \cdot (\nabla \times \boldsymbol{z}) &= 0 \text{ on } \Gamma. \end{aligned}$$

Thus again by Proposition 3.7 of [2],

$$\nabla \times \boldsymbol{z} \in \left(H^{1/2+\delta}(\Omega)\right)^3$$

with the norm bound

$$||\nabla \times \boldsymbol{z}||_{H^{1/2+\delta}(\Omega)} \leq C||\boldsymbol{e}_0^h||.$$

We conclude that $\boldsymbol{z} \in H^{1/2+\delta}(\text{curl};\Omega)$. Hence the interpolant $r_h \boldsymbol{z}$ is well defined, and we can use Theorem 26 to obtain the error estimate

$$||\boldsymbol{z} - P_h \boldsymbol{z}||_{curl} \leq ||\boldsymbol{z} - r_h \boldsymbol{z}||_{curl} \leq C h^{1/2+\delta} ||\boldsymbol{e}_0^h||.$$

Now using (17) and the fact that $\boldsymbol{z}$ is divergence free we have

$$||\boldsymbol{e}_0^h||^2 = a(\boldsymbol{e}_0^h, \boldsymbol{z}) = a(\boldsymbol{e}_0^h + \nabla p^h, \boldsymbol{z}) = a(\boldsymbol{e}_h, \boldsymbol{z}).$$

Hence by the Galerkin condition (9)

$$||\boldsymbol{e}_0^h||^2 = a(\boldsymbol{e}_h, \boldsymbol{z} - P_h \boldsymbol{z}) \leq C||\boldsymbol{e}_h||_{curl} ||\boldsymbol{z} - P_h \boldsymbol{z}||_{curl} \leq C h^{1/2+\delta} ||\boldsymbol{e}_0^h|| \, ||\boldsymbol{e}_h||_{curl}.$$

We have thus proved that

$$\|\boldsymbol{e}_0^h\| \le Ch^{1/2+\delta}\|\boldsymbol{e}_h\|_{curl} \tag{18}$$

Now we estimate the term $(\nabla p^h, \boldsymbol{v}_{0,h})$ in (15). Since $\boldsymbol{v}_{0,h}$ is discrete divergence free Lemma 5 implies that there is a divergence free function $\boldsymbol{v}_0^h \in H(\mathrm{curl};\Omega)$ with

$$\|\boldsymbol{v}_0^h - \boldsymbol{v}_{0,h}\| \le Ch^{1/2+\delta}\|\nabla\times\boldsymbol{v}_{0,h}\| = Ch^{1/2+\delta}\|\nabla\times\boldsymbol{v}_h\|.$$

Now using the fact that $\boldsymbol{v}_0^h$ is divergence free, and the error estimate above, we have

$$\begin{aligned}(\nabla p^h, \boldsymbol{v}_{0,h}) &= (\nabla p^h, \boldsymbol{v}_{0,h} - \boldsymbol{v}_0^h)\\ &\le Ch^{1/2+\delta}\|\nabla p^h\|\,\|\nabla\times\boldsymbol{v}_h\|.\end{aligned} \tag{19}$$

Using (18) in (16) and using the resulting estimate together with (19) in (15) proves the desired result.

We now state and prove our main theorem:

Theorem 27. *Let Ω be a simply connected Lipschitz polyhedron with connected boundary Γ. Suppose k is not a Maxwell eigenvalue for Ω. Then if $\boldsymbol{E}$ satisfies (2) and $\boldsymbol{E}_h \in X_h$ satisfies (5) there is a constant C independent of h, $\boldsymbol{E}$ and $\boldsymbol{E}_h$ and a constant $h_0 > 0$ independent of $\boldsymbol{E}$ and $\boldsymbol{E}_h$ such that for all $0 < h < h_0$,*

$$\|\boldsymbol{E} - \boldsymbol{E}_h\|_{curl} \le \frac{1}{1 - Ch^{1/2+\delta}} \inf_{\boldsymbol{v}_h \in X_h} \|\boldsymbol{E} - \boldsymbol{v}_h\|_{curl}.$$

Here $\delta > 0$ is the exponent in Lemma 5.

Remarks:

1. Choosing h small enough that (for example) $Ch^{1/2+\delta} < 1/2$ proves quasi-optimal convergence of the edge element approximation. Furthermore the constant $1/(1 - Ch^{1/2+\delta})$ can be made arbitrarily close to unity. This seems to me to be a surprising result given that the norm involved in this estimate is the standard $H(\mathrm{curl};\Omega)$ norm and is not k dependent (of course the constant C does depend on k via the a priori estimate for the dual problem).
2. If $\boldsymbol{u} \in H^s(\mathrm{curl};\Omega)$ for some s with $1/2 < s \le l$, then Theorems 26 and 27 show that for all sufficiently small h there is a constant C such that

$$\|\boldsymbol{E} - \boldsymbol{E}_h\|_{curl} \le Ch^s.$$

In general the polyhedral boundary Γ causes singularities in the solution that prevent high global regularity. Nevertheless, as we have seen, we can expect sufficient regularity to guarantee a convergence rate of better than $O(h^{1/2})$.

Proof. Lemma 6 shows that

$$\sup_{\boldsymbol{v}_h \in X_h} \frac{|(\boldsymbol{e}_h, \boldsymbol{v}_h)|}{\|\boldsymbol{v}_h\|_{curl}} \leq C h^{1/2+\delta} \|\boldsymbol{e}_h\|_{curl}.$$

Putting this together with (10) shows that

$$\|\boldsymbol{e}_h\|_{curl} \leq \|\boldsymbol{E} - P_h \boldsymbol{E}\|_{curl} + C h^{1/2+\delta} \|\boldsymbol{e}_h\|_{curl}.$$

Choosing h small enough that $1 - Ch^{1/2+\delta} > 0$ proves the result.

Corollary 3. *For any $\boldsymbol{F} \in H_0(curl\ ;\Omega)'$, there is an $h_0 > 0$ such that for all $h < h_0$, equation (5) has a unique solution.*

Proof. It suffices to prove uniqueness. Let $\boldsymbol{F} = 0$, then since k is not a Maxwell eigenvalue, $\boldsymbol{E} = 0$ in (2) and $\boldsymbol{E}_h = 0$ is one solution of the discrete problem. By the above error estimate, for any solution $\boldsymbol{E}_h$ of the discrete problem $\|\boldsymbol{E}_h\|_{curl} \leq C \inf_{\boldsymbol{v}_h \in X_h} \|\boldsymbol{v}_h\| = 0$. Hence $\boldsymbol{E}_h = 0$, and uniqueness is proved.

4 Conclusion

The proof we have given rests critically on regularity results for the dual problem, and on the estimate for in Lemma 3 for the approximation of a discrete divergence free function by a divergence free function. For smooth coefficients these results still hold. But for general coefficients both results might be difficult to obtain, however it is possible that using arguments like those in [7] the explicit estimates used here could be replaced by uniform convergence estimates based on compactness arguments of the type used by Schatz and Wang [26].

From the point of view of analyzing other elements, the proof here is also valid for second family edge elements on tetrahedra [24] and for the edge finite elements on parallelepipeds in [23]. Extending these results to $h-p$ elements such as those in [27, 15] would require the low regularity interpolation results in Theorem 26 and the estimate in Lemma 2. As far as I am aware, these are not yet proved.

References

1. A. Alonso and A. Valli, *An optimal domain decomposition preconditioner for low-frequency time-harmonic Maxwell equations*, Mathematics of Computations, 68 (1999), pp. 607–631.
2. C. Amrouche, C. Bernardi, M. Dauge, and V. Girault, *Vector potentials in three-dimensional nonsmooth domains*, Math. Meth. Appl. Sci., 21 (1998), pp. 823–864.

3. D. Arnold, R. Falk, and R. Winthur, *Multigrid in H(div) and H(curl)*, Numerische Mathematik, 85 (2000), pp. 197–217.
4. D. Boffi, *Fortin operators and discrete compactness for edge elements*, Numer. Math., 87 (2000), pp. 229–246.
5. ——, *A note on the de Rham complex and a discrete compactness property*, Applied Mathematics Letters, 14 (2001), pp. 33–38.
6. D. Boffi and L. Gastaldi, *Edge finite elements for the approximation of Maxwell resolvent operators*. Preprint.
7. S. Caorsi, P. Fernandes, and M. Raffetto, *On the convergence of Galerkin finite element approximations of electromagnetic eigenproblems*, SIAM. J. Numer. Anal., 38 (2000), pp. 580–607.
8. P. Ciarlet, *The Finite Element Method for Elliptic Problems*, vol. 4 of Studies In Mathematics and It's Applications, Elsevier North-Holland, New York, 1978.
9. D. Colton and R. Kress, *Inverse Acoustic and Electromagnetic Scattering Theory*, no. 93 in Applied Mathematical Sciences, Springer-Verlag, New York, second ed., 1998.
10. M. Costabel, *A remark on the regularity of solutions of Maxwell's equations on Lipschitz domains*, Math. Meth. Appl. Sci., 12 (1990), pp. 365–368.
11. M. Costabel and M. Dauge, *Maxwell and Lamé eigenvalues on polyhedra*, Mathematical Methods in the Applied Sciences, 22 (1999), pp. 243–258.
12. M. Dauge, M. Costabel, and D. Martin, *Numerical investigation of a boundary penalization method for Maxwell equations.* Report available at `http://www.maths.univ-rennes1.fr/~dauge/`, 1999.
13. ——, *Weighted regularization of Maxwell equations in polyhedral domains.* Report available at `http://www.maths.univ-rennes1.fr/~dauge/`, 2001.
14. L. Demkowicz and P. Monk, *Discrete compactness and the approximation of Maxwell's equations in* $\mathbb{R}^3$, Math. Comp., 70 (2001), pp. 507–523.
15. L. Demkowicz, P. Monk, and L. Vardapetyan, *de Rham diagram for hp finite element spaces*, Comput. Math. Appl., 39 (2000), pp. 29–38.
16. L. Demkowicz and L. Vardapetyan, *Modelling electromagnetic absorbtion/scattering problems using hp-adaptive finite elements*, Compu. Methods Appl. Mech. Engrg., 152 (1998), pp. 103–124.
17. J. Descloux, N. Nassif, and J. Rappaz, *On spectral approximation. I. The problem of convergence*, RAIRO Anal. Númer., 12 (1978), pp. 97–112.
18. V. Girault, *Incompressible finite element methods for Navier–Stokes equations with nonstandard boundary conditions in* $\mathbb{R}^3$, Math. Comp., 51 (1988), pp. 53–58.
19. V. Girault and P.-A. Raviart, *Finite Element Approximation of the Navier-Stokes Equations*, vol. 749 of Lecture Notes in Mathematics, Springer-Verlag, Berlin, 1979.
20. J. Gopalakrishnan and J. Pasciak, *Overlapping Schwarz preconditioners for indefinite Maxwell equations.* To appear in *Math. Comp.*
21. F. Kikuchi, *On a discrete compactness property for the Nedelec finite elements*, J. Fac. Sci. Univ. Tokyo, Sect. 1A, Math, 36 (1989), pp. 479–490.
22. P. Monk, *A finite element method for approximating the time-harmonic Maxwell equations*, Numer. Math., 63 (1992), pp. 243–261.
23. J. Nédélec, *Mixed finite elements in* $\mathbb{R}^3$, Numer. Math., 35 (1980), pp. 315–341.
24. ——, *A new family of mixed finite elements in* $\mathbb{R}^3$, Numer. Math., 50 (1986), pp. 57–81.

25. A. Schatz, *An observation concerning Ritz-Galerkin methods with indefinite bilinear forms*, Math. Comp., 28 (1974), pp. 959–962.
26. A. Schatz and J. Wang, *Some new error estimates for Ritz-Galerkin methods with minimal regularity assumptions*, Mathematics of Computation, 65 (1996), p. 19.
27. L. Vardapetyan and L. Demkowicz, *hp-Adaptive finite elements in electromagnetics*, Computers Methods in Applied Mechanics and Engineering, 169 (1999), pp. 331–344.
28. C. Weber, *A local compactness theorem for Maxwell's equations*, Math. Meth. in the Appl. Sci., 2 (1980), pp. 12–25.

The Time-Harmonic Eddy-Current Problem in General Domains: Solvability via Scalar Potentials

Ana Alonso Rodríguez[1], Paolo Fernandes[2], and Alberto Valli[3]

[1] Dipartimento di Matematica, Università di Milano, via Saldini 50, 20133 Milano, Italy.
[2] Istituto per la Matematica Applicata del C.N.R., via De Marini 6, Torre di Francia, 16149 Genova, Italy.
[3] Dipartimento di Matematica, Università di Trento, 38050 Povo (Trento), Italy.

Summary. The eddy-current problem for the time-harmonic Maxwell equations in domains of general topology is solved by introducing a scalar "potential" for the magnetic field in the insulator part of the domain. Indeed, since in general the insulator Ω_I is multiply-connected, the magnetic field differs from the gradient of a potential by a harmonic field. We rewrite the problem in a two-domain formulation, in term of a scalar magnetic potential and a harmonic field in Ω_I. Then the finite element numerical approximation based on this two-domain formulation is presented, using edge elements in the conductor and nodal elements in the insulator, and an optimal error estimate is proved. An iteration-by-subdomain procedure for the solution of the problem is also proposed.

1 Introduction

Let us consider a bounded connected open set $\Omega \subset \mathbb{R}^3$, with boundary $\partial\Omega$. The unit outward normal vector on $\partial\Omega$ will be denoted by $\mathbf{n}$. We assume that $\overline{\Omega}$ is split into two parts, $\overline{\Omega} = \overline{\Omega_C} \cup \overline{\Omega_I}$, where Ω_C (a non-homogeneous non-isotropic conductor) and Ω_I (a perfect insulator) are open disjoint subsets, such that $\overline{\Omega_C} \subset \Omega$. For the sake of simplicity, we also suppose that Ω_I is connected (the general case can be treated in a similar way, focusing on each connected component of Ω_I, but some technical modifications are needed when the boundary of a connected component of Ω_I has empty intersection with $\partial\Omega$).

We denote by $\Gamma := \partial\Omega_I \cap \partial\Omega_C$ the interface between the two subdomains; note that, in the present situation, $\partial\Omega_C = \Gamma$ and $\partial\Omega_I = \partial\Omega \cup \Gamma$. Moreover, we indicate by Γ_j, $j = 1, \ldots, p_\Gamma$, the connected components of Γ, and by $(\partial\Omega)_r$, $r = 0, 1, \ldots, p_{\partial\Omega}$, the connected components of $\partial\Omega$ (in particular, we have denoted by $(\partial\Omega)_0$ the external one). Finally, we indicate by $n^*_{\partial\Omega}$ the number of cycles on $\partial\Omega$ non-homotopic to 0 in Ω_I, and by n^*_Γ the number of cycles on Γ non-homotopic to 0 in Ω_I.

In this paper we study the time-harmonic *eddy-current problem* for the Maxwell equations (see, e.g., Bossavit [7]). For the sake of definiteness, we will consider the *magnetic* boundary value problem, in which the tangential component $\mathbf{H} \times \mathbf{n}$ of the magnetic field is assumed to vanish on $\partial\Omega$ (for the *electric* boundary value problem, in which the tangential component $\mathbf{E} \times \mathbf{n}$ of the electric field is assumed to vanish on $\partial\Omega$, see Section 7).

In Alonso Rodríguez, Fernandes and Valli [1] it has been proved that the complete system of equations describing the eddy-current problem is:

$$\begin{cases} \operatorname{rot} \mathbf{H} - \sigma \mathbf{E} = \mathbf{J}_e & \text{in } \Omega \\ \operatorname{rot} \mathbf{E} + i\alpha\mu \mathbf{H} = \mathbf{0} & \text{in } \Omega \\ \operatorname{div}(\varepsilon \mathbf{E})_{|\Omega_I} = 0 & \text{in } \Omega_I \\ \mathbf{H} \times \mathbf{n} = \mathbf{0} & \text{on } \partial\Omega \\ \varepsilon \mathbf{E} \cdot \mathbf{n} = 0 & \text{on } \partial\Omega \\ \langle (\varepsilon \mathbf{E})_{|\Omega_I} \cdot \mathbf{n}, 1 \rangle_{\Gamma_j} = 0 & \text{for all } j = 1, \dots, p_\Gamma - 1 \\ \int_{\Omega_I} (\varepsilon \mathbf{E})_{|\Omega_I} \cdot \boldsymbol{\pi}_k = 0 & \text{for all } k = 1, \dots, n_{\partial\Omega} \,, \end{cases} \tag{1}$$

where $\langle \cdot, \cdot \rangle_{\Gamma_j}$ denotes the duality pairing between $H^{-1/2}(\Gamma_j)$ and $H^{1/2}(\Gamma_j)$ (in a formal way, one could interpret it as the integral over Γ_j), and $\boldsymbol{\pi}_k$ are the basis functions of the space of harmonic fields $\mathcal{H}_{\varepsilon_I}(\Gamma; \partial\Omega)$ (for precise definitions and notation, see Section 2).

The magnetic permeability μ is assumed to be a symmetric matrix, uniformly positive definite in Ω, with entries in $L^\infty(\Omega)$. The same assumption holds for the dielectric coefficient ε in Ω_I. Since Ω_I is a perfect insulator, we require that $\sigma_{|\Omega_I} \equiv 0$; moreover, as Ω_C is a non-homogeneous non-isotropic conductor, $\sigma_{|\Omega_C}$ is assumed to be a symmetric matrix, uniformly positive definite in Ω_C, with entries in $L^\infty(\Omega_C)$. The applied current density $\mathbf{J}_e$ is not assumed to vanish in Ω_C, so that also the skin effect in current driven massive conductors can be modelled.

In Alonso Rodríguez, Fernandes and Valli [1] we have given an existence and uniqueness result for problem (1) without assuming topological restrictions neither on the domain Ω nor on the conductor Ω_C (note that, in engineering practice, it is rather common that Ω_C is multiply-connected).

In this paper, that is a continuation of [1], we want to solve (1) by introducing a scalar "potential" for the magnetic field $\mathbf{H}$ in Ω_I (indeed, since in general Ω_I is multiply-connected, the magnetic field $\mathbf{H}$ differs from the gradient of a potential by a harmonic field: see (29)). The starting point in the proof is the $(L^2(\Omega_I))^3$-orthogonal decomposition result in Section 3, which, in addition to the weak formulation presented in [1] (see (28)), leads to the two-domain weak formulation in term of the scalar magnetic "potential" described in Theorem 29. An alternative two-domain formulation, that does not require the knowledge of the harmonic fields, but only uses some easily computed interpolants, is also proposed in Section 4; this latter formulation is the most indicated for numerical approximation.

We want to underline that results similar to those here contained have been obtained in the interesting paper by Bermúdez, Rodríguez and Salgado [6], that we have known only recently. One of the main features in their paper is the treatment, via Lagrangian multipliers, of non-homogeneous boundary conditions for the tangential components of the magnetic field **H**. However, our weak formulations (31) and (38) are somehow different from their, even for homogeneous boundary conditions, as we explicitly construct the part of the magnetic field that is not a gradient (see (29) and (36)), while Bermúdez, Rodríguez and Salgado [6] work with a multivalued potential. For the usual way in which the magnetic scalar potential in multiply-connected regions is dealt with, see Leonard and Rodger [18], [19] and the references therein.

The paper is organized as follows. For the ease of the reader, Section 2 is devoted to the notation and the preliminary results that are needed to formulate and solve the problem (a more complete presentation can be found in Alonso Rodríguez, Fernandes and Valli [1]). In Section 3 we will prove the $(L^2(\Omega_I))^3$-orthogonal decomposition result, and in Section 4 we will rewrite the problem in a two-domain formulation, in term of a scalar magnetic potential and a harmonic field in Ω_I. Numerical approximation based on the two-domain formulation is presented in Section 5. Finally, in Section 6 we will propose an iteration-by-subdomain procedure for the solution. Section 7 is devoted to present the modifications that are needed to treat the case in which the boundary condition is $\mathbf{E} \times \mathbf{n} = \mathbf{0}$ on $\partial\Omega$.

2 Notation and preliminaries

We indicate by $H^s(\Omega)$, $H^s(\partial\Omega)$, $s \in \mathbb{R}$, $H(\mathrm{rot};\Omega)$ and $H(\mathrm{div};\mu;\Omega)$ the usual Sobolev spaces of real or complex functions. With $H_0(\mathrm{rot};\Omega)$ we indicate the subspace of $H(\mathrm{rot};\Omega)$ constituted by those functions $\boldsymbol{\phi}$ satisfying $(\boldsymbol{\phi}\times\mathbf{n})_{|\partial\Omega} = \mathbf{0}$.

Let us assume that

(*H*1) either $\partial\Omega \in C^{1,1}$, or else Ω is a Lipschitz polyhedron.

From Cessenat [12] (see also Buffa and Ciarlet [10], [11] and Buffa [9]) one knows that for a vector field $\boldsymbol{\lambda}$ satisfying $(\boldsymbol{\lambda}\cdot\mathbf{n})_{|\partial\Omega} = 0$ it is possible to define the tangential divergence $\mathrm{div}_\tau\,\boldsymbol{\lambda}$ of $\boldsymbol{\lambda}$ and the tangential curl $\mathrm{rot}_\tau\,\boldsymbol{\lambda}$ of $\boldsymbol{\lambda}$. One also has that, for all $\mathbf{v} \in H(\mathrm{rot};\Omega)$,

$$\mathrm{div}_\tau((\mathbf{v}\times\mathbf{n})_{|\partial\Omega}) = (\mathrm{rot}\,\mathbf{v}\cdot\mathbf{n})_{|\partial\Omega} \,. \tag{2}$$

Moreover, the spaces

$$\begin{aligned} H^{-1/2}_{\mathrm{div},\tau}(\partial\Omega) := \{\boldsymbol{\lambda} \in (H^{-1/2}(\partial\Omega))^3 \mid \\ \mid \boldsymbol{\lambda}\cdot\mathbf{n} = 0 \text{ on } \partial\Omega \text{ and } \mathrm{div}_\tau\boldsymbol{\lambda} \in H^{-1/2}(\partial\Omega)\} \end{aligned} \tag{3}$$

$$\begin{aligned} H^{-1/2}_{\mathrm{rot},\tau}(\partial\Omega) := \{\boldsymbol{\lambda} \in (H^{-1/2}(\partial\Omega))^3 \mid \\ \mid \boldsymbol{\lambda}\cdot\mathbf{n} = 0 \text{ on } \partial\Omega \text{ and } \mathrm{rot}_\tau\boldsymbol{\lambda} \in H^{-1/2}(\partial\Omega)\} \end{aligned} \tag{4}$$

are in duality, and the following Green formula holds:

$$\begin{aligned}&\int_\Omega (\mathbf{v} \cdot \operatorname{rot} \mathbf{w} - \operatorname{rot} \mathbf{v} \cdot \mathbf{w}) \\ &\quad = \langle\langle \mathbf{v} \times \mathbf{n}, \mathbf{n} \times (\mathbf{w} \times \mathbf{n}) \rangle\rangle_{\partial\Omega} \qquad \forall\, \mathbf{v}, \mathbf{w} \in H(\operatorname{rot}; \Omega) \ ,\end{aligned} \tag{5}$$

where we have denoted by $\langle\langle \cdot, \cdot \rangle\rangle_{\partial\Omega}$ the duality pairing between $H^{-1/2}_{\operatorname{div},\tau}(\partial\Omega)$ and $H^{-1/2}_{\operatorname{rot},\tau}(\partial\Omega)$.

Another relation which will be useful is, for all $\boldsymbol{\lambda} \in H^{-1/2}_{\operatorname{div},\tau}(\partial\Omega)$ and all $\varphi \in H^1(\Omega)$,

$$\langle \operatorname{div}_\tau \boldsymbol{\lambda}, \varphi_{|\partial\Omega} \rangle_{\partial\Omega} = -\langle\langle \boldsymbol{\lambda}, \mathbf{n} \times (\nabla\varphi \times \mathbf{n}) \rangle\rangle_{\partial\Omega} \ , \tag{6}$$

where we have denoted by $\langle \cdot, \cdot \rangle_{\partial\Omega}$ the duality pairing between $H^{-1/2}(\partial\Omega)$ and $H^{1/2}(\partial\Omega)$.

Finally, we are interested in introducing a couple of linear spaces of harmonic vector fields. As for each of them we want to construct an explicit basis, it is useful to make some other geometrical assumptions on Ω_I (on the other hand, from here to the end of the Section it is enough to assume that $\partial\Omega$ and Γ are Lipschitz manifolds). Let us denote by n_Γ the number of cycles on Γ that are neither homotopic in Ω_I to 0, nor homotopic in Ω_I to a cycle on $\partial\Omega$ (therefore, in general we have $n_\Gamma \leq n^*_\Gamma$; for example, if Ω and Ω_C are two coaxial tori, we have $n^*_\Gamma = 2$ and $n_\Gamma = 0$). These cycles will be called the *singular* cycles on Γ. We assume that (see, e.g., Foias and Temam [15], Picard [23], Amrouche, Bernardi, Dauge and Girault [5], Fernandes and Gilardi [14]):

(*H*2) there exist n_Γ "cuts" Σ_m, which are the interior of two-dimensional, mutually disjoint, compact and connected Lipschitz manifolds $\overline{\Sigma_m}$ with boundary $\partial\Sigma_m$, such that $\Sigma_m \subset \Omega_I$ and $\partial\Sigma_m \subset \Gamma$, and such that in the open set $\widehat{\Omega_I} := \Omega_I \setminus \cup_m \Sigma_m$, assumed to be connected, every curl-free vector field with vanishing tangential component on $\partial\Omega$ has a global potential.

In other words, each surface Σ_m, $m = 1, \ldots, n_\Gamma$, "cuts" a singular cycle on Γ. These cuts are named "essential" cuts in Fernandes and Gilardi [14].

In a similar way, denoting by $n_{\partial\Omega}$ the number of singular cycles on $\partial\Omega$ (i.e., cycles that are neither homotopic in Ω_I to 0, nor homotopic in Ω_I to a cycle on Γ), we assume that

(*H*3) there exist $n_{\partial\Omega}$ "cuts" Σ^*_i, which are the interior of two-dimensional, mutually disjoint, compact and connected Lipschitz manifolds $\overline{\Sigma^*_i}$ with boundary $\partial\Sigma^*_i$, such that $\Sigma^*_i \subset \Omega_I$ and $\partial\Sigma^*_i \subset \partial\Omega$, and such that in the open set $\widetilde{\Omega_I} := \Omega_I \setminus \cup_i \Sigma^*_i$, assumed to be connected, every curl-free vector field with vanishing tangential component on Γ has a global potential.

In this case, each surface Σ_i^*, $i = 1, \ldots, n_{\partial\Omega}$, "cuts" a singular cycle on $\partial\Omega$.

Let us introduce now the following space of harmonic fields:

$$\begin{aligned}\mathcal{H}_{\mu_I}(\partial\Omega;\Gamma) := \{\mathbf{v}_I \in (L^2(\Omega_I))^3 \mid \operatorname{rot}\mathbf{v}_I = \mathbf{0}, \operatorname{div}(\mu_I\mathbf{v}_I) = 0, \\ \mathbf{v}_I \times \mathbf{n} = \mathbf{0} \ \text{ on } \partial\Omega, \mu_I\mathbf{v}_I \cdot \mathbf{n}_I = 0 \ \text{ on } \Gamma\}\ ,\end{aligned} \tag{7}$$

where we have set $\mu_I := \mu_{|\Omega_I}$. It is well-known that this space has finite dimension, that is equal to $n_\Gamma + p_{\partial\Omega}$ (see Fernandes and Gilardi [14], Proposition 5.6; see also Alonso and Valli [2], Kress [17], Picard [22] [23]). A basis for this space is given by ∇z_r, $r = 1, \ldots, p_{\partial\Omega}$, and $\boldsymbol{\rho}_l$, $l = 1, \ldots, n_\Gamma$, where

$$\begin{cases}\operatorname{div}(\mu_I\nabla z_r) = 0 & \text{in } \Omega_I \\ \mu_I\nabla z_r \cdot \mathbf{n}_I = 0 & \text{on } \Gamma \\ z_r = 0 & \text{on } \partial\Omega \setminus (\partial\Omega)_r \\ z_r = 1 & \text{on } (\partial\Omega)_r\ ,\end{cases} \tag{8}$$

and $\boldsymbol{\rho}_l \in \mathcal{H}_{\mu_I}(\partial\Omega;\Gamma)$ satisfy

$$\int_{\varphi_m} \boldsymbol{\rho}_l \cdot d\boldsymbol{\varphi} = \delta_{lm} \tag{9}$$

for each singular cycle φ_m on Γ. It is worthy to recall that for an irrotational vector field belonging to $(L^2(\Omega_I))^3$ the definition of the line integrals in (9) has a meaning (see, e.g., Dautray and Lions [13]).

In an analogous way one can define another space of harmonic fields

$$\begin{aligned}\mathcal{H}_{\varepsilon_I}(\Gamma;\partial\Omega) := \{\mathbf{v}_I \in (L^2(\Omega_I))^3 \mid \operatorname{rot}\mathbf{v}_I = \mathbf{0}, \operatorname{div}(\varepsilon_I\mathbf{v}_I) = 0, \\ \mathbf{v}_I \times \mathbf{n}_I = \mathbf{0} \ \text{ on } \Gamma, \varepsilon_I\mathbf{v}_I \cdot \mathbf{n} = 0 \ \text{ on } \partial\Omega\}\ ,\end{aligned} \tag{10}$$

having set $\varepsilon_I := \varepsilon_{|\Omega_I}$. The dimension of $\mathcal{H}_{\varepsilon_I}(\Gamma;\partial\Omega)$ is $n_{\partial\Omega} + p_\Gamma - 1$. The construction of a basis ∇w_j, $j = 1, \ldots, p_\Gamma - 1$, and $\boldsymbol{\pi}_k$, $k = 1, \ldots, n_{\partial\Omega}$, can be done similarly as before, interchanging the role of Γ and $\partial\Omega$.

The basis functions $\boldsymbol{\rho}_l \in \mathcal{H}_{\mu_I}(\partial\Omega;\Gamma)$ and $\boldsymbol{\pi}_k \in \mathcal{H}_{\varepsilon_I}(\Gamma;\partial\Omega)$ can be more explicitly expressed by resorting to the solution of a suitable elliptic problem in $\Omega_I \setminus \Sigma_l$ and $\Omega_I \setminus \Sigma_k^*$, respectively. For example, it is well known that in $\Omega_I \setminus \Sigma_l$ the basis function $\boldsymbol{\rho}_l$ is the gradient of p_l, the solution of

$$\begin{cases}\operatorname{div}(\mu_I\nabla p_l) = 0 & \text{in } \Omega_I \setminus \Sigma_l \\ \mu_I\nabla p_l \cdot \mathbf{n}_I = 0 & \text{on } \Gamma \setminus \partial\Sigma_l \\ p_l = 0 & \text{on } \partial\Omega \\ [\mu_I\nabla p_l \cdot \mathbf{n}_\Sigma]_{\Sigma_l} = 0 & \\ [p_l]_{\Sigma_l} = 1\ , & \end{cases} \tag{11}$$

where $[\cdot]_{\Sigma_l}$ denotes the jump across the surface Σ_l and $\mathbf{n}_\Sigma$ denotes the unit normal vector on Σ_l (see, e.g., Foias and Temam [15], where this construction is used for the space of tangential harmonic fields).

3 The orthogonal decomposition of $(L^2(\Omega_I))^3$

The results in this Section are essentially contained in Fernandes and Gilardi [14]. Here we present a more explicit construction of the orthogonal decomposition that will be useful in the sequel.

Instead of $(H1)$ for the subset Ω_I, let us only assume in this Section that the boundary $\partial\Omega$ and the interface Γ are Lipschitz manifolds. Assume moreover that $(H2)$ and $(H3)$ are satisfied.

For any (real valued) vector function $\mathbf{u}_I \in (L^2(\Omega_I))^3$, let us construct the vector function $\mathbf{q}_I$, the scalar function ψ_I and the constants $\alpha_{I,l}$, $\beta_{I,r}$, $l = 1, \ldots, n_\Gamma$, $r = 1, \ldots, p_{\partial\Omega}$, in the following way.

The vector function $\mathbf{q}_I \in H(\mathrm{rot}; \Omega_I)$ is the (weak) solution to

$$\begin{cases} \mathrm{rot}(\mu_I^{-1} \,\mathrm{rot}\, \mathbf{q}_I) = \mathrm{rot}\, \mathbf{u}_I & \text{in } \Omega_I \\ \mathrm{div}\, \mathbf{q}_I = 0 & \text{in } \Omega_I \\ \mathbf{q}_I \times \mathbf{n}_I = \mathbf{0} & \text{on } \Gamma \\ \mu_I^{-1} \,\mathrm{rot}\, \mathbf{q}_I \times \mathbf{n} = \mathbf{u}_I \times \mathbf{n} & \text{on } \partial\Omega \\ \mathbf{q}_I \cdot \mathbf{n} = 0 & \text{on } \partial\Omega \\ \mathbf{q}_I \perp \mathcal{H}(\Gamma; \partial\Omega) \,, \end{cases} \tag{12}$$

where

$$\mathcal{H}(\Gamma; \partial\Omega) := \{\mathbf{v}_I \in (L^2(\Omega_I))^3 \mid \mathrm{rot}\, \mathbf{v}_I = \mathbf{0}, \mathrm{div}\, \mathbf{v}_I = 0, \\ \mathbf{v}_I \times \mathbf{n}_I = \mathbf{0} \ \text{ on } \Gamma, \mathbf{v}_I \cdot \mathbf{n} = 0 \ \text{ on } \partial\Omega\} \,.$$

Let us note that the existence and uniqueness of the solution $\mathbf{q}_I$ can be proved as in Alonso and Valli [2], (33), where at the right hand side one takes $\int_{\Omega_I} \mathbf{u}_I \cdot \boldsymbol{\phi}$ and at the left hand side the term $\mathrm{rot}\, \mathbf{q}_I$ has been substituted by $\mu_I \,\mathrm{rot}\, \mathbf{q}_I$. It is worthy to underline that the existence result reported in [2], (33), is true not only for a domain Ω_I with a $C^{1,1}$ boundary, but also for a Lipschitz domain. In fact, it is only based on the compactness of the immersion of X_I in $L^2(\Omega_I)$, where

$$X_I := \{\mathbf{v}_I \in H(\mathrm{rot}; \Omega_I) \cap H(\mathrm{div}; \Omega_I) \,|\, \mathbf{v}_I \times \mathbf{n} = \mathbf{0} \text{ on } \Gamma, \mathbf{v}_I \cdot \mathbf{n}_I = 0 \text{ on } \partial\Omega\} \,.$$

For this compactness result, one can see Fernandes and Gilardi [14], Proposition 7.3, or else use the regularity results in Alonso and Valli [4], Theorems 4.3 and 4.4.

The scalar function $\psi_I \in H^1(\Omega_I)$ is the (weak) solution to the elliptic mixed boundary value problem

$$\begin{cases} \mathrm{div}(\mu_I \nabla \psi_I) = \mathrm{div}(\mu_I \mathbf{u}_I) & \text{in } \Omega_I \\ \mu_I \nabla \psi_I \cdot \mathbf{n}_I = \mu_I \mathbf{u}_I \cdot \mathbf{n}_I & \text{on } \Gamma \\ \psi_I = 0 & \text{on } \partial\Omega \,. \end{cases} \tag{13}$$

The existence and uniqueness of the solution ψ_I is well-known from the classical theory on elliptic boundary value problems.

Finally, the vector $(\alpha_{I,l}, \beta_{I,r})$, $l = 1, \ldots, n_\Gamma$, $r = 1, \ldots, p_{\partial\Omega}$, is the solution of the linear system

$$A \begin{pmatrix} \alpha_{I,l} \\ \beta_{I,r} \end{pmatrix} = \begin{pmatrix} \int_{\Omega_I} \mu_I \mathbf{u}_I \cdot \boldsymbol{\rho}_m \\ \int_{\Omega_I} \mu_I \mathbf{u}_I \cdot \nabla z_s \end{pmatrix} , \tag{14}$$

$m = 1, \ldots, n_\Gamma$, $s = 1, \ldots, p_{\partial\Omega}$, where $A := \begin{pmatrix} D & B \\ B^T & C \end{pmatrix}$ and

$$\begin{aligned} D_{ml} &:= \int_{\Omega_I} \mu_I \boldsymbol{\rho}_l \cdot \boldsymbol{\rho}_m \ , \quad B_{mr} := \int_{\Omega_I} \mu_I \nabla z_r \cdot \boldsymbol{\rho}_m \ , \\ C_{rs} &:= \int_{\Omega_I} \mu_I \nabla z_r \cdot \nabla z_s \ , \end{aligned} \tag{15}$$

and the harmonic vector fields ∇z_r and $\boldsymbol{\rho}_l$ are the basis functions of the space $\mathcal{H}_{\mu_I}(\partial\Omega; \Gamma)$ introduced in (7).

It is easily proved that the matrix A is symmetric and positive definite, as the matrix $\mu_I(\mathbf{x})$ is symmetric and positive definite, uniformly with respect to $\mathbf{x}$, and the functions $\boldsymbol{\rho}_l$ and ∇z_r are linearly independent.

The following proposition will be useful in the sequel (the proof is an easy consequence of the orthogonality results in Fernandes and Gilardi [14], and can be obtained by straightforward computations).

Proposition 2. *The vector functions $\mu_I^{-1} \operatorname{rot} \mathbf{q}_I$ and $\nabla \psi_I$ are orthogonal, and are both orthogonal to $\mathcal{H}_{\mu_I}(\partial\Omega; \Gamma)$, with respect to the scalar product*

$$(\mathbf{v}_I, \mathbf{w}_I)_{\mu_I, \Omega_I} := \int_{\Omega_I} \mu_I \mathbf{v}_I \cdot \mathbf{w}_I \ . \tag{16}$$

We are now in a position to prove the main result of this Section: an orthogonal decomposition result for $(L^2(\Omega_I))^3$, with respect to the scalar product introduced in (16). Other decomposition results, in which the role of the boundary conditions is different, are well-known (see, for instance, Dautray and Lions [13], Saranen [24], [25], Valli [26]). For the sake of completeness, here below we present the proof, which could be also derived by the results in Fernandes and Gilardi [14].

Theorem 28. *Any given (real valued) vector function $\mathbf{u}_I \in (L^2(\Omega_I))^3$ can be decomposed into the following sum*

$$\mathbf{u}_I = \mu_I^{-1} \operatorname{rot} \mathbf{q}_I + \nabla \psi_I + \sum_{r=1}^{p_{\partial\Omega}} \beta_{I,r} \nabla z_r + \sum_{l=1}^{n_\Gamma} \alpha_{I,l} \boldsymbol{\rho}_l \ , \tag{17}$$

where $\mathbf{q}_I$ has been introduced in (12), *ψ_I in* (13) *and $\alpha_{I,l}$, $\beta_{I,r}$, $l = 1, \ldots, n_\Gamma$, $r = 1, \ldots, p_{\partial\Omega}$, in* (14).

Proof. Define

$$\mathbf{V}_I := \mathbf{u}_I - \mu_I^{-1} \operatorname{rot} \mathbf{q}_I - \nabla \psi_I - \sum_{r=1}^{p_{\partial\Omega}} \beta_{I,r} \nabla z_r - \sum_{l=1}^{n_\Gamma} \alpha_{I,l} \boldsymbol{\rho}_l \ . \tag{18}$$

The thesis follows from the fact that

$$\mathbf{V}_I \in \mathcal{H}_{\mu_I}(\partial\Omega;\Gamma) \cap \mathcal{H}_{\mu_I}(\partial\Omega;\Gamma)^\perp \ .$$

In fact, the verification of $\operatorname{rot}\mathbf{V}_I = \mathbf{0}$ and $\operatorname{div}(\mu_I \mathbf{V}_I) = 0$ is trivial. From $\psi_I = 0$ on $\partial\Omega$ one has $\nabla\psi_I \times \mathbf{n} = \mathbf{0}$ on $\partial\Omega$, and consequently $\mathbf{V}_I \times \mathbf{n} = \mathbf{0}$ on $\partial\Omega$. From $\mathbf{q}_I \times \mathbf{n}_I = \mathbf{0}$ on Γ and (2) it follows that $\operatorname{rot}\mathbf{q}_I \cdot \mathbf{n}_I = 0$ on Γ, hence $\mu_I \mathbf{V}_I \cdot \mathbf{n}_I = 0$ on Γ. Finally, $\mathbf{V}_I$ is orthogonal to $\mathcal{H}_{\mu_I}(\partial\Omega;\Gamma)$ as a direct consequence of the orthogonality properties described in Proposition 2. □

4 The introduction of the magnetic potential and the two-domain weak formulation

From now on we will assume that either $\partial\Omega \in C^{1,1}$ and $\Gamma \in C^{1,1}$, or that Ω, Ω_I and Ω_C are Lipschitz polyhedra, and that assumptions $(H2)$ and $(H3)$, introduced in Section 2, are satisfied.

Let us also recall the following necessary assumptions on the current density $\mathbf{J}_e \in (L^2(\Omega_I))^3$ (see also Alonso Rodríguez, Fernandes and Valli [1]):

$$\operatorname{div}(\mathbf{J}_{e|\Omega_I}) = 0 \quad \text{in } \Omega_I \ , \ \mathbf{J}_{e|\Omega_I} \cdot \mathbf{n} = \mathbf{0} \quad \text{on } \partial\Omega \tag{19}$$

$$\begin{aligned} &\langle \mathbf{J}_{e|\Omega_I} \cdot \mathbf{n}_I, 1\rangle_{\Gamma_j} = 0 \quad \forall\, j = 1,\dots,p_\Gamma - 1 \ , \\ &\textstyle\int_{\Omega_I} \mathbf{J}_{e|\Omega_I} \cdot \boldsymbol{\pi}_k = 0 \quad \forall\, k = 1,\dots,n_{\partial\Omega} \end{aligned} \tag{20}$$

(see Section 2 for notation; in particular, we have denoted by $\langle\cdot,\cdot\rangle_{\Gamma_j}$ the duality pairing between $H^{-1/2}(\Gamma_j)$ and $H^{1/2}(\Gamma_j)$).

As a consequence of these assumptions, there exist two vector fields $\mathbf{H}_{e,I} \in H(\mathrm{rot};\Omega_I)$ and $\mathbf{H}_{e,C} \in H(\mathrm{rot};\Omega_C)$ satisfying

$$\begin{cases} \operatorname{rot}\mathbf{H}_{e,I} = \mathbf{J}_{e|\Omega_I} & \text{in } \Omega_I \\ \mathbf{H}_{e,I} \times \mathbf{n} = \mathbf{0} & \text{on } \partial\Omega \end{cases} \tag{21}$$

and

$$\mathbf{H}_{e,C} \times \mathbf{n}_C + \mathbf{H}_{e,I} \times \mathbf{n}_I = \mathbf{0} \quad \text{on } \Gamma \ . \tag{22}$$

We will denote by $\mathbf{H}^* \in H_0(\mathrm{rot};\Omega)$ the vector field defined as

$$\mathbf{H}^* := \begin{cases} \mathbf{H}_{e,I} & \text{in } \Omega_I \\ \mathbf{H}_{e,C} & \text{in } \Omega_C \ . \end{cases} \tag{23}$$

In Alonso Rodríguez, Fernandes and Valli [1] it has been proved that the strong problem (1) is equivalent to the following weak problem. Introduce the Hilbert space of *complex-valued* vector functions

$$V := \{\mathbf{v} \in H_0(\mathrm{rot};\Omega) \,|\, \operatorname{rot}\mathbf{v}_I = \mathbf{0} \ \text{ in } \Omega_I\} \ , \tag{24}$$

where we have denoted by $\mathbf{v}_I := \mathbf{v}_{|\Omega_I}$ (and similarly it will be done in the sequel for any other vector function). Then define in $H(\mathrm{rot};\Omega) \times H(\mathrm{rot};\Omega)$ the bilinear form $\mathcal{A}(\cdot,\cdot)$ as

$$\mathcal{A}(\mathbf{w},\mathbf{v}) := \mathcal{A}_C(\mathbf{w}_C,\mathbf{v}_C) + \mathcal{A}_I(\mathbf{w}_I,\mathbf{v}_I) \ , \tag{25}$$

where

$$\mathcal{A}_C(\mathbf{w}_C,\mathbf{v}_C) := \int_{\Omega_C} (\sigma^{-1} \operatorname{rot} \mathbf{w}_C \cdot \operatorname{rot} \overline{\mathbf{v}_C} + i\alpha\mu_C \mathbf{w}_C \cdot \overline{\mathbf{v}_C}) \tag{26}$$

and

$$\mathcal{A}_I(\mathbf{w}_I,\mathbf{v}_I) := i\alpha \int_{\Omega_I} \mu_I \mathbf{w}_I \cdot \overline{\mathbf{v}_I} \ . \tag{27}$$

(The bilinear form $\mathcal{A}_I(\cdot,\cdot)$ is essentially the scalar product $(\cdot,\cdot)_{\mu_I,\Omega_I}$ for complex valued vector functions, see (16).)

Setting $\mathbf{Z} := \mathbf{H} - \mathbf{H}^*$, the weak problem equivalent to (1) is given by:

$$\begin{aligned} \text{find } \mathbf{Z} \in V \ : \ \mathcal{A}(\mathbf{Z},\mathbf{v}) = & -\mathcal{A}(\mathbf{H}^*,\mathbf{v}) \\ & + \textstyle\int_{\Omega_C} \sigma^{-1} \mathbf{J}_{e,C} \cdot \operatorname{rot} \overline{\mathbf{v}_C} \quad \forall\, \mathbf{v} \in V \ . \end{aligned} \tag{28}$$

We want to rewrite this problem in a two-domain formulation, using the orthogonal decomposition (17). First of all, the restriction $\mathbf{Z}_I$ of the solution $\mathbf{Z}$ to (28) can be written as

$$\mathbf{H}_I - \mathbf{H}_{e,I} = \mathbf{Z}_I = \nabla\psi_I + \sum_{r=1}^{p_{\partial\Omega}} \beta_{I,r} \nabla z_r + \sum_{l=1}^{n_\Gamma} \alpha_{I,l} \boldsymbol{\rho}_l \ , \tag{29}$$

where the complex valued function ψ_I and the complex constants $\beta_{I,r}$, $\alpha_{I,l}$ are defined in (13) and (14), respectively (more precisely, the real part and the imaginary part of $\beta_{I,r}$, $\alpha_{I,l}$ are obtained as in (14), substituting $\mathbf{u}_I$ with the real part and the imaginary part of $\mathbf{Z}_I$, respectively). In a similar way, for any $\mathbf{v} \in V$ we have

$$\mathbf{v}_I = \nabla\phi_I + \sum_{r=1}^{p_{\partial\Omega}} \beta_{v,r} \nabla z_r + \sum_{l=1}^{n_\Gamma} \alpha_{v,l} \boldsymbol{\rho}_l \ . \tag{30}$$

To simplify the notation, we will denote by $\boldsymbol{\eta}_I \in \mathbb{C}^N$, $N = n_\Gamma + p_{\partial\Omega}$, the complex vector with components $\eta_{I,l} = \alpha_{I,l}$, $l = 1,\ldots,n_\Gamma$, $\eta_{I,n_\Gamma+r} = \beta_{I,r}$, $r = 1,\ldots,p_{\partial\Omega}$, and similarly denote by $\boldsymbol{\theta}_I \in \mathbb{C}^N$ the complex vector with components $\theta_{I,l} = \alpha_{v,l}$, $l = 1,\ldots,n_\Gamma$, $\theta_{I,n_\Gamma+r} = \beta_{v,r}$, $r = 1,\ldots,p_{\partial\Omega}$. Moreover, the basis functions of the space $\mathcal{H}_{\mu_I}(\partial\Omega;\Gamma)$ will be indicated by $\boldsymbol{\omega}_q$, $q = 1,\ldots,N$, where $\boldsymbol{\omega}_l = \boldsymbol{\rho}_l$, $l = 1,\ldots,n_\Gamma$, and $\boldsymbol{\omega}_{n_\Gamma+r} = \nabla z_r$, $r = 1,\ldots,p_{\partial\Omega}$. Finally, we will denote by $[\cdot,\cdot]$ the scalar product in $\mathbb{C}^N$.

We have:

Theorem 29. *The weak problem* (28) *is equivalent to the following one: find* $(\widetilde{\mathbf{Z}}_C, \widetilde{\psi}_I, \widetilde{\boldsymbol{\eta}}_I) \in W$ *such that*

$$
\begin{aligned}
&\mathcal{A}_C(\widetilde{\mathbf{Z}}_C, \mathbf{v}_C) + \mathcal{A}_I(\nabla\widetilde{\psi}_I, \nabla\phi_I) + i\alpha[A\widetilde{\boldsymbol{\eta}}_I, \boldsymbol{\theta}_I] \\
&\quad = -\mathcal{A}_C(\mathbf{H}_{e,C}, \mathbf{v}_C) - \mathcal{A}_I(\mathbf{H}_{e,I}, \nabla\phi_I) - \mathcal{A}_I\Big(\mathbf{H}_{e,I}, \textstyle\sum_{q=1}^N \theta_{I,q}\,\boldsymbol{\omega}_q\Big) \\
&\qquad + \textstyle\int_{\Omega_C} \sigma^{-1}\mathbf{J}_{e,C} \cdot \operatorname{rot}\overline{\mathbf{v}_C} \quad \forall\, (\mathbf{v}_C, \phi_I, \boldsymbol{\theta}_I) \in W \ ,
\end{aligned}
\tag{31}
$$

where the matrix A *is defined in* (15) *and*

$$
\begin{aligned}
W := \{(\mathbf{v}_C, \phi_I, \boldsymbol{\theta}_I) \in H(\mathrm{rot}; \Omega_C) \times H^1(\Omega_I) \times \mathbb{C}^N \mid \phi_{I|\partial\Omega} = 0, \\
\mathbf{v}_C \times \mathbf{n}_C + \nabla\phi_I \times \mathbf{n}_I + \textstyle\sum_{q=1}^N \theta_{I,q}\,\boldsymbol{\omega}_q \times \mathbf{n}_I = \mathbf{0} \ \text{on}\ \Gamma\} \ .
\end{aligned}
\tag{32}
$$

In particular, the solution to (31) *is given by* $(\mathbf{Z}_C, \psi_I, \boldsymbol{\eta}_I)$, *where* $\mathbf{Z}_C$ *is the restriction of the solution* $\mathbf{Z}$ *to* (28) *to* Ω_C, *and the restriction* $\mathbf{Z}_I$ *to* Ω_I *is decomposed as* $\mathbf{Z}_I = \nabla\psi_I + \sum_{q=1}^N \eta_{I,q}\,\boldsymbol{\omega}_q$.

Proof. Let $\mathbf{Z}$ be the solution to (28). From (29) and (30) we have $\mathbf{Z}_I = \nabla\psi_I + \sum_{q=1}^N \eta_{I,q}\boldsymbol{\omega}_q$ and $\mathbf{v}_I = \nabla\phi_I + \sum_{q=1}^N \theta_{I,q}\boldsymbol{\omega}_q$. Therefore, recalling the orthogonality result given in Proposition 2, it is easily seen that the solution to problem (28) gives a solution $(\mathbf{Z}_C, \psi_I, \boldsymbol{\eta}_I)$ to (31). On the other hand, the bilinear form at the left hand side of (31) is clearly continuous and coercive in $W \times W$, endowed with the natural norm. Therefore, from Lax–Milgram lemma, there exists a unique solution $(\widetilde{\mathbf{Z}}_C, \widetilde{\psi}_I, \widetilde{\boldsymbol{\eta}}_I) \in W$ to (31), and, due to the uniqueness property, this solution has to coincide with $(\mathbf{Z}_C, \psi_I, \boldsymbol{\eta}_I)$. □

A different two-domain weak formulation, which could be useful in view of numerical approximation, can be derived in the following way. Let us suppose that Ω is a polyhedral domain, and that there is a triangulation $\mathcal{T}_{h^*}$ such that $\overline{\Omega} = \cup_{K \in \mathcal{T}_{h^*}} K$, where K is a tetrahedron or a parallelepiped and $h^* > 0$ is the (fixed) mesh size. Let us denote by I_q^* the piecewise polynomial function taking value 1 at the nodes on one side of Σ_q, say Σ_q^+, and 0 at all the other nodes (including those on Σ_q^-, the other side of Σ_q) for $q = 1, \ldots, n_\Gamma$, and taking value 1 at the nodes on $(\partial\Omega)_{q-n_\Gamma}$ and 0 at all the other nodes for $q = n_\Gamma + 1, \ldots, n_\Gamma + p_{\partial\Omega}$. Set also

$$
\boldsymbol{\lambda}_q := \begin{cases} \widetilde{\nabla} I_q^* & \text{for } q = 1, \ldots, n_\Gamma \\ \nabla I_q^* & \text{for } q = n_\Gamma + 1, \ldots, n_\Gamma + p_{\partial\Omega} \ , \end{cases}
\tag{33}
$$

where $\widetilde{\nabla} I_q^*$ denotes the $(L^2(\Omega_I))^3$-extension of ∇I_q^* computed in $\Omega_I \setminus \Sigma_q$. It is verified at once that for each $q = 1, \ldots, N$ the vector function $\boldsymbol{\lambda}_q$ belongs to $(L^2(\Omega_I))^3$ and satisfies

$$
\begin{cases} \operatorname{rot}\boldsymbol{\lambda}_q = \mathbf{0} & \text{in } \Omega_I \\ \boldsymbol{\lambda}_q \times \mathbf{n} = \mathbf{0} & \text{on } \partial\Omega \ . \end{cases}
$$

Denoting by $g_q \in H^1(\Omega_I)$ the (weak) solution to

$$\begin{cases} \operatorname{div}(\mu_I \nabla g_q) = \operatorname{div}(\mu_I \boldsymbol{\lambda}_q) & \text{in } \Omega_I \\ \mu_I \nabla g_q \cdot \mathbf{n} = \mu_I \boldsymbol{\lambda}_q \cdot \mathbf{n} & \text{on } \Gamma \\ g_q = 0 & \text{on } \partial\Omega \ , \end{cases} \tag{34}$$

one can easily check that the basis functions $\boldsymbol{\omega}_q$ of the space $\mathcal{H}_{\mu_I}(\partial\Omega;\Gamma)$ introduced in (8) and (9) can be written as

$$\boldsymbol{\omega}_q = \boldsymbol{\lambda}_q - \nabla g_q \ . \tag{35}$$

Using this result in the representation formula (29) one finds

$$\begin{aligned} \mathbf{Z}_I &= \nabla\psi_I + \textstyle\sum_{q=1}^N \eta_{I,q}\boldsymbol{\omega}_q = \nabla\psi_I + \sum_{q=1}^N \eta_{I,q}(\boldsymbol{\lambda}_q - \nabla g_q) \\ &= \nabla\psi_I^* + \textstyle\sum_{q=1}^N \eta_{I,q}\boldsymbol{\lambda}_q \ , \end{aligned} \tag{36}$$

having set

$$\psi_I^* := \psi_I - \sum_{q=1}^N \eta_{I,q} \nabla g_q \ . \tag{37}$$

If we set

$$\mathcal{B}_I(\psi_I^*, \boldsymbol{\eta}_I; \phi_I^*, \boldsymbol{\theta}_I) := \mathcal{A}_I\Big(\nabla\psi_I^* + \sum_{q=1}^N \eta_{I,q}\boldsymbol{\lambda}_q, \nabla\phi_I^* + \sum_{q=1}^N \theta_{I,q}\boldsymbol{\lambda}_q\Big) \ ,$$

the two-domain weak problem (31) can thus be rewritten as follows: find $(\mathbf{Z}_C, \psi_I^*, \boldsymbol{\eta}_I) \in W^*$ such that

$$\begin{aligned} \mathcal{A}_C(\mathbf{Z}_C, \mathbf{v}_C) + \mathcal{B}_I(\psi_I^*, \boldsymbol{\eta}_I; \phi_I^*, \boldsymbol{\theta}_I) &= -\mathcal{A}_C(\mathbf{H}_{e,C}, \mathbf{v}_C) - \mathcal{A}_I(\mathbf{H}_{e,I}, \nabla\phi_I^*) \\ &\quad -\mathcal{A}_I\Big(\mathbf{H}_{e,I}, \textstyle\sum_{q=1}^N \theta_{I,q}\,\boldsymbol{\lambda}_q\Big) + \int_{\Omega_C} \sigma^{-1}\mathbf{J}_{e,C} \cdot \operatorname{rot}\overline{\mathbf{v}_C}, \end{aligned} \tag{38}$$

for all $(\mathbf{v}_C, \phi_I^*, \boldsymbol{\theta}_I) \in W^*$, where

$$\begin{aligned} W^* := \{&(\mathbf{v}_C, \phi_I^*, \boldsymbol{\theta}_I) \in H(\operatorname{rot};\Omega_C) \times H^1(\Omega_I) \times \mathbb{C}^N \mid \phi^*_{I|\partial\Omega} = 0, \\ &\mathbf{v}_C \times \mathbf{n}_C + \nabla\phi_I^* \times \mathbf{n}_I + \textstyle\sum_{q=1}^N \theta_{I,q}\,\boldsymbol{\lambda}_q \times \mathbf{n}_I = \mathbf{0} \text{ on } \Gamma\} \ . \end{aligned} \tag{39}$$

It is clear that the two problems (31) and (38) are equivalent, therefore (38) is well-posed. Moreover, it is worthy of noting that the bilinear form at the left hand side in (38) is continuous in $W^* \times W^*$, and it can also be seen that it is coercive in $W^* \times W^*$. In fact, taking into account (35) and the orthogonality properties presented in Proposition 2, one has

$$\begin{aligned} \mathcal{B}_I(\phi_I^*, \boldsymbol{\theta}_I; \phi_I^*, \boldsymbol{\theta}_I) &= \mathcal{A}_I\Big(\nabla\phi_I^* + \textstyle\sum_{q=1}^N \theta_{I,q}\nabla g_q, \nabla\phi_I^* + \sum_{q=1}^N \theta_{I,q}\nabla g_q\Big) \\ &\quad +\mathcal{A}_I\Big(\textstyle\sum_{q=1}^N \theta_{I,q}\boldsymbol{\omega}_q, \sum_{q=1}^N \theta_{I,q}\boldsymbol{\omega}_q\Big) \\ &= \mathcal{A}_I\Big(\nabla\phi_I^* + \textstyle\sum_{q=1}^N \theta_{I,q}\nabla g_q, \nabla\phi_I^* + \sum_{q=1}^N \theta_{I,q}\nabla g_q\Big) \\ &\quad +i\alpha[A\boldsymbol{\theta}_I, \boldsymbol{\theta}_I] \ . \end{aligned}$$

Therefore,

$$
\begin{aligned}
&|\mathcal{A}_C(\mathbf{v}_C,\mathbf{v}_C) + \mathcal{B}_I(\phi_I^*,\boldsymbol{\theta}_I;\phi_I^*,\boldsymbol{\theta}_I)| \\
&\quad \geq c_1 \Big[\int_{\Omega_C} (|\operatorname{rot}\mathbf{v}_C|^2 + |\mathbf{v}_C|^2) + \int_{\Omega_I} |\nabla\phi_I^* + \textstyle\sum_{q=1}^N \theta_{I,q}\nabla g_q|^2 + |\boldsymbol{\theta}_I|^2 \Big] \\
&\quad \geq c_1 \Big[\int_{\Omega_C} (|\operatorname{rot}\mathbf{v}_C|^2 + |\mathbf{v}_C|^2) + (1-\delta)\int_{\Omega_I} |\nabla\phi_I^*|^2 \\
&\qquad\qquad -(1-\delta)\delta^{-1}\int_{\Omega_I} |\textstyle\sum_{q=1}^N \theta_{I,q}\nabla g_q|^2 + |\boldsymbol{\theta}_I|^2 \Big] \\
&\quad \geq c_1 \Big[\int_{\Omega_C} (|\operatorname{rot}\mathbf{v}_C|^2 + |\mathbf{v}_C|^2) + (1-\delta)\int_{\Omega_I} |\nabla\phi_I^*|^2 \\
&\qquad\qquad -c_2(1-\delta)\delta^{-1}|\boldsymbol{\theta}_I|^2 + |\boldsymbol{\theta}_I|^2 \Big] ,
\end{aligned}
$$

for each $0 < \delta < 1$. Choosing $c_2/(1+c_2) < \delta < 1$, the proof of the coerciveness is complete.

5 Numerical approximation

Concerning numerical approximation, the use of finite elements in (31) must be clearly based on nodal elements in Ω_I and edge elements in Ω_C. However, any approximation procedure starting from the weak formulation (31) has some drawbacks, as it is also necessary either compute explicitly $\boldsymbol{\omega}_q$ (and in general this is not feasible), or to approximate them through (8) and (11) (and this is rather expensive, especially for large N). Moreover, the matching condition on Γ, that is included in the definition of the space W, cannot be satisfied at the discrete level, as $\boldsymbol{\omega}_q \times \mathbf{n}_I$ is not a discrete function on Γ. Therefore one should consider the matching condition for a suitable interpolant (or projection) of $\boldsymbol{\omega}_q$ on Γ, thus resulting in a non-conforming finite element approximation.

Instead, the numerical approximation based on (38) is much easier to implement, as the construction of each $\boldsymbol{\lambda}_q$ is straightforward; moreover, if the finite element triangulation $\mathcal{T}_h$ is a refinement of the triangulation $\mathcal{T}_{h^*}$ used in the construction of $\boldsymbol{\lambda}_q$, the matching condition on Γ can be imposed directly, as in this case $\boldsymbol{\lambda}_q \times \mathbf{n}$ is a discrete function also in any finer mesh.

In the sequel we assume that Ω_I and Ω_C are polyhedrical domains, and that $\mathcal{T}_{I,h}$ and $\mathcal{T}_{C,h}$ are two regular families of triangulations of Ω_I and Ω_C, respectively; for the sake of simplicity, we suppose that each element K of $\mathcal{T}_{I,h}$ and $\mathcal{T}_{C,h}$ is a tetrahedron. We assume that these triangulations match on Γ.

Let $\mathbf{P}_k$, $k \geq 1$, be the space of polynomials of degree less than or equal to k, and denote by $\mathbf{P}_k^*$ the space of homogeneous polynomials of degree k. We set

$$
S_k := \{\mathbf{p} \in (\mathbf{P}_k^*)^3 |\ \mathbf{p}(\mathbf{x}) \cdot \mathbf{x} = 0\} \quad , \quad R_k := (\mathbf{P}_{k-1})^3 \oplus S_k \ .
$$

Notice that $(\mathbf{P}_{k-1})^3 \subset R_k \subset (\mathbf{P}_k)^3$. We will employ the finite element spaces

$$N^k_{C,h} := \{\mathbf{v}_{C,h} \in H(\mathrm{rot};\Omega_C) \mid \mathbf{v}_{C,h|K} \in R_k \ \forall\ K \in \mathcal{T}_{C,h}\}$$

$$\Sigma^k_{I,h} := \{\psi_{I,h} \in H^1(\Omega_I) \mid \psi_{I,h|K} \in \mathbf{P}_k \ \forall\ K \in \mathcal{T}_{I,h}\} \ .$$

The space $N^k_{C,h}$ is the Nédélec edge finite element space related to Ω_C (see Nédélec [20]); other choices would be possible, for instance the finite element space introduced and analyzed by Nédélec in [21].

We also consider the spaces

$$N^k_{I,h} := \{\mathbf{v}_{I,h} \in H(\mathrm{rot};\Omega_I) \mid \mathbf{v}_{I,h|K} \in R_k \ \forall\ K \in \mathcal{T}_{I,h}\}$$

and

$$\mathcal{X}_{\Gamma,h} := \{(\mathbf{v}_{I,h} \times \mathbf{n}_I)_{|\Gamma} \mid \mathbf{v}_{I,h} \in N^k_{I,h}\} = \{(\mathbf{v}_{C,h} \times \mathbf{n}_C)_{|\Gamma} \mid \mathbf{v}_{C,h} \in N^k_{C,h}\} \ .$$

Notice that for all $\psi_h \in \Sigma^k_{I,h}$ one has $\nabla\psi_h \in N^k_{I,h}$, and that $\boldsymbol{\lambda}_q \in N^k_{I,h}$ for $q = 1, \ldots, N$. Hence the space

$$\begin{aligned} W^*_h := \{(\mathbf{v}_{C,h}, \phi^*_{I,h}, \boldsymbol{\theta}_I) \in N^k_{C,h} \times \Sigma^k_{I,h} \times \mathbb{C}^N \mid \phi^*_{I,h|\partial\Omega} = 0, \\ \mathbf{v}_{C,h} \times \mathbf{n}_C + \nabla\phi^*_{I,h} \times \mathbf{n}_I + \textstyle\sum_{q=1}^N \theta_{I,q}\, \boldsymbol{\lambda}_q \times \mathbf{n}_I = \mathbf{0} \text{ on } \Gamma\} \end{aligned}$$

is well defined, since $\mathbf{v}_{C,h} \times \mathbf{n}_C$, $\nabla\phi_{I,h} \times \mathbf{n}_I$ and $\sum_{q=1}^N \theta_{I,q}\, \boldsymbol{\lambda}_q \times \mathbf{n}_I$ are in the same space $\mathcal{X}_{\Gamma,h}$. (We note that, for $k = 1$, a simple way of imposing the matching condition on Γ is proposed in Bermúdez, Rodríguez and Salgado [6], Lemma 6.1.) Moreover, W^*_h is clearly contained in W^*.

For the numerical approximation based on (38), it is now easy to obtain an optimal error estimate. Indeed, by using the Céa lemma, due to the fact that the bilinear form $\mathcal{A}_C(\cdot,\cdot)+\mathcal{B}_I(\cdot;\cdot)$ is continuous and coercive in $W^* \times W^*$ one finds at once

$$\begin{aligned} |\mathbf{Z}_C - \mathbf{Z}_{C,h}||_{H(\mathrm{rot};\Omega_C)} + ||\psi^*_I - \psi^*_{I,h}||_{H^1(\Omega_I)} + |\boldsymbol{\eta}_I - \boldsymbol{\eta}_{I.h}| \\ \leq c^*(||\mathbf{Z}_C - \mathbf{v}_{C,h}||_{H(\mathrm{rot};\Omega_C)} + ||\psi^*_I - \phi^*_{I,h}||_{H^1(\Omega_I)} + |\boldsymbol{\eta}_I - \boldsymbol{\theta}_{I.h}|) \end{aligned} \tag{40}$$

for any choice of $(\mathbf{v}_{C,h}, \phi^*_{I,h}, \boldsymbol{\theta}_{I.h})$ in W^*_h.

We note that, if ϕ is so regular that the interpolant $\mathbf{r}_{I,h}(\nabla\phi) \in N^k_{I,h}$ is well defined, then

$$\mathbf{r}_{I,h}(\nabla\phi) = \nabla\phi_{I,h} \ \text{ for some } \phi_{I,h} \in \Sigma^k_{I,h}.$$

In fact, proceeding as in the proof of Lemma 5.10, Chapter III in Girault and Raviart [16] we see that

$$\mathrm{rot}\, \mathbf{r}_{I,h}(\nabla\phi) = \mathbf{0} \text{ in } \Omega_I \quad \text{and} \quad \mathbf{r}_{I,h}(\nabla\phi) \times \mathbf{n} = \mathbf{0} \text{ on } \partial\Omega.$$

Moreover we know that for each edge e of $\mathcal{T}_{I,h}$ one has $\int_e \mathbf{r}_{I,h}(\nabla\phi) \cdot \tau\, de = \int_e \nabla\phi \cdot \tau\, de$, therefore

$$\int_{\varphi_m} \mathbf{r}_{I,h}(\nabla\phi) \cdot d\boldsymbol{\varphi} = 0$$

for each singular cycle φ_m on Γ. Consequently, $\mathbf{r}_{I,h}(\nabla\phi) = \nabla\phi_I$ for some $\phi_I \in H^1(\Omega_I)$ such that $\phi_{I|\partial\Omega} = 0$. Moreover $\phi_{I|K} \in \mathbf{P}_k$ for each $K \in \mathcal{T}_{I,h}$, therefore $\phi_I \in \Sigma^k_{I,h}$.

In conclusion, in (40) one can choose as $\mathbf{v}_{C,h}$ the interpolant of $\mathbf{Z}_C$, $\phi^*_{I,h}$ such that $\nabla\phi^*_{I,h} = \mathbf{r}_{I,h}(\nabla\psi^*_I)$ (provided that these solutions are regular enough), and $\boldsymbol{\theta}_{I.h} = \boldsymbol{\eta}_I$, as with this choice the matching condition on Γ is satisfied.

As a consequence, if for $r > 1/2$ the solutions satisfy $\mathbf{Z}_C \in H^r(\mathrm{rot};\Omega_C) := \{\mathbf{v}_C \in (H^r(\Omega_C))^3 \mid \mathrm{rot}\,\mathbf{v}_C \in (H^r(\Omega_C))^3\}$ and $\psi^*_I \in H^{1+r}(\Omega_I)$, the following error estimate holds:

$$\begin{aligned} &||\mathbf{Z}_C - \mathbf{Z}_{C,h}||_{H(\mathrm{rot};\Omega_C)} + ||\psi^*_I - \psi^*_{I,h}||_{H^1(\Omega_I)} + |\boldsymbol{\eta}_I - \boldsymbol{\eta}_{I.h}| \\ &\quad \le c_* h^{\min(r,k)}(||\mathbf{Z}_C||_{H^r(\mathrm{rot};\Omega_C)} + ||\psi^*_I||_{H^{1+r}(\Omega_I)}) \end{aligned} \tag{41}$$

(see Alonso and Valli [4]).

6 The two-domain strong formulation and a domain decomposition approach

In Alonso Rodríguez, Fernandes and Valli [1] it has been proved that the strong form of (28) is

$$\mathrm{rot}(\sigma^{-1}\,\mathrm{rot}\,\mathbf{H}_C) + i\alpha\mu_C\mathbf{H}_C = \mathrm{rot}(\sigma^{-1}\mathbf{J}_{e,C}) \quad \text{in } \Omega_C \tag{42a}$$

$$\mathrm{rot}\,\mathbf{H}_I = \mathbf{J}_{e,I} \quad \text{in } \Omega_I \tag{42b}$$

$$\mathrm{div}(\mu_I\mathbf{H}_I) = 0 \quad \text{in } \Omega_I \tag{42c}$$

$$\langle \mu_I\mathbf{H}_I\cdot\mathbf{n}, 1\rangle_{(\partial\Omega)_r} = 0 \quad \forall\, r = 1,\dots,p_{\partial\Omega} \tag{42d}$$

$$\begin{aligned} &i\alpha \int_{\Omega_I} \mu_I \mathbf{H}_I\cdot\boldsymbol{\rho}_l \\ &\quad + \langle\langle[\sigma^{-1}(\mathrm{rot}\,\mathbf{H}_C - \mathbf{J}_{e,C})]\times\mathbf{n}_C, \mathbf{n}_I\times(\boldsymbol{\rho}_l\times\mathbf{n}_I)\rangle\rangle_\Gamma = 0 \\ &\qquad\qquad\qquad\qquad \forall\, l = 1,\dots,n_\Gamma \end{aligned} \tag{42e}$$

$$\mathbf{H}_I\times\mathbf{n} = \mathbf{0} \quad \text{on } \partial\Omega \tag{42f}$$

$$\mu_I\mathbf{H}_I\cdot\mathbf{n}_I + \mu_C\mathbf{H}_C\cdot\mathbf{n}_C = 0 \quad \text{on } \Gamma \tag{42g}$$

$$\mathbf{H}_I\times\mathbf{n}_I + \mathbf{H}_C\times\mathbf{n}_C = \mathbf{0} \quad \text{on } \Gamma\,. \tag{42h}$$

It must be noted that the conditions (42d) and (42e) are necessary for determining the correct projection of the solution $\mathbf{Z}_I = \mathbf{H}_I - \mathbf{H}_{e,I}$ over the space of harmonic fields $\mathcal{H}_{\mu_I}(\partial\Omega;\Gamma)$ (see the orthogonal decomposition formula (29)). It is also useful to recall that, if the solution is regular enough, the duality $\langle\langle\mathbf{v},\mathbf{w}\rangle\rangle_\Gamma$ can be interpreted as the integral over Γ of the scalar product $\mathbf{v}\cdot\overline{\mathbf{w}}$.

We are now interested in rewriting this problem in an equivalent form, in term of $\mathbf{H}_C$, ψ_I and $\boldsymbol{\eta}_I = (\boldsymbol{\alpha}_I, \boldsymbol{\beta}_I)$ (see (29)).

Theorem 30. *Problem* (42*a*)–(42*h*) *is equivalent to the following one:*

$$\begin{cases} \operatorname{rot}(\sigma^{-1} \operatorname{rot} \mathbf{H}_C) + i\alpha\mu_C \mathbf{H}_C = \operatorname{rot}(\sigma^{-1}\mathbf{J}_{e,C}) & \text{in } \Omega_C \\ \mathbf{H}_C \times \mathbf{n}_C = -\nabla\psi_I \times \mathbf{n}_I - \sum_{q=1}^{N} \eta_{I,q}\, \boldsymbol{\omega}_q \times \mathbf{n}_I - \mathbf{H}_{e,I} \times \mathbf{n}_I & \text{on } \Gamma \end{cases} \tag{43}$$

$$\begin{cases} \operatorname{div}(\mu_I \nabla\psi_I) = -\operatorname{div}(\mu_I \mathbf{H}_{e,I}) & \text{in } \Omega_I \\ \mu_I \nabla\psi_I \cdot \mathbf{n}_I = -\mu_C \mathbf{H}_C \cdot \mathbf{n}_C - \mu_I \mathbf{H}_{e,I} \cdot \mathbf{n}_I & \text{on } \Gamma \\ \psi_I = 0 & \text{on } \partial\Omega \end{cases} \tag{44}$$

$$\begin{aligned} i\alpha(A\boldsymbol{\eta}_I)_q = &-\langle\langle[\sigma^{-1}(\operatorname{rot}\mathbf{H}_C - \mathbf{J}_{e,C})] \times \mathbf{n}_C, \mathbf{n}_I \times (\boldsymbol{\omega}_q \times \mathbf{n}_I)\rangle\rangle_\Gamma \\ &-\mathcal{A}_I(\mathbf{H}_{e,I}, \boldsymbol{\omega}_q) \quad \forall\, q = 1, \ldots, N\ . \end{aligned} \tag{45}$$

Proof. Equations (43) and (44) are easily obtained from (42a) and (42h), (42c) and (42g), respectively, and the representation formula (29). We are therefore left with the proof of (45).

Recalling that $z_r = 1$ on $(\partial\Omega)_r$ and $z_r = 0$ on $\partial\Omega \setminus (\partial\Omega)_r$, $r = 1, \ldots, p_{\partial\Omega}$, and taking into account (42c), (42d), (42g) and (42a), by integrating by parts we find that

$$\begin{aligned} i\alpha \int_{\Omega_I} \mu_I \mathbf{H}_I \cdot \nabla z_s &= i\alpha\langle\mu_I \mathbf{H}_I \cdot \mathbf{n}_I, z_{s|\Gamma}\rangle_\Gamma = -i\alpha\langle\mu_C \mathbf{H}_C \cdot \mathbf{n}_C, z_{s|\Gamma}\rangle_\Gamma \\ &= \langle\operatorname{rot}[\sigma^{-1}(\operatorname{rot}\mathbf{H}_C - \mathbf{J}_{e,C})] \cdot \mathbf{n}_C, z_{s|\Gamma}\rangle_\Gamma \\ &= \langle\operatorname{div}_\tau([\sigma^{-1}(\operatorname{rot}\mathbf{H}_C - \mathbf{J}_{e,C})] \times \mathbf{n}_C), z_{s|\Gamma}\rangle_\Gamma \\ &= -\langle\langle[\sigma^{-1}(\operatorname{rot}\mathbf{H}_C - \mathbf{J}_{e,C})] \times \mathbf{n}_C, \mathbf{n}_I \times (\nabla z_s \times \mathbf{n}_I)\rangle\rangle_\Gamma\ , \end{aligned} \tag{46}$$

having used (2) and (6).

In other words, putting together (42e) and (46), we have

$$i\alpha \int_{\Omega_I} \mu_I \mathbf{H}_I \cdot \boldsymbol{\omega}_q = -\langle\langle[\sigma^{-1}(\operatorname{rot}\mathbf{H}_C - \mathbf{J}_{e,C})] \times \mathbf{n}_C, \mathbf{n}_I \times (\boldsymbol{\omega}_q \times \mathbf{n}_I)\rangle\rangle_\Gamma \tag{47}$$

for each $q = 1, \ldots, N$.

Recalling the definition of the matrix A in (15), the representation formula (29) and the orthogonality of $\nabla\psi_I$ and $\boldsymbol{\omega}_q$ with respect to (16), we finally find

$$\begin{aligned} i\alpha(A\boldsymbol{\eta}_I)_q &= i\alpha \int_{\Omega_I} \mu_I \sum_{p=1}^{N} \eta_{I,p}\, \boldsymbol{\omega}_p \cdot \boldsymbol{\omega}_q \\ &= i\alpha \int_{\Omega_I} \mu_I (\mathbf{H}_I - \mathbf{H}_{e,I}) \cdot \boldsymbol{\omega}_q \\ &= -\langle\langle[\sigma^{-1}(\operatorname{rot}\mathbf{H}_C - \mathbf{J}_{e,C})] \times \mathbf{n}_C, \mathbf{n}_I \times (\boldsymbol{\omega}_q \times \mathbf{n}_I)\rangle\rangle_\Gamma \\ &\qquad -i\alpha \int_{\Omega_I} \mu_I \mathbf{H}_{e,I} \cdot \boldsymbol{\omega}_q\ , \end{aligned} \tag{48}$$

which ends the proof. □

An alternative two-domain strong formulation derives from the week problem (38). First of all, from (34) we have

$$\begin{aligned} i\alpha \int_{\Omega_I} \mu_I \mathbf{H}_I \cdot \nabla g_q &= -i\alpha \int_{\Omega_I} \operatorname{div}(\mu_I \mathbf{H}_I)\, g_q + i\alpha\langle\mu_I \mathbf{H}_I \cdot \mathbf{n}_I, g_{q|\Gamma}\rangle_\Gamma \\ &= -i\alpha\langle\mu_C \mathbf{H}_C \cdot \mathbf{n}_C, g_{q|\Gamma}\rangle_\Gamma\ , \end{aligned}$$

having used (42c) and (42g). Moreover, from (6) and (2)

$$\begin{aligned}&\langle\langle[\sigma^{-1}(\operatorname{rot}\mathbf{H}_C - \mathbf{J}_{e,C})] \times \mathbf{n}_C, \mathbf{n}_I \times (\nabla g_q \times \mathbf{n}_I)\rangle\rangle_\Gamma \\ &\quad = -\langle \operatorname{div}_\tau([\sigma^{-1}(\operatorname{rot}\mathbf{H}_C - \mathbf{J}_{e,C})] \times \mathbf{n}_C), g_{q|\Gamma}\rangle_\Gamma \\ &\quad = -\langle \operatorname{rot}[\sigma^{-1}(\operatorname{rot}\mathbf{H}_C - \mathbf{J}_{e,C})] \cdot \mathbf{n}_C, g_{q|\Gamma}\rangle_\Gamma \\ &\quad = i\alpha\langle \mu_C \mathbf{H}_C \cdot \mathbf{n}_C, g_{q|\Gamma}\rangle_\Gamma \ ,\end{aligned}$$

having used (42a). Therefore, putting (35) in (47) we find

$$i\alpha \int_{\Omega_I} \mu_I \mathbf{H}_I \cdot \boldsymbol{\lambda}_q = -\langle\langle[\sigma^{-1}(\operatorname{rot}\mathbf{H}_C - \mathbf{J}_{e,C})] \times \mathbf{n}_C, \mathbf{n}_I \times (\boldsymbol{\lambda}_q \times \mathbf{n}_I)\rangle\rangle_\Gamma \quad (49)$$

for each $q = 1, \ldots, N$.

Inserting now the relation (35) in (43) and (44), and repeating the procedure used for obtaining (48), with (49) as a starting point, we find

$$\begin{cases} \operatorname{rot}(\sigma^{-1}\operatorname{rot}\mathbf{H}_C) + i\alpha\mu_C\mathbf{H}_C = \operatorname{rot}(\sigma^{-1}\mathbf{J}_{e,C}) & \text{in } \Omega_C \\ \mathbf{H}_C \times \mathbf{n}_C = -\nabla\psi_I^* \times \mathbf{n}_I - \sum_{q=1}^N \eta_{I,q}\,\boldsymbol{\lambda}_q \times \mathbf{n}_I - \mathbf{H}_{e,I} \times \mathbf{n}_I & \text{on } \Gamma \end{cases} \quad (50)$$

$$\begin{cases} \operatorname{div}(\mu_I\nabla\psi_I^*) = -\operatorname{div}(\mu_I\mathbf{H}_{e,I}) - \sum_{q=1}^N \eta_{I,q}\operatorname{div}(\mu_I\boldsymbol{\lambda}_q) & \text{in } \Omega_I \\ \mu_I\nabla\psi_I^* \cdot \mathbf{n}_I = -\mu_C\mathbf{H}_C \cdot \mathbf{n}_C - \mu_I\mathbf{H}_{e,I}\cdot\mathbf{n}_I & \\ \qquad\qquad - \sum_{q=1}^N \eta_{I,q}\,\mu_I\boldsymbol{\lambda}_q\cdot\mathbf{n}_I & \text{on } \Gamma \\ \psi_I^* = 0 & \text{on } \partial\Omega \end{cases} \quad (51)$$

$$\begin{aligned} i\alpha(A^*\boldsymbol{\eta}_I)_q = &-\langle\langle[\sigma^{-1}(\operatorname{rot}\mathbf{H}_C - \mathbf{J}_{e,C})] \times \mathbf{n}_C, \mathbf{n}_I \times (\boldsymbol{\lambda}_q \times \mathbf{n}_I)\rangle\rangle_\Gamma \\ &-\mathcal{A}_I(\mathbf{H}_{e,I}, \boldsymbol{\lambda}_q) - \mathcal{A}_I(\nabla\psi_I^*, \boldsymbol{\lambda}_q) \quad \forall\, q = 1, \ldots, N \ , \end{aligned} \quad (52)$$

where the matrix A^* is defined as

$$A^*_{pq} := \int_{\Omega_I} \mu_I \boldsymbol{\lambda}_p \cdot \boldsymbol{\lambda}_q \quad , \quad p, q = 1, \ldots, N \ . \quad (53)$$

It is easily seen that this matrix is symmetric and positive definite, as the vectors $\boldsymbol{\lambda}_q$ are linearly independent.

Remark 11. We have already seen in the proof of Theorem 30 that conditions (49) for $q = n_\Gamma + 1, \ldots, N$ can be reduced to (42d).

On the other hand, it is worthy to note that, proceeding formally, conditions (49) for $q = 1, \ldots, n_\Gamma$ can be read as

$$\begin{aligned} &i\alpha \textstyle\int_{\Sigma_q} \mu_I \mathbf{H}_I \cdot \mathbf{n}_\Sigma \\ &\quad = \textstyle\int_{\partial\Sigma_q} (\mathbf{n}_\Sigma \times \mathbf{n}_C) \cdot [\sigma^{-1}(\operatorname{rot}\mathbf{H}_C - \mathbf{J}_{e,C})] \qquad \forall\, q = 1, \ldots, n_\Gamma \ . \end{aligned} \quad (54)$$

In fact, from (42c) it holds

$$\begin{aligned} i\alpha \textstyle\int_{\Omega_I} \mu_I \mathbf{H}_I \cdot \boldsymbol{\lambda}_q &= i\alpha \textstyle\int_{\Omega_I \setminus \Sigma_q} \mu_I \mathbf{H}_I \cdot \nabla I_q^* = i\alpha \int_{\partial(\Omega_I \setminus \Sigma_q)} \mu_I \mathbf{H}_I \cdot \mathbf{n}_I \; I_q^* \\ &= i\alpha \textstyle\int_{\Gamma \setminus \partial\Sigma_q} \mu_I \mathbf{H}_I \cdot \mathbf{n}_I \; I_q^* + i\alpha \int_{\Sigma_q} \mu_I \mathbf{H}_I \cdot \mathbf{n}_\Sigma \; [I_q^*] \\ &= i\alpha \textstyle\int_\Gamma \mu_I \mathbf{H}_I \cdot \mathbf{n}_I \; I_q^* + i\alpha \int_{\Sigma_q} \mu_I \mathbf{H}_I \cdot \mathbf{n}_\Sigma \ , \end{aligned}$$

and, from (6), (2) and (42a),

$$
\begin{aligned}
-\langle\langle[\sigma^{-1}(\operatorname{rot}\mathbf{H}_C - \mathbf{J}_{e,C})] \times \mathbf{n}_C, \mathbf{n}_I \times (\boldsymbol{\lambda}_q \times \mathbf{n}_I)\rangle\rangle_\Gamma \\
= -\int_{\Gamma\setminus\partial\Sigma_q} \{[\sigma^{-1}(\operatorname{rot}\mathbf{H}_C - \mathbf{J}_{e,C})] \times \mathbf{n}_C\} \cdot \nabla I_q^* \\
= \int_{\Gamma\setminus\partial\Sigma_q} \operatorname{div}_\tau\{[\sigma^{-1}(\operatorname{rot}\mathbf{H}_C - \mathbf{J}_{e,C})] \times \mathbf{n}_C\}\, I_q^* \\
- \int_{\partial\Sigma_q} \{[\sigma^{-1}(\operatorname{rot}\mathbf{H}_C - \mathbf{J}_{e,C})] \times \mathbf{n}_C\} \cdot \mathbf{n}_\Sigma\, [I_q^*] \\
= -i\alpha \int_\Gamma \mu_C \mathbf{H}_C \cdot \mathbf{n}_C\, I_q^* - \int_{\partial\Sigma_q} \{[\sigma^{-1}(\operatorname{rot}\mathbf{H}_C - \mathbf{J}_{e,C})] \times \mathbf{n}_C\} \cdot \mathbf{n}_\Sigma\ .
\end{aligned}
$$

Therefore, using (42g) we have

$$
i\alpha \int_{\Sigma_q} \mu_I \mathbf{H}_I \cdot \mathbf{n}_\Sigma = -\int_{\partial\Sigma_q} \{[\sigma^{-1}(\operatorname{rot}\mathbf{H}_C - \mathbf{J}_{e,C})] \times \mathbf{n}_C\} \cdot \mathbf{n}_\Sigma\ ,
$$

which is equivalent to (54).

Taking into account that $\operatorname{rot}\mathbf{H}_C - \sigma\mathbf{E}_C = \mathbf{J}_{e,C}$ and that the matching condition $\mathbf{E}_C \times \mathbf{n}_C + \mathbf{E}_I \times \mathbf{n}_I = \mathbf{0}$ on Γ has to be satisfied, one sees that (54) is equivalent to the Faraday law imposed on the cut Σ_q. □

The strong formulation presented in Theorem 30 suggests a possible iterative procedure for solving (43)–(45). Given $\boldsymbol{\lambda}^{\text{old}} \in H^{-1/2}_{\operatorname{div},\tau}(\Gamma)$, solve in parallel

$$
\begin{cases}
\operatorname{div}(\mu_I \nabla\psi_I) = -\operatorname{div}(\mu_I \mathbf{H}_{e,I}) & \text{in } \Omega_I \\
\mu_I \nabla\psi_I \cdot \mathbf{n}_I = (i\alpha)^{-1}\operatorname{div}_\tau \boldsymbol{\lambda}^{\text{old}} - \mu_I \mathbf{H}_{e,I} \cdot \mathbf{n}_I & \text{on } \Gamma \\
\psi_I = 0 & \text{on } \partial\Omega
\end{cases}
\tag{55}
$$

and

$$
\begin{aligned}
i\alpha(A\boldsymbol{\eta}_I)_q = -\langle\langle\boldsymbol{\lambda}^{\text{old}}, \mathbf{n}_I \times (\boldsymbol{\omega}_q \times \mathbf{n}_I)\rangle\rangle_\Gamma \\
-\mathcal{A}_I(\mathbf{H}_{e,I}, \boldsymbol{\omega}_q)\ \forall\, q = 1, \dots, N,
\end{aligned}
\tag{56}
$$

then solve (43), namely

$$
\begin{cases}
\operatorname{rot}(\sigma^{-1}\operatorname{rot}\mathbf{H}_C) + i\alpha\mu_C \mathbf{H}_C = \operatorname{rot}(\sigma^{-1}\mathbf{J}_{e,C}) & \text{in } \Omega_C \\
\mathbf{H}_C \times \mathbf{n}_C = -\nabla\psi_I \times \mathbf{n}_I - \sum_{q=1}^N \eta_{I,q}\boldsymbol{\omega}_q \times \mathbf{n}_I - \mathbf{H}_{e,I} \times \mathbf{n}_I & \text{on } \Gamma\ ,
\end{cases}
\tag{57}
$$

finally set

$$
\boldsymbol{\lambda}^{\text{new}} = (1-\delta)\boldsymbol{\lambda}^{\text{old}} + \delta\,[\sigma^{-1}(\operatorname{rot}\mathbf{H}_C - \mathbf{J}_{e,C})] \times \mathbf{n}_C \quad \text{on } \Gamma\ ,
\tag{58}
$$

and iterate until convergence (here $\delta > 0$ is an acceleration parameter).

An analogous iterative scheme can be proposed for formulation (50)–(52). In this case, however, problems (51) and (52) cannot be solved in parallel, as they are not independent.

Similar iterative procedures have been proposed in the framework of domain decomposition methods, and have been given the name of Dirichlet/Neumann iterative procedures. For instance, for the eddy-current problem with the *electric* boundary condition $\mathbf{E} \times \mathbf{n} = \mathbf{0}$ on $\partial\Omega$, one can see

Alonso and Valli [3], where the convergence of the iterative scheme is proved in the case of the finite dimensional approximation by means of edge finite elements in Ω_C and nodal finite elements in Ω_I.

The convergence and the numerical performance of the iterative scheme (55)–(58) (and of that derived from (50)–(52)) will be studied elsewhere.

7 Appendix

In this Section we shortly present how the results of Sections 3–6 have to be modified if we consider the *electric* boundary value problem $\mathbf{E} \times \mathbf{n} = \mathbf{0}$ instead of the *magnetic* boundary value problem $\mathbf{H} \times \mathbf{n} = \mathbf{0}$ (for the same problem, see Alonso and Valli [3], where the solution has be done in terms of the electric field $\mathbf{E}$; see also Bossavit [8]).

For the sake of simplicity, we will focus only on the problems expressed in terms of the harmonic fields $\widehat{\boldsymbol{\rho}}_s$ (see (63) here below), leaving to the interested reader the extension of the results to the case in which, instead of $\widehat{\boldsymbol{\rho}}_s$, one uses interpolants like those introduced in (33).

In Alonso Rodríguez, Fernandes and Valli [1], assuming that $\operatorname{div} \mathbf{J}_{e,I} = 0$ in Ω_I and $\langle \mathbf{J}_{e|\Omega_I} \cdot \mathbf{n}_I, 1\rangle_{\Gamma_j} = 0$ for all $j = 1, \ldots, p_\Gamma - 1$, $\langle \mathbf{J}_{e|\Omega_I} \cdot \mathbf{n}, 1\rangle_{\partial\Omega_r} = 0$ for all $r = 0, \ldots, p_{\partial\Omega}$, it has been proved that the complete system of equations describing the eddy-current problem

$$\begin{cases} \operatorname{rot} \mathbf{H} - \sigma \mathbf{E} = \mathbf{J}_e & \text{in } \Omega \\ \operatorname{rot} \mathbf{E} + i\alpha\mu \mathbf{H} = \mathbf{0} & \text{in } \Omega \\ \operatorname{div}(\varepsilon \mathbf{E})_{|\Omega_I} = 0 & \text{in } \Omega_I \\ \mathbf{E} \times \mathbf{n} = \mathbf{0} & \text{on } \partial\Omega \\ \langle (\varepsilon \mathbf{E})_{|\Omega_I} \cdot \mathbf{n}, 1\rangle_{\Gamma_j} = 0 & \text{for all } j = 1, \ldots, p_\Gamma - 1 \\ \langle (\varepsilon \mathbf{E})_{|\Omega_I} \cdot \mathbf{n}, 1\rangle_{(\partial\Omega)_r} = 0 & \text{for all } r = 0, \ldots, p_{\partial\Omega} \end{cases} \tag{59}$$

has an unique solution.

The weak problem for the magnetic field $\mathbf{H}$ is

$$\mathcal{A}(\mathbf{H}, \widehat{\mathbf{v}}) = \int_{\Omega_C} \sigma^{-1} \mathbf{J}_{e,C} \cdot \operatorname{rot} \overline{\widehat{\mathbf{v}}_C} \quad \forall\, \widehat{\mathbf{v}} \in \widehat{V} , \tag{60}$$

where

$$\widehat{V} := \{ \widehat{\mathbf{v}} \in H(\operatorname{rot}; \Omega) \mid \operatorname{rot} \widehat{\mathbf{v}}_I = \mathbf{0} \ \text{ in } \Omega_I \} . \tag{61}$$

The orthogonal decomposition result of Section 3 now reads:

$$\mathbf{u}_I = \mu_I^{-1} \operatorname{rot} \widehat{\mathbf{q}}_I + \nabla \widehat{\psi}_I + \sum_{s=1}^{n^*} \widehat{\alpha}_{I,s} \widehat{\boldsymbol{\rho}}_s , \tag{62}$$

where $\widehat{\boldsymbol{\rho}}_s$, $s = 1, \ldots, n^*$, are the basis functions of the space of harmonic fields

$$\mathcal{H}_{\mu_I}(m) := \{\mathbf{v}_I \in (L^2(\Omega_I))^3 \mid \operatorname{rot} \mathbf{v}_I = \mathbf{0}, \operatorname{div}(\mu_I \mathbf{v}_I) = 0, \\ \mu_I \mathbf{v}_I \cdot \mathbf{n}_I = 0 \text{ on } \partial\Omega \cup \Gamma\} . \tag{63}$$

(We have set $n^* = n^*_{\partial\Omega_I}$, the total number of cycles on $\partial\Omega_I = \partial\Omega \cup \Gamma$ that are not homotopic to 0 in Ω_I.)

The functions $\widehat{\mathbf{q}}_I$ and $\widehat{\psi}_I$ and the constants $\widehat{\alpha}_{I,s}$ are solutions of problems like (12), (13) and (14), respectively, where one has to substitute Γ with $\Gamma \cup \partial\Omega$ (and $\widehat{\psi}_I$ has also to satisfy $\int_{\Omega_I} \widehat{\psi}_I = 0$).

Therefore, using the representation formula (62) for the solution $\mathbf{H}_I - \mathbf{H}_{e,I}$ and the test functions $\widehat{\mathbf{v}}_I$ (in both cases it turns out that $\widehat{\mathbf{q}}_I = \mathbf{0}$, as $\operatorname{rot}(\mathbf{H}_I - \mathbf{H}_{e,I}) = \operatorname{rot} \widehat{\mathbf{v}}_I = \mathbf{0}$), one finds the two-domain weak formulation

$$\mathcal{A}_C(\mathbf{H}_C, \mathbf{v}_C) + \mathcal{A}_I(\nabla\widehat{\psi}_I, \nabla\widehat{\phi}_I) + i\alpha[\widehat{A}\widehat{\boldsymbol{\alpha}}_I, \widehat{\boldsymbol{\theta}}_I] \\ = -\mathcal{A}_I(\mathbf{H}_{e,I}, \nabla\widehat{\phi}_I) - \mathcal{A}_I\Big(\mathbf{H}_{e,I}, \textstyle\sum_{s=1}^{n^*} \widehat{\theta}_{I,s} \widehat{\boldsymbol{\rho}}_s\Big) \\ + \int_{\Omega_C} \sigma^{-1} \mathbf{J}_{e,C} \cdot \operatorname{rot} \overline{\mathbf{v}_C} \quad \forall\, (\mathbf{v}_C, \widehat{\phi}_I, \widehat{\boldsymbol{\theta}}_I) \in \widehat{W} , \tag{64}$$

where

$$\widehat{W} := \{(\mathbf{v}_C, \widehat{\phi}_I, \widehat{\boldsymbol{\theta}}_I) \in H(\operatorname{rot}; \Omega_C) \times H^1(\Omega_I) \times \mathbb{C}^{n^*} \mid \int_{\Omega_I} \widehat{\phi}_I = 0, \\ \mathbf{v}_C \times \mathbf{n}_C + \nabla\widehat{\phi}_I \times \mathbf{n}_I + \textstyle\sum_{s=1}^{n^*} \widehat{\theta}_{I,s} \widehat{\boldsymbol{\rho}}_s \times \mathbf{n}_I = \mathbf{0} \text{ on } \Gamma\} . \tag{65}$$

The domain decomposition iterations in this case read: given $\boldsymbol{\lambda}^{\text{old}} \in H^{-1/2}_{\operatorname{div},\tau}(\Gamma)$, solve in parallel

$$\begin{cases} \operatorname{div}(\mu_I \nabla\widehat{\psi}_I) = -\operatorname{div}(\mu_I \mathbf{H}_{e,I}) & \text{in } \Omega_I \\ \mu_I \nabla\widehat{\psi}_I \cdot \mathbf{n}_I = (i\alpha)^{-1} \operatorname{div}_\tau \boldsymbol{\lambda}^{\text{old}} - \mu_I \mathbf{H}_{e,I} \cdot \mathbf{n}_I & \text{on } \partial\Omega \cup \Gamma \\ \int_{\Omega_I} \widehat{\psi}_I = 0 \end{cases} \tag{66}$$

and

$$i\alpha(\widehat{A}\widehat{\boldsymbol{\eta}}_I)_s = -\langle\langle \boldsymbol{\lambda}^{\text{old}}, \mathbf{n}_I \times (\widehat{\boldsymbol{\rho}}_s \times \mathbf{n}_I)\rangle\rangle_\Gamma \\ -\mathcal{A}_I(\mathbf{H}_{e,I}, \widehat{\boldsymbol{\rho}}_s) \quad \forall\, s = 1, \dots, n^* , \tag{67}$$

then solve

$$\begin{cases} \operatorname{rot}(\sigma^{-1} \operatorname{rot} \mathbf{H}_C) + i\alpha\mu_C \mathbf{H}_C = \operatorname{rot}(\sigma^{-1} \mathbf{J}_{e,C}) & \text{in } \Omega_C \\ \mathbf{H}_C \times \mathbf{n}_C = -\nabla\widehat{\psi}_I \times \mathbf{n}_I - \sum_{s=1}^{n^*} \widehat{\eta}_{I,s} \widehat{\boldsymbol{\rho}}_s \times \mathbf{n}_I - \mathbf{H}_{e,I} \times \mathbf{n}_I & \text{on } \Gamma , \end{cases} \tag{68}$$

finally set

$$\boldsymbol{\lambda}^{\text{new}} = (1 - \delta)\boldsymbol{\lambda}^{\text{old}} + \delta\, [\sigma^{-1}(\operatorname{rot} \mathbf{H}_C - \mathbf{J}_{e,C})] \times \mathbf{n}_C \quad \text{on } \Gamma , \tag{69}$$

and iterate until convergence.

References

1. Alonso Rodríguez, A., Fernandes, P. & Valli, A. (2001) Weak and strong formulations for the time-harmonic eddy-current problem in general domains, preprint UTM 603, Department of Mathematics, University of Trento, submitted to European J. Appl. Math.
2. Alonso, A. & Valli, A. (1996) Some remarks on the characterization of the space of tangential traces of $H(\text{rot}; \Omega)$ and the construction of an extension operator. Manuscr. Math. **89**, 159–178
3. Alonso, A. & Valli, A. (1997) A domain decomposition approach for heterogeneous time-harmonic Maxwell equations. Comput. Meth. Appl. Mech. Engrg. **143**, 97–112
4. Alonso, A. & Valli, A. (1999) An optimal domain decomposition preconditioner for low-frequency time-harmonic Maxwell Equations. Math. Comput. **68**, 607–631
5. Amrouche, C., Bernardi, C., Dauge, M. & Girault, V. (1998) Vector potentials in three-dimensional non-smooth domains. Math. Methods Appl. Sci. **21**, 823–864
6. Bermúdez, A., Rodríguez, R. & Salgado, P. (2001) A finite element method with Lagrange multipliers for low-frequency harmonic Maxwell equations. Preprint
7. Bossavit, A. (1993) Électromagnétisme, en Vue de la Modélisation. Springer–Verlag, Paris
8. Bossavit, A. (1999) "Hybrid" electric–magnetic methods in eddy-current problems. Comput. Meth. Appl. Mech. Engrg. **178**, 383–391
9. Buffa, A. (2001) Hodge decomposition on the boundary of non-smooth domains: the multi-connected case. Math. Models Meth. Appl. Sci. **11**, 1491–1504
10. Buffa, A. & Ciarlet, P. Jr. (2001) On traces for functional spaces related to Maxwell's equations. I. An integration by parts formula in Lipschitz polyhedra. Math. Methods Appl. Sci. **24**, 9–30
11. Buffa, A. & Ciarlet, P. Jr. (2001) On traces for functional spaces related to Maxwell's equations. II. Hodge decompositions on the boundary of Lipschitz polyhedra and applications. Math. Methods Appl. Sci. **24**, 31–48
12. Cessenat, M. (1996) Mathematical Methods in Electromagnetism: Linear Theory and Applications. World Scientific Pub. Co., Singapore
13. Dautray, R. & Lions, J.-L. (1992) Mathematical Analysis and Numerical Methods for Science and Technology. Volume 5: Evolution Problems I, Springer–Verlag, Berlin
14. Fernandes, P. & Gilardi, G. (1997) Magnetostatic and electrostatic problems in inhomogeneous anisotropic media with irregular boundary and mixed boundary conditions. Math. Models Meth. Appl. Sci. **7**, 957–991
15. Foias, C. & Temam, R. (1978) Remarques sur les équations de Navier-Stokes stationnaires et les phénomènes successifs de bifurcations. Ann. Scuola Norm. Sup. Pisa **5** (IV), 29–63
16. Girault, V. & Raviart, P.-A. (1986) Finite Element Methods for Navier–Stokes Equations. Theory and algorithms. Springer–Verlag, Berlin
17. Kress, R. (1971) Ein kombiniertes Dirichlet-Neumannsches Randwertproblem bei harmonischen Vektorfeldern. Arch. Rational Mech. Anal. **42**, 40–49
18. Leonard, P.J. & D. Rodger (1989) A new method for cutting the magnetic scalar potential in multiply connected eddy currents problems. IEEE Trans. Magnetics **25**, 4132–4134

19. Leonard, P.J. & D. Rodger (1990) Comparison of methods for dealing with multiply connected regions in 3D eddy current problems using the A-ψ method. COMPEL **9** Supplement A, 55–57
20. Nédélec, J.-C. (1980) Mixed finite elements in $\mathbb{R}^3$. Numer. Math. **35**, 315–341
21. Nédélec, J.-C. (1986) A new family of mixed finite elements in $\mathbb{R}^3$. Numer. Math. **50**, 57–81
22. Picard, R. (1979) Zur Theorie der harmonischen Differentialformen. Manuscr. Math. **27**, 31–45
23. Picard, R. (1982) On the boundary value problems of electro- and magnetostatics. Proc. Royal Soc. Edinburgh **92 A**, 165–174
24. Saranen, J. (1982) On generalized harmonic fields in domains with anisotropic nonhomogeneous media. J. Math. Anal. Appl. **88**, 104–115; Erratum: J. Math. Anal. Appl. **91** (1983) 300
25. Saranen, J. (1983) On electric and magnetic problems for vector fields in anisotropic nonhomogeneous media. J. Math. Anal. Appl. **91**, 254–275
26. Valli, A. (1996) Orthogonal decompositions of $(L^2(\Omega))^3$. preprint UTM 493, Department of Mathematics, University of Trento

Finite Element Micromagnetics

Thomas Schrefl, Dieter Suess, Werner Scholz, Hermann Forster, Vassilios Tsiantos, and Josef Fidler

Vienna University of Technology, Institute of Applied and Technical Physics, Wiedner Hauptstr. 8-10, A-1040 Vienna, Austria

Summary. The development of advanced magnetic materials such as magnetic sensors, recording heads, and magneto-mechanic devices requires a precise understanding of the magnetic behavior. As the size of the magnetic components approach the nanometer regime, detailed predictions of the magnetic properties becomes possible using micromagnetic simulations. Micromagnetics combines Maxwell's equations for the magnetic field with an equation of motion describing the time evolution of the magnetization. The local arrangement of the magnetic moments follows from the complex interaction between intrinsic magnetic properties such as the magnetocrystalline anisotropy and the physical/chemical microstructure of the material. This paper reviews the basic numerical methods used in finite element micromagnetic simulations and presents numerical examples in the field of soft magnetic sensor elements, polycrystalline thin film elements, and magnetic nanowires.

1 Introduction

Micromagnetism is a continuum theory to describe magnetization processes on a significant length scale which is large enough to replace atomistic magnetic moments by a continuous function of position and small enough to reveal the transitions between magnetic domains [1]. With the rapid increase in computer power, numerical micromagnetics has become an important tool to characterize magnetic materials as used in high density magnetic recording and magneto-electronics [2]. The development of ultrahigh density storage media [3] and magneto-electronic devices [4] requires a precise understanding of the magnetization reversal process. The numerical integration of the equation of motions which describe the dynamic response of a magnetic system under the influence of an external field provides a detailed understanding of the microscopic processes that determine the macroscopic magnetic properties like switching time and switching field. In addition to external parameters like the applied magnetic field and the temperature, the magnetization reversal process significantly depends on the interplay between the physical/chemical microstructure of a magnet and the local arrangement of the magnetic moments.

The finite element method is a highly flexible tool to describe magnetization processes, since it is possible to incorporate the physical grain structure and to adjust the finite element mesh according to the local magnetization.

An efficient a posteriori error indicator can be defined making use of a conservation law inherent to the physics of the problem. In order to treat the magnetostatic interactions of distinct magnetic parts, the finite element method can be combined with a boundary element method. The space discretization of the partial differential equations which describe the magnetization dynamics leads to a stiff system of ordinary differential equations. Preconditioned backward differentiation methods significantly reduce the CPU time as compared to Adams or Runge-Kutta methods for time integration.

The paper is organized as follows. Section 2 introduces the basic set of partial differential equations that describes the time evolution of a magnetic system. Section 3 presents basic numerical techniques used in the simulation of magnetic microstructures. Section 3.1 presents the hybrid finite element / boundary element methods for the calculation of the magnetostatic field. Section 3.2 deals with the time integration of the equation of motion, and section 3.3 briefly discusses an adaptive refinement scheme. Section 4 presents some recent examples of micromagnetic simulations. Section 4a presents magnetostatically driven reversal processes in magnetic nano-dots. Section 4b shows the influence of surface roughness on the magnetization reversal of magnetic nano-elements. Section 4c treats the motion of domains walls in magnetic nano-wires using adaptive mesh refinement.

2 Micromagnetics

2.1 Basic principles of micromagnetism

The basic concept of micromagnetism is to replace the atomic magnetic moments by a continuous function of position. In a continuum theory the local direction of the magnetic moments may be described by the magnetic polarization vector

$$\mathbf{J}(\mathbf{r}) = \mu_0 \mathbf{M}(\mathbf{r}) = \mu_0 \mathbf{m}/V. \tag{1}$$

The magnetic polarization $\mathbf{J}$ is proportional to the magnetization, which is given by the magnetic moment, $\mathbf{m}$, per unit volume, V. μ_0 is the magnetic permeability of vacuum. The second principle of micromagnetism treats the magnitude of the magnetization as a function of temperature only. The modulus of $\mathbf{J}$,

$$|\mathbf{J}| = J_s(T), \tag{2}$$

is assumed to be a function of temperature and to be independent of the local magnetic field. Thus the magnetic state of the system can be uniquely described by the directions cosines $b_i(\mathbf{r})$ of the magnetic polarization, $\mathbf{J} = \mathbf{b}J_s$. In a metastable equilibrium state, $\mathbf{b}(\mathbf{r})$ minimizes the the total Gibbs free energy energy of the system.

2.2 Total magnetic Gibbs free energy

The contributions to the total magnetic Gibbs free energy are derived from classical electrodynamics, condensed matter physics, and quantum mechanics so that the continuous expressions for the energy describe the interactions of the spins with the external field, the crystal lattice, and the interactions of the spins with one another. The latter consists of long-range magnetostatic interactions and short-range quantum-mechanical exchange interactions. The competitive effects of the micromagnetic energy contributions upon minimization determine the equilibrium distribution of the magnetization. The minimization of the ferromagnetic exchange energy aligns the magnetic moments parallel to each other, whereas the minimization of the magnetostatic energy favors the existence of magnetic domains. The magnetocrystalline anisotropy energy describes the interaction of the magnetization with the crystal lattice. Its minimization orients the magnetization preferably along certain crystallographic directions. The minimization of the Zeeman energy of the magnetization in an external field rotates the magnetization parallel to the applied field.

The total magnetic Gibbs free energy, E_{t} may be written in the following form [5]

$$E_{\mathrm{t}} = \int_{\Omega_{\mathrm{int}}} \left\{ -\frac{1}{2}\mathbf{H}_{\mathrm{d}} \cdot \mathbf{J} + A\sum_{i=1}^{3}(\nabla b_i)^2 + f_{\mathrm{k}}(\mathbf{b}) - \mathbf{H}_{\mathrm{app}} \cdot \mathbf{J} \right\} \mathrm{d}^3 r. \qquad (3)$$

In (3) the first term of the integrand is the magnetostatic energy density, the second term is the exchange energy density, the third term denotes the magnetocrystalline anisotropy density, and the last term is the Zeeman energy. The integral extends over the total volume of all magnetic particles, Ω_{int}. $\mathbf{H}_{\mathrm{d}}$, A, and $\mathbf{H}_{\mathrm{app}}$ denote the demagnetizing field, the exchange constant, and the applied magnetic field, respectively. For uniaxial materials the magnetocrystalline anisotropy density may be written as

$$f_{\mathrm{k}} = -K_{\mathrm{u}}(\mathbf{u} \cdot \mathbf{b})^2, \qquad (4)$$

where K_{u} is the anisotropy constant and $\mathbf{u}$ is the unit vector along the easy axis. The intrinsic magnetic properties A, K_{u}, and J_{s} and the spatial distribution of the easy axes can be determined experimentally. In a polycrystalline material the direction of the easy axis changes from grain to grain. In addition, the intrinsic magnetic properties may be space dependent.

The demagnetizing field follows from a magnetic scalar potential, U,

$$\mathbf{H}_{\mathrm{d}} = -\nabla U. \qquad (5)$$

The scalar potential solves the Poisson equation

$$\Delta U(\mathbf{r}) = \frac{\nabla \mathbf{J}(\mathbf{r})}{\mu_0} \qquad \text{for } \mathbf{r} \in \Omega_{\text{int}} \tag{6}$$

within Ω_{int}, the space occupied by the magnetic particles. Outside the magnetic particles, Ω_{ext}, (6) reduces to the Laplace equation

$$\Delta U(\mathbf{r}) = 0 \qquad \text{for } \mathbf{r} \in \Omega_{\text{ext}}. \tag{7}$$

At the boundary of the magnet Γ the boundary conditions

$$U^{\text{int}} = U^{\text{ext}}, \quad \left(\nabla U^{\text{int}} - \nabla U^{\text{ext}}\right) \cdot \mathbf{n} = \frac{\mathbf{J} \cdot \mathbf{n}}{\mu_0}, \tag{8}$$

hold. Here $\mathbf{n}$ denotes the outward pointing normal unit vector on Γ. The magnetic scalar potential is regular at infinity

$$U \propto 1/\mathbf{r} \quad \text{for } \mathbf{r} \to \infty. \tag{9}$$

The minimization of (3) subjects to the constraints $|\mathbf{J}| = J_s$, and (6) to (9) provides a metastable equilibrium state of the magnetic system. The subsequent minimization of (3) for different applied field gives the hysteresis curve. Within this static approach, hysteresis is the way the system follows its path through the local minima of the energy landscape [6].

2.3 The Landau-Lifshitz-Gilbert equation

The coercive field of a magnetic is a dynamic property. The measured coercivity significantly depends on the rate of change of the external field. Several experiments show an enhancement of the coercive field with decreasing pulse width of the external field [7]. The dynamic coercivity becomes important in ultra-high density and high data rate magnetic storage [8]. In addition to thermal effects, the gyromagnetic precession causes the increase of the coercive field at short times [9]. The precessional motion of a magnetic moment in the absence of damping is described by the torque equation. According to quantum theory the angular momentum associated with a magnetic moment $\mathbf{m}$ is

$$\mathbf{L} = \mathbf{m}/\gamma, \tag{10}$$

where γ is the gyromagnetic ratio. The torque on the magnetic moment, $\mathbf{m}$, exerted by an effective magnetic field, $\mathbf{H}_{\text{eff}}$,

$$\mathbf{T} = \mathbf{m} \times \mathbf{H}_{\text{eff}}. \tag{11}$$

The change of the angular momentum with time equals the torque,

$$\frac{\partial}{\partial t}\left(\frac{\mathbf{m}}{\gamma}\right) = \mathbf{m} \times \mathbf{H}_{\text{eff}}, \tag{12}$$

which describes the precession of the magnetic moment around the effective field. In equilibrium the change of the angular momentum with time is zero and thus the torque is zero. In order to describe the motion of the magnetic moment towards equilibrium a viscous damping term can be included. A dissipative term proportional to the generalized velocity, $(\partial \mathbf{m}/\partial t)$, is added to the effective field. With η being a positive constant, the dissipative term $-\eta(\partial \mathbf{m}/\partial t)$ slows down the motion of the magnetic moment and aligns $\mathbf{m}$ parallel to $\mathbf{H}_{\text{eff}}$. This gives the Gilbert equation of motion [10]

$$\frac{\partial \mathbf{m}}{\partial t} = -|\gamma|\mathbf{m} \times \left(\mathbf{H}_{\text{eff}} - \eta\frac{\partial \mathbf{m}}{\partial t}\right). \tag{13}$$

Within the framework of a continuum theory (13) has to hold in every point within a ferromagnetic material. Thus we can replace the magnetic moment, $\mathbf{m}$, with the magnetic polarization vector, $\mathbf{J}$, and write the equation of motion in continuous form

$$\frac{\partial \mathbf{J}}{\partial t} = -|\gamma|\mathbf{J} \times \mathbf{H}_{\text{eff}} + \frac{\alpha}{J_{\text{s}}}\mathbf{J} \times \frac{\partial \mathbf{J}}{\partial t}. \tag{14}$$

In (14) the dimensionless Gilbert damping constant $\alpha = \gamma\eta J_{\text{s}}$ was introduced. Multiplying (14) with $\mathbf{J}\cdot$ shows that the equation of motion conserves the norm of the magnetic polarization vector, since the right hand side vanishes: $\partial\left(\mathbf{J}\cdot\mathbf{J}\right)/\partial t = \partial J_{\text{s}}^2/\partial t = 0$. Multiplying both sides of (14) with $\mathbf{J}\times$ gives

$$\mathbf{J} \times \frac{\partial \mathbf{J}}{\partial t} = -|\gamma|\mathbf{J} \times (\mathbf{J} \times \mathbf{H}_{\text{eff}}) + \frac{\alpha}{J_{\text{s}}}\mathbf{J} \times \left(\mathbf{J} \times \frac{\partial \mathbf{J}}{\partial t}\right). \tag{15}$$

Using the $\mathbf{a} \times (\mathbf{b} \times \mathbf{c}) = (\mathbf{a}\cdot\mathbf{c})\mathbf{b} - (\mathbf{a}\cdot\mathbf{b})\mathbf{c}$, we can rewrite (15)

$$\begin{aligned}
\mathbf{J} \times \frac{\partial \mathbf{J}}{\partial t} &= -|\gamma|\mathbf{J} \times (\mathbf{J} \times \mathbf{H}_{\text{eff}}) + \frac{\alpha}{J_{\text{s}}}\left(\mathbf{J}\cdot\frac{\partial \mathbf{J}}{\partial t}\right)\mathbf{J} - \frac{\alpha}{J_{\text{s}}}(\mathbf{J}\cdot\mathbf{J})\frac{\partial \mathbf{J}}{\partial t}, \\
\mathbf{J} \times \frac{\partial \mathbf{J}}{\partial t} &= -|\gamma|\mathbf{J} \times (\mathbf{J} \times \mathbf{H}_{\text{eff}}) - \alpha\frac{\partial \mathbf{J}}{\partial t}.
\end{aligned} \tag{16}$$

If we substitute this result into (14), we obtain the Landau-Lifshitz-Gilbert equation

$$\frac{\partial \mathbf{J}}{\partial t} = -\frac{|\gamma|}{1+\alpha^2}\mathbf{J} \times \mathbf{H}_{\text{eff}} - \frac{|\gamma|\alpha}{(1+\alpha^2)J_{\text{s}}}\mathbf{J} \times (\mathbf{J} \times \mathbf{H}_{\text{eff}}). \tag{17}$$

The effective field,

$$\mathbf{H}_{\text{eff}} = -\frac{\delta E_{\text{t}}}{\delta \mathbf{J}}, \tag{18}$$

is the negative variational derivative of the total magnetic Gibbs free energy. Each energy term contributes to the effective field. The different contributions to the effective field are the demagnetzing field, $\mathbf{H}_{\text{d}}$, the exchange field, $\mathbf{H}_{\text{ex}}$

the anisotropy field, $\mathbf{H}_\mathrm{K}$, and the applied field, $\mathbf{H}_\mathrm{app}$. Whereas the anisotropy field depends only locally on the magnetic polarization, the exchange field and the demagnetizing field account for interactions. The exchange interactions are short range. The magnetostatic interactions are long range, since the magnetic potential U depends on the magnetic volume charges, $\nabla \cdot \mathbf{J}/\mu_0$, and magnetic surface charges, $\mathbf{J} \cdot \mathbf{n}/\mu_0$, over all magnetic particles. The variation of the exchange energy gives the exchange field

$$\mathbf{H}_\mathrm{ex} = \frac{2A}{J_\mathrm{s}} \Delta \mathbf{b}. \tag{19}$$

The LLG equation (17) is a partial differential equation which is coupled to the magnetostatic boundary value problem (6) to (9).

3 Numerical methods

3.1 Magnetostatics

A key part in micromagnetic simulations is the calculation of the magnetic field which arises from the interaction of the magnetization with the element geometry. This so-called demagnetizing field is crucial for the formation of the magnetic domain structure in large elements and determines the external field required to reverse the magnetization of small elements. The magnetostatic interactions between distinct magnetic elements become important in magnetic multilayers or arrays of magnetic dots used for sensor applications, and magnetic storage.

The partial differential equations for the magnetic scalar potential (6) to (9) define an open boundary problem. The potential or its normal derivative are only know at infinity. In principle one has to mesh a wide region outside the magnetic particles in order to account for the boundary conditions at infinity (9). In order to overcome these problems various techniques to treat open boundaries in finite element simulations have been proposed [11]. A hybrid finite element/boundary element method originally introduced by Fredkin and Koehler [12] is very suitable for micromagnetic simulations, as it treats the magnetostatic interactions between distinct magnetic parts without the need to mesh the space outside the magnetic bodies. This feature becomes important simulating the magnetostatic interactions between sensors elements or the writing process in magnetic recording.

The basic concept of this approach is to split the calculations into to parts using the superposition principle. First a potential, U_1, which arises from the magnetic charges within the individual magnetic bodies is calculated. In a second step, a potential, U_2, which accounts for the magnetostatic interactions between distinct bodies and the boundary conditions at infinity, is calculated. The potential U_1 is assumed to solve a closed boundary value problem. Then the equations for U_2 can be derived from (6)–(9), which hold for the total potential $U = U_1 + U_2$. The potential U_1 can be computed from the closed boundary value problem,

$$\Delta U_1(\mathbf{r}) = \frac{\nabla \cdot \mathbf{J}(\mathbf{r})}{\mu_0} \quad \text{for } \mathbf{r} \in \Omega_{\text{int}} \tag{20}$$

$$U_1 = 0 \quad \text{for } \mathbf{r} \in \Omega_{\text{ext}} \tag{21}$$

$$\nabla U_1 \cdot \mathbf{n} = \frac{\mathbf{J} \cdot \mathbf{n}}{\mu_0} \quad \text{for } \mathbf{r} \in \Gamma. \tag{22}$$

The potential U_1 is the solution of the Poisson equation within the magnetic particles and equals zero outside the magnets. At the surface of the magnets natural boundary conditions hold. The potential U_2 satisfies the Laplace equation everywhere

$$\Delta U(\mathbf{r}) = 0 \quad \text{for } \mathbf{r} \in \Omega_{\text{int}} \cup \Omega_{\text{ext}}, \tag{23}$$

with the following boundary conditions for $\mathbf{r} \in \Gamma$

$$U_2^{\text{int}} - U_2^{\text{ext}} = U_1, \tag{24}$$

$$\left(\nabla U_2^{\text{int}} - \nabla U_2^{\text{ext}}\right) \cdot \mathbf{n} = 0. \tag{25}$$

The potential U_2 shows a jump at the surfaces of the magnetic bodies.

The closed boundary value problem (20)–(22) can be solved using a standard finite element method. Both the potential U_1 and the direction cosines of the magnetic polarization, b_i, are interpolated by piecewise linear functions on a tetrahedral finite element grid. The resulting linear equation is solved using a conjugate gradient method with relaxed incomplete factorization (RILU) preconditioning [13]. During time integration of the LLG equation, the iterative solver can be started with the previous solution for U_1 as initial guess. Typically about 25 iterations are required in a system with 2×10^4 nodes. The equations (23)–(25) define a double layer potential

$$U_2(\mathbf{r}) = \frac{1}{4\pi} \int_\Gamma U_1(\mathbf{r}') \nabla' \frac{1}{|\mathbf{r} - \mathbf{r}'|} \cdot \mathbf{n}' \mathrm{d}r'^2 \tag{26}$$

which is created by a dipole sheet with magnitude U_1. In principle U_2 can be evaluated everywhere within the magnetic bodies using (26). However, instead of the direct computation of U_2 discretizing (26), we evaluate U_2 at the boundary and then we solve (23) within Ω_{int} using the known boundary values as Dirichlet conditions. To compute U_2 on Γ, we have to take the limit $\mathbf{r} \to \Gamma$ of the surface integral from inside Ω_{int}

$$U_2(\mathbf{r}) = \frac{1}{4\pi} \int_\Gamma U_1(\mathbf{r}') \nabla' \frac{1}{|\mathbf{r} - \mathbf{r}'|} \cdot \mathbf{n}' \mathrm{d}r'^2 + \left(\frac{\Omega(\mathbf{r})}{4\pi} - 1\right) U_1(\mathbf{r}). \tag{27}$$

We discretize (27) using piecewise linear functions to interpolate U_1 on a triangular surface mesh. U_2 follows from a matrix vector product $\underline{U_2} = \mathbf{B}\underline{U_1}$.

The boundary element matrix, **B**, depends only on the geometry of the problem and has to be computed only once. **B** is a fully populated $m \times m$ matrix which relates the m boundary nodes with each other.

In summary we have to perform the following procedure to compute the demagnetizing field. Prior to the time integration of the LLG equation we assemble the system matrices and compute an incomplete ILU factorization of the linear systems corresponding to equations (20) and (23). The setup phase also involves the computation of the boundary matrix **B**. Generally we are interested on the dynamic response of a system over a time span which is about two orders of magnitude larger than intrinsic precession time. Thus the CPU time of the setup phase is only a small fraction of the total CPU time. At each iteration during the time integration we have to perform the following steps to update the magnetostatic field:

1. Iterative solution of a linear system for U_1 (equation 20).
2. Matrix vector product to obtain U_2 at the boundary of the magnetic bodies (equation 27).
3. Iterative solution of a linear sytem for U_2 within the magnetic bodies (equation 23).
4. Sum U_1 and U_2 and build the gradient (equation 5).

3.2 Time integration

The precise understanding of the switching process of thin film nanomagnets is important for sensor and spin electronic applications. Surface irregularities and grain structure drastically change the reversal mechanism of thin film elements [14]. Taking into account surface roughness and grain structures requires an inhomogeneous computational grid which in turn causes very small time steps for time integration. Toussaint and co-workers [15] showed that the time step required to obtain a stable solution of the LLG equation with an explicit time integration scheme has to be proportional to $1/h^2$, where h is the size of the spatial grid. Edge roughness and an irregular grain structure may force small finite elements which leads to a small time step when an explicit time integration method is applied to solve the LLG equation. Yang and Fredkin [16] originally applied a BDF method in dynamic micromagnetic simulations. They apply the Galerkin variant of the finite element method for space discretization and a generalized minimum residual method (GMRES) to solve the linear systems involved in the solution process.

We use a collocation method to integrate the LLG equation (17) and assume that the equation is fulfilled at the nodes of the finite element mesh. Using spatial averaging we assign a magnetic moment

$$\mathbf{m}_i = \frac{1}{\mu_0} \int_{V_i} \mathbf{J}(\mathbf{r}) \mathrm{d}^3 r \tag{28}$$

to node i of the finite element mesh. The box volumes V_i have the following properties

$$\sum_{i=1}^{N} V_i = \int_{\Omega_{\text{int}}} \mathrm{d}^3 r, \tag{29}$$

$$V_i \cap V_i = 0 \quad \text{for} \quad i \neq j, \tag{30}$$

where the sum in (29) runs from 1 to the total number of nodes of the finite element mesh, N. The effective field at node i can be approximated as

$$\mathbf{H}_{\text{eff}}^{i} = -\frac{1}{\mu_0}\frac{\partial E_{\text{t}}}{\partial \mathbf{m}^i}. \tag{31}$$

Again a piecewise linear interpolation of the magnetic polarization vector on a tetrahedral finite element mesh is used to discretize the total magnetic Gibbs free energy, E_{t}.

Using (28) and (31) we can define a magnetic moment vector and an effective field vector at each node of the finite element mesh which leads to a system of $3N$ ordinary differential equations. It is solved using a BDF method [17]. Within the framework of this software package, the linear system at each Newton iteration is solved using a GMRES method. The GMRES method is a matrix free iterative method to solve a linear system of equations. Within the time integration package, the product of the Jacobian matrix times a vector is approximated using finite differences. Preconditioning partly replaces the finite difference approximation with exact curvature information. We provide the parts of the Jacobian matrix which are associated with the magnetocrystalline anisotropy and the ferromagnetic exchange interactions. As the short range interactions are the major source of stiffness in micromagnetic simulations, we obtain a significant speed up while keeping the system matrix sparse. A nonlinear system of equations has to be solved at each time step which can be effectively solved using the Newton method. Typically only 1–2 Newton steps are required to obtain convergence. However, the linear system to be solved at each Newton step may be ill-conditioned so that most of the total CPU time is spent in solving this system. A twofold procedure helps to speed up the calculation by more a factor of 40.

1. We provide an approximate Jacobian containing the short range interactions. This information is used to apply a left preconditioner to the internal matrix free GMRES solver of the time integration software.
2. The auxiliary linear equation which has to be solved for preconditioning of the internal GMRES solver is solved iteratively using a biconjugate gradient stabilized (BICSTAB) algorithm [18]. Among various preconditioners the imcomplete factorization (ILU) preconditioning proved to be most efficent for this auxiliary systems of linear equations.

3.3 Adaptive meshing

The numerical treatment of magnetization processes involves a wide range of length scale. A sufficiently fine finite element mesh is required to accurately predict the switching field of magnetic particles. Numerical experiments by Rave and co-workers [20] showed that the accurate simulation of the nucleation of reversed domains requires a mesh size comparable with the characteristic length of the material. As the mesh size reaches the critical length the exchange energy density and the magnetostatic energy density balance each other and the nucleation field becomes independent of the grid spacing. The critical length scale depends on the relative strength of the exchange energy density with respect to the other micromagnetic energy terms. In soft magnetic materials the most dominant energy contribution is the magnetostatic energy. The significant length is given by the exchange length

$$l_{\mathrm{ex}} = \sqrt{\frac{2\mu_0 A}{J_{\mathrm{s}}^2}}. \tag{32}$$

In hard magnetic materials the most dominant energy contribution is the magneto-crystalline anisotropy energy. The significant length is given by the Bloch parameter

$$\delta_0 = \sqrt{\frac{A}{K_{\mathrm{u}}}}. \tag{33}$$

The typical length scales involved are the following:

- The typical sample size is in the range of micrometer range. Examples are the length of magnetic-nanowires or the lateral extension of magnetic thin film elements as used for sensor or storage elements [4].
- Most magnetic materials exhibit a polycrystalline structure. The grain size of magnetic thin film elements is in order of 10 nm [19].
- The characterstic length is an intrinsic property an is typically in the range from 3 nm – 5 nm. It is the length scale on which the magnetic polarization vector changes its direction [20].

Adaptive refinement and coarsening schemes are applied in order to scope with these different length scale in micromagnetic simulations. Hertel and Kronmüller [21] introduced an adaptive refinement scheme in static micromagnetic simulations, to calculate domain configurations in thin film elements. Here present a scheme to refine and coarse the mesh during the time integration of the LLG equation.

The general outline of adaptive algorithms is as follows. Starting from an initial triangulation $\mathcal{T}_0$, we produce a sequence of refined grids $\mathcal{T}_{\mathrm{k}}$, until the estimated error is below a given tolerance ε. The nature of the micromagnetic problem allows to define a cheap a posteriori error indicator. Using piecewise linear function to represent the direction cosines of the magnetic polarization, the constraint (2) can only be hold at the nodal points of the finite element mesh. Thus an error indicator for the element e may be constructed by

$$\eta_e = \int_{\Omega_e} \left| \frac{\mathbf{J} \cdot \mathbf{J}}{J_{\mathrm{s}}^2} - 1 \right| \mathrm{d}^3 r / V_e. \tag{34}$$

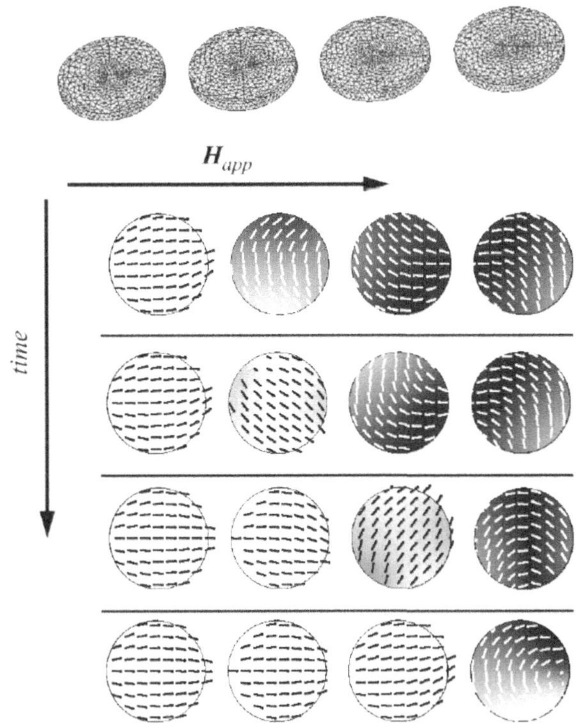

Fig. 1. Top: Surface of the finite element mesh of interacting NiFe nano-dots. Bottom: Time evolution of the magnetization pattern under the influence of an applied field

The mesh is adjusted to the current magnetization distrubution during the solution of the LLG equation. We start with an initial triangulation τ_0. The mesh is refined in regions with non-uniform magnetization, whereas elements are taken out where the magnetization is uniform. After each time step we compute the error indicators. Depending on the distribution of the error indicators one of the following three procedures are applied:

Refinement. The error indicator of at least one element exceeds the global error tolerance ε. The time step is rejected. The mesh is refined in regions with high η_e. The previously accepted magnetization distribution is interpolated on the new nodes.

Coarsening. A certain percentage of elements shows an error indicator below a certain threshold, $\sigma\varepsilon$, with $\sigma < 1$. The time step is rejected and the current magnetization distribution is interpolated on the initial mesh, τ_0.

Proceed in time. Otherwise the time step is accepted and the time integration continues with the given grid.

The above algorithm guarantees that the simulation proceeds in time only if the space discretization error is below a certain threshold. Simulations of wall motion in nanowires show [22] showed that this adaptive scheme reduces the total CPU time by more than a factor of 4 as compared to a uniform mesh.

4 Examples

4.1 Magnetostatic interactions between circular nanodots

Circular magnetic nano-dots may be the basic structural units of future magnetic logic devices [23]. A critical parameter of these magnetic structures is the switching speed. As magnetization switching is induced by the magnetostatic interaction field of switched neighboring dots, the switching behavior is governed by the strength of the magnetostatic interactions and the nonuniform magnetization distribution during within the individual dots during the magnetization reversal process. Fig. 1 shows the finite element mesh (top) and the time evolution of the magnetization pattern (bottom) of interacting magnetic nano-dots. The diameter of the dots is 110 nm and their thickness is 10 nm. The intrinsic magnetic properties of permalloy were used for the simulations. The reversal time per dot is about 0.5 ns for an applied field of 5.6 kA/m using a Gilbert damping constant $\alpha = 0.1$. The magnetostatic interactions between the dots cause a dot by dot reversal of the chain. The magnetization of neighboring dots rotates in opposite directions, forming partial flux closure structures during the reversal process.

4.2 Surface roughness in magnetic nanoelements

Magnetic nano-elements have important applications as magnetic field sensors and might be used in future discrete storage media. A well defined switching field and a predictable domain structure are important prerequisites for the application of thin film elements. However, both the switching field and the switching time were found to depend strongly on the physical structure of the elements such as the surface roughness and the polycrystalline grain structure. In the following the switching process is compared for three different Co elements. One element denoted by (A) consists of a perfect microstructure. The surface is flat, no grains are assumed within the particle and the crystalline anisotropy is zero. Element (B) takes account of surface roughness. The notches are in average 8 nm. Element (C) consists of 500 columnar grains (diameter is 8 nm) with random distribution of the magnetocrystalline anisotropy directions. All the elements are 400 nm long, 80 nm wide and 25 nm thick. The granular element (C) has the largest coercive field, $H_c = 72$ kA/m. The coercive field decreases by less than 10 % for the perfect Co-element without crystalline anisotropy. Surface roughness leads to

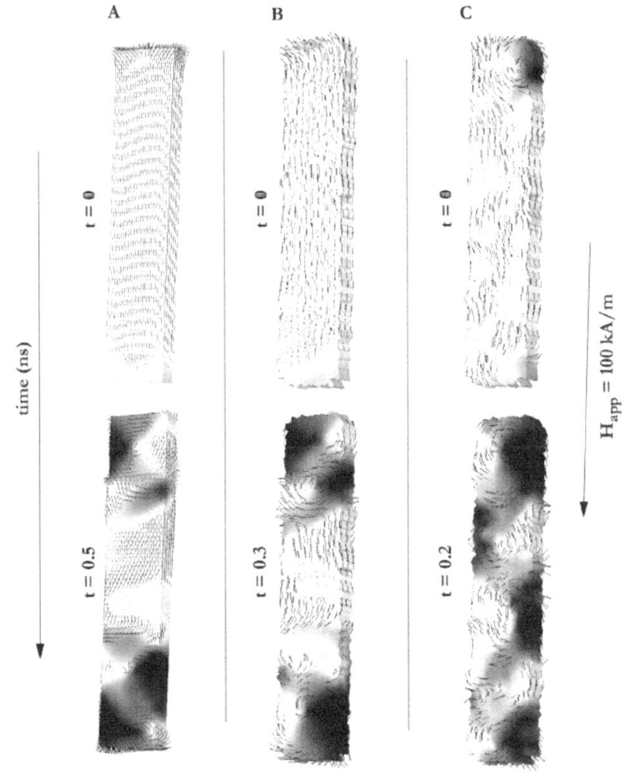

Fig. 2. Influence of the physical structure on the magnetization reversal of Co nano-elements. The plots compare the time evolution of (A) a flat element, (B) an element with surface roughness, and (C) an element with surface roughness and polycrystalline grain structure. Top row: Remanent magnetization distribution for zero applied field. Bottom row: Transient magnetic states during irreversible switching under the influence of a field of 100 kA/m . The component of the magnetic polarization is color coded. White: $\mathbf{J}$ antiparallel to $\mathbf{H}_{\text{app}}$, black: $\mathbf{J}$ parallel to $\mathbf{H}_{\text{app}}$

a reduction of the coercive field of about 20%. Fig. 2 shows the onset of magnetization reversal for the different elements under the influence of a reverse field of 100 kA/m. The switching time decreases from 0.75 ns for element (A) to 0.5 ns for element (C). Again a Gilbert damping constant $\alpha = 0.1$ was used. Fig. 3 shows the effect of preconditioning on the time integration error,

$$\eta_{\text{time}} = max_i |1 - \mathbf{J} \cdot \mathbf{J} / J_s^2|, \tag{35}$$

where i runs over all the nodes of the finite element mesh. The BDF solver without preconditioning does not preserve the norm of the magnetic polarization vector at the nodes of the mesh. Several renormalization steps of $\mathbf{J}$ are required during the time integration of the LLG equation. Preconditioning keeps the time discretization error small.

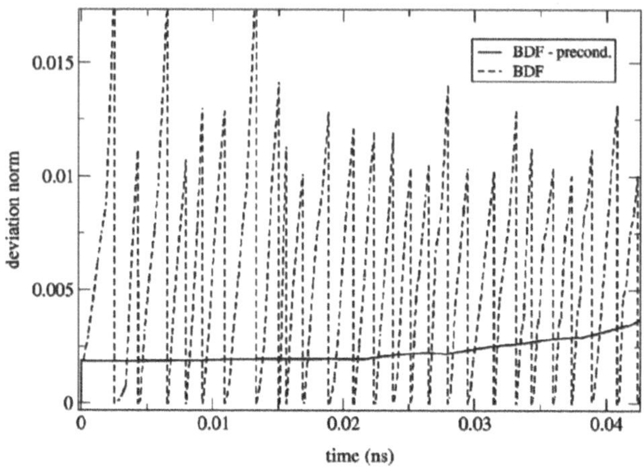

Fig. 3. Comparison of the time discretization error, η_{time}, for the integration of the LLG equation for sample (C) using a BDF method and a preconditioned BDF method

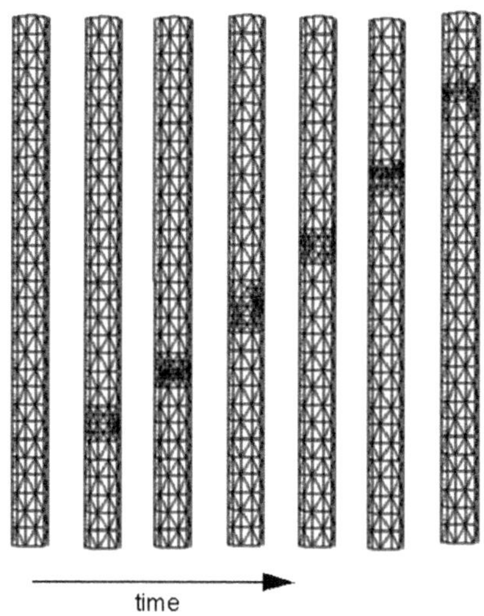

Fig. 4. Sequence of meshes during the motion of a domain wall in a magnetic nanowire. The coarse mesh has a size of 20 nm. At the wall position the mesh size is about 4 nm

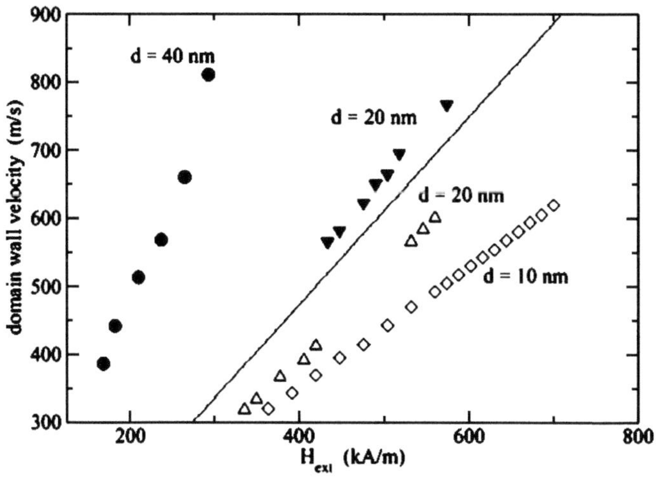

Fig. 5. Domain wall velocity as a function of the applied field for different diameters, d of the Co wire. For $d < 20$ nm transverse walls occur, for $d > 20$ nm vortex walls are energetically more favorable. Both types are found at $d = 20$ nm

4.3 Domain wall motion magnetic nano-wires

The domain wall motion has been calculated in Co nano-wires as a function of the wire thickness. The thickness is varied in the range from 10 nm to 40 nm. The length of the wire is 600 nm. Initially, a reversed domain is created in one end. Under the influence of an applied field the domain with the magnetization parallel to the field direction expands and the domain wall moves through the wire. During this process the magnetization remains nearly uniform within the core of both domains. Thus it is sufficient to resolve only the magnetization transition in the domain wall and use a coarse finite element mesh within the domains. As the wall moves, the finite element mesh is adjusted to the current wall position. Fig. 4 gives a sequence of finite element meshes during wall motion. The structure of the domain wall has a strong effect on the wall velocity. Thin wires show a transverse wall which move slowly. In thick wires, a vortex can form within the wall which causes an increase in the wall velocity. Fig. 5 shows the calculated domain wall velocities as a function of the applied field.

Acknowledgement

Work supported by the Austrian Science fund (Y-132PHY,13260TEC).

References

1. Aharoni A. (1996) Introduction to the Theory of Ferromagnetism. Oxford University Press, New York
2. Dahlberg, E. D., Zhu, J. G. (1995) Micromagnetic Microscopy and Modeling. Physics Today **48**, 34–40
3. Schabes, M. E., Fullerton, E. E., Margulies, D. T. (2001) Theory of Antiferromagnetically Coupled Magnetic Recording Media. J. Appl. Phys., in press
4. Johnson, M. (2000) Magnetoelectronic memories last and last ... . IEEE Spectrum **37**, 33–40
5. Brown Jr., W. F. (1963) Micromagnetics, Interscience, New York
6. Kinderlehrer, D., Ma, L. (1994) Computational Hysteresis in Modeling Magnetic Systems. IEEE Trans. Magn. **30**, 4380–4382
7. He L., Doyle W. D. et al. (1996) High-speed switching in magnetic recording media. J. Magn. Magn. Mat. **155**, 6–12
8. Akagi F., Nakamura A. et al. (2000) Computer Simulation of Magnetization Switching Behavior in High-Data-Rate Hard-Disk Media Masukazu Igarashi, IEEE. Trans. Magn. **36**, 154–158
9. Harrell, R. W. (2001) Orientation dependence of the dynamic coercivity of Stoner-Wohlfarth particles. IEEE Trans. Magn. **37**, 533–537
10. Gilbert, T. L. (1955)A Lagrangian formulation of gyromagnetic equation of the magnetization field, Phys. Rev. **100**, 1243
11. Chen, Q., Konrad, A. (1997) A review of finite element open boundary techniques for static and quasi-static electromagnetic field problems. IEEE Trans. Magn. **33**, 663–676
12. Fredkin, D. R., Koehler, T. R. (1990) Hybrid method for computing demagnetizing fields. IEEE Trans. Magn. **26**, 415–417
13. Bruaset, A. M. (1997) Krylov subspace iterations for sparse linear systems. In: Morten Daehlen, M., Tveito, A. (Eds.) Numerical Meth and Software Tools in Industrial Mathematics. Birkhauser, Boston, 21
14. Gadbois, J., and Zhu, J. G. (1995) Effect of Edge Roughness in Nano-Scale Magnetic Bar Switching. IEEE Trans. Magn. **31**, 3802–3804
15. Toussaint, J. C., Kevorkian, B., Givord, D., and Rossignol, M. F. (1996) Micromagnetic Modeling of Magnetization Reversal in Permanent Magnets. In: Proceedings of the 9th International Symposium Magnetic Anisotropy and Coercivity In Rare-Earth Transition Metal Alloys, World Scientific, Singapore, 59–68
16. Yang, B., Fredkin, D. R. (1998) Dynamical micromagnetics by the finite element method. IEEE Trans. Magn. **34**, 3842–3852
17. Cohen, S. D., and Hindmarsh, A. C. (1996) CVODE, A Stiff/Nonstiff ODE Solver in C. Computers in Physics, **10** 138–143.
18. Saad, Y. (1996) Iterative methods for sparse linear systems, PWS Publishing Company, Boston
19. Kirk, K. J., Chapman, J. N., Wilkinson, C. D. W. (1997) Switching fields and magnetostatic interactions of thin film magnetic nanoelements. Appl. Phys. Lett. **71**, 539–541
20. Rave, W., Ramstock, K., Hubert, A. (1998) Corners and Nucleation in Micromagnetics. J. Magn. Magn. Mater. **183**, 329–333

21. Hertel, R., Kronmuller, H. (1998) Adaptive finite element mesh refinement techniques in three-dimensional micromagnetic modeling. IEEE Trans. Magn. **34**, 3922–3930
22. Schrefl T., Forster H. et al. (2001) Micromagnetic Simulation of Switching Events, In: Kramer, B. (Ed.) Advances in Solid State Physics 41. Springer, Berlin Heidelberg, 623-635
23. Cowburn, R. P., Welland, M. E. (2000) Room temperature magnetic quantum cellular automata, Science **287**, 1466–1458

Finite Integration Method and Discrete Electromagnetism

Thomas Weiland

Technische Universitt Darmstadt
Faculty of Electrical Engineering and Information Technology
Schlossgartenstr. 8, D-64289 Germany

Summary. We review some basic properties of the Finite Integration Technique (FIT), a generalized finite difference scheme for the solution of Maxwell's equations. Special emphasis is put on its relations to the Finite Difference Time Domain (FDTD) method, as both algorithms are found to be computationally equivalent for the special case of an explicit time-stepping scheme with Cartesian grids. The more general discretization approach of the FIT, however, inherently includes an elegant matrix-vector notation, which enables the application of powerful tools for the analysis of consistency, stability, and other issues. On the implementation side this leads to many important consequences concerning the basic method as well as all kinds of extensions.

1 Introduction

Aiming at a general discretization method for the whole range of electromagnetic phenomena, the best choice for a numerical analysis is to start from the fundamental equations rather than from derived or specialized ones. We will thus, use Maxwell's equations without any a-priori restriction, given here in their integral form for non moving materials:

$$\oint_{\partial A} \mathbf{E} \cdot d\mathbf{s} = - \iint_{A} \frac{\partial \mathbf{B}}{\partial t} \cdot d\mathbf{A} \quad (1) \qquad \oint_{\partial A} \mathbf{H} \cdot d\mathbf{s} = \iint_{A} \left(\frac{\partial \mathbf{D}}{\partial t} + \mathbf{J} \right) \cdot d\mathbf{A} \quad (3)$$

$$\oint_{\partial V} \mathbf{B} \cdot d\mathbf{A} = 0 \, , \quad (2) \qquad \oint_{\partial V} \mathbf{D} \cdot d\mathbf{A} = \iiint_{V} \varrho dV \quad (4)$$

$\forall A \in \mathbb{R}^2$ and $\forall V \in \mathbb{R}^3$. The field and flux vectors are related by the material equations

$$\mathbf{B} = \mu \mathbf{H} + \mathbf{M} \, , \qquad \mathbf{D} = \varepsilon \mathbf{E} + \mathbf{P} \, , \qquad \mathbf{J} = \kappa \mathbf{E} \, . \tag{5}$$

For the solution of the above equations we present in this paper a generalized FDTD scheme, known as Finite Integration Technique (FIT). This method has originally been developed independently of the well known FDTD method by [1] for frequency domain cases [2] and later completed to a generalized FDTD scheme for the entire application range of Maxwell's equations [3, 4].

The paper is organized as follows: In the next Section we introduce the notation of FIT and derive the basic equations by means of a two-dimensional triangular computational grid (as one example out of the numerous types of computational grids usable within this method). In Section 3 we proceed with the discrete material relations of FIT, before some of its most important properties are shortly reviewed (Section 4). In Section 5, the special choice of a 3D Cartesian grid allows a direct comparison between FIT, supplemented by the explicit leapfrog time integration scheme, and the FDTD method. From a stability analysis of this time-domain case we finally draw some important consequences concerning any extensions of both FIT and FDTD (Section 6).

2 The Finite Integration Technique (FIT)

Like the FDTD method, FIT uses a pair of staggered grids, the primary grid G and the dual grid $\widetilde{G}$, which however can have a more general structure as the standard "Yee cell" of FDTD. Actually, there are only very few restricting prerequisites for the grids to be used, and so far the method has been implemented into computer codes for 2D- and 3D-Cartesian and cylindrical coordinate grids [4], for 2D triangular [5, 6] and (under a different name) 3D-tetrahedral grids [7, 8], as well as for generalized nonorthogonal coordinate grids [9]. The primary grid defines a decomposition of the entire computation domain into a collection of cells, denoted as $V_i(i = 1..N_V)$, each of which is surrounded by facets $A_i(i = 1 \dots N_A)$, which are again surrounded by edges $L_i(i = 1 \dots N_L)$. For an approximation of a real material distribution we assume that in each cell the material may be different but homogenous (for a even more general approach including arbitrary inhomogeneous material fillings inside single mesh cells cf. [10]). The required duality-property of the grids assures that, each edge of one grid penetrates a surface of the other grid and that each mesh point of one grid lies at the center of a cell volume of the other grid. (Thus, $\widetilde{N}_A = N_L$, $\widetilde{N}_L = N_A$, etc.). All quantities (edges, facets, etc.) referring to the dual grid are denoted by a *tilde*. As a simple example for the derivation of the method we consider the two cells of a two-dimensional triangular grid[1] and the related dual grid as shown in Fig. 1.

A key issue in the construction of any numerical method is to make a proper choice for the unknown quantities to be computed (the state variables of the approach). Depending on the formulation this could be vector potentials, Hertz potentials, scalar potentials, field vectors or any derived quantities. An obvious choice are (the components of) field vectors as they

[1] As Maxwell's equations are inherently three-dimensional: If necessary, 2D-grids are implicitly extended in the third dimension by a unit step size (see Fig. 2).

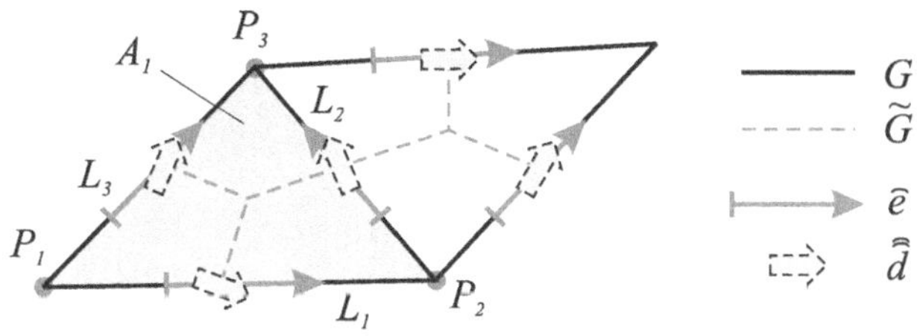

Fig. 1. Triangular grid and dual grid, elementary geometric objects (edges L_i, facets A_i) of a primary grid cell, electric state variables

appear in the fundamental equations (1)-(4); and the distributed allocation of Cartesian field components in the middle of grid edges has been one of the basic ideas of Yee's FDTD method. In the FIT we use a closely related, but different type of state variables, the so-called grid voltages and grid fluxes. These scalar quantities are defined as integrals of the electric and magnetic field vectors along the geometric objects of the grids (cf. Fig. 1):

$$\begin{aligned} \widehat{\mathbf{e}}_i &= \int_{L_i} \mathbf{E} \cdot d\mathbf{s} \,, \quad \widehat{\widehat{\mathbf{b}}}_i = \int_{A_i} \mathbf{B} \cdot d\mathbf{A} \,, \\ \widehat{\mathbf{h}}_i &= \int_{\widetilde{L}_i} \mathbf{H} \cdot d\mathbf{s} \,, \quad \widehat{\widehat{\mathbf{d}}}_i = \int_{\widetilde{A}_i} \mathbf{D} \cdot d\mathbf{A} \,, \quad \widehat{\widehat{\mathbf{j}}}_i = \int_{\widetilde{A}_i} \mathbf{J} \cdot d\mathbf{A} \,. \end{aligned} \tag{6}$$

The integral quantities are denoted using the above special symbols and notifying the dimension of the integrations[2].

To transfer Maxwell's equations into the discrete space of these scalar state variables, we consider e.g. Faraday's law in its integral form (hence the name "Finite Integration Technique"), applied to one of the triangular facets in Fig. 1. With the definitions in (6) and using the notations of Fig. 1 we find the simple, but exact discrete representation

$$\oint_{\partial A_1} \mathbf{E} \cdot d\mathbf{s} = -\frac{d}{dt} \int_{A_1} \mathbf{B} \cdot d\mathbf{A} \quad \Leftrightarrow \quad \widehat{\mathbf{e}}_1 + \widehat{\mathbf{e}}_2 - \widehat{\mathbf{e}}_3 = \frac{d}{dt} \widehat{\widehat{\mathbf{b}}}_1 \,. \tag{7}$$

The left hand side contains all voltage components related to edges adjacent to the considered facet, combined with proper signs ± 1 according to their

[2] Besides some important consequences for the theoretical analysis of the method which we will focus on later, it is remarkable that also in the continuous world only integral quantities (and not the fields themselves) are accessible to measurements. This feature of the FIT formulation has recently been recognized in the Finite Element literature as well as in some publications proposing a reformulation of Maxwell's equations in terms of the language of differential algebra (cf. [11, 12] and the references therein).

orientation. This process of picking the right components out of the set of all grid voltages and assigning ± 1 can be summarized for all facets in a matrix-vector notation with a sparse $N_A \times N_E$-matrix $\mathbb{C}$, containing only $\{\pm 1, 0\}$:

$$\mathbb{C}\widehat{\mathbf{e}} = -\frac{d}{dt}\widehat{\widehat{\mathbf{b}}} \,. \tag{8}$$

In a topological sense, $\mathbb{C}$ describes the facet-to-edge incidence relation of the grid. As in our context it is used to make up finite curl-operations for the discrete fields, $\mathbb{C}$ is referred to as the discrete curl-operator of the primary grid. A further analysis of its sub-structure shows that for Cartesian and other coordinate grids (using a special indexing scheme) is composed of even more basic, two-banded matrices $\mathbf{P}_u$,$\mathbf{P}_v$ and $\mathbf{P}_w$, each of which represents a discrete partial differentiation operator [3].

A completely analogous derivation can be performed for Ampere's law, applied to one facet of the dual grid, where now magnetic voltages and electric fluxes have to used. This leads, for all dual facets, to the discrete form of Ampere's law

$$\mathcal{T}\widehat{\mathbf{h}} = \frac{d}{dt}\widehat{\widehat{\mathbf{d}}} + \widehat{\widehat{\mathbf{j}}} \,, \tag{9}$$

with the *dual*-curl-*matrix* $\mathcal{T}$. From the grid topology we have the important relation:

$$\mathcal{T} = \mathbb{C}^{\mathrm{T}} \,. \tag{10}$$

In a similar way we can transform the remaining Maxwell's equations into the discrete space. As integration domains in (2) and (4) we choose one cell $\{V_i, \widetilde{V}_i\}$ of the primary or the dual grid, respectively, and all we have to do is adding up the signed fluxes on the cell surface:

$$\sum_{A_j \subset \partial V_i} \pm \widehat{\widehat{\mathbf{b}}}_j = 0 \,, \quad \sum_{\widetilde{A}_j \subset \partial \widetilde{V}_i} \pm \widehat{\widehat{\mathbf{d}}}_j = \mathbf{q}_i = \int_{\widetilde{V}_i} \varrho dV \,. \tag{11}$$

Similarly to the definition of $\mathbb{C}$ we can define a new $\widetilde{N}_V \times \widetilde{N}_A$ - matrix that also contains only ± 1 and 0 and constructs the corresponding equation for each of the N_V cells of G (or $\widetilde{N}_V$ cells of $\widetilde{G}$ respectively):

$$\mathbf{S}\widehat{\widehat{\mathbf{b}}} = \mathbf{0} \,, \quad \widetilde{\mathbf{S}}\widehat{\widehat{\mathbf{d}}} = \mathbf{q} \,. \tag{12}$$

These matrices are referred to as *discrete divergence operators* of the primary or the dual grid. Again, grid topology supplies an important relation between the operator matrices:

$$\mathbf{S}\mathbb{C} = \mathbf{0} \,, \quad \widetilde{\mathbf{S}}\mathcal{T} = \mathbf{0} \,. \tag{13}$$

Additionally we can also derive discrete gradient-matrices for each of the grids [13] and find the duality properties

$$\mathbf{G} = -\widetilde{\mathbf{S}} \,, \quad \widetilde{G} = -\mathbf{S}^{\mathrm{T}} \,. \tag{14}$$

As a result of this derivation we obtain a complete set of algebraic equations, transforming Maxwell's equations one-to-one into the discrete space of FIT's state variables. These so-called *Maxwell's Grid Equations* read:

$$\begin{aligned}
\oint_{\partial A} \mathbf{E}\cdot d\mathbf{s} = -\iint_{A} \frac{\partial \mathbf{B}}{\partial t}\cdot d\mathbf{A} \quad &\xleftrightarrow{\forall A\in\{A_i\}\subset G} \quad \mathbb{C}\widehat{\mathbf{e}} = -\frac{d}{dt}\widehat{\widehat{\mathbf{b}}}\,,\\
\oint_{\partial A} \mathbf{H}\cdot d\mathbf{s} = \iint_{A} \left(\frac{\partial \mathbf{D}}{\partial t} + \mathbf{J}\right)\cdot d\mathbf{A} \quad &\xleftrightarrow{\forall A\in\{\widetilde{A}_i\}\subset G} \quad \widetilde{\mathbb{C}}\widehat{\mathbf{h}} = \frac{d}{dt}\widehat{\widehat{\mathbf{d}}} + \widehat{\widehat{\mathbf{j}}}\,,\\
\oint_{\partial V} \mathbf{B}\cdot d\mathbf{A} = 0 \quad &\xleftrightarrow{\forall V\in\{V_i\}\subset G} \quad \mathbf{S}\widehat{\widehat{\mathbf{b}}} = \mathbf{0}\,,\\
\oint_{\partial V} \mathbf{D}\cdot d\mathbf{A} = \iiint_{V} \varrho\, dV \quad &\xleftrightarrow{\forall V\in\{\widetilde{V}_i\}\subset G} \quad \widetilde{\mathbf{S}}\widehat{\widehat{\mathbf{d}}} = \boldsymbol{q}\,.
\end{aligned} \tag{15}$$

Note that this transformation is still exact, as e.g. the electric voltage related to one edge of the surface represents - by definition in (6) - the exact value of the integral of the electric field strength along that path, and $\widehat{\widehat{\mathbf{b}}}$ represents the exact value of the magnetic flux density integral over a cell surface. A discretization takes place only in the sense that geometrical grid objects are chosen as special integration domains in Maxwell's equations, and thus the complexity of continuous electric and magnetic field quantities is reduced to a finite number of well-defined discrete unknowns.

3 Discrete Material Relations

To complete the discrete system in (13), we have to supply operators for the constitutive relations to transform electric and magnetic voltages into fluxes and vice versa. In the simplest case[3] these operators can be postulated to be square matrices, so-called generalized material matrices, which are denoted by $\mathbf{M}_\varepsilon$, $\mathbf{M}_\mu$ and $\mathbf{M}_\kappa$. Depending on the type of grid, there are many different possibilities to derive these material relations within the FIT-framework. For simplicity - and to compare our results with the FDTD approach - we stay here with the important class of orthogonal grid doublets, where we will find that the definition (6) of FIT's integral state variables suggests an inherent way to derive the corresponding material operators. More advanced material operators have been developed for nonorthogonal grids [9], and a recent approach even utilizes Whitney elements known from Finite Element methods [6], thus bridging the gap between FIT, FDTD and FE approaches.

Orthogonal grid doublets $\{G\perp\widetilde{G}\}$ are characterized by an orthogonal intersection of primary edges and dual facets (and vice versa), whereas the

[3] For simplicity, we will only describe here non-dispersive, isotropic and linear material relations. For more details on discrete models for more complicated material phenomena cf. [14, 15, 16, 17].

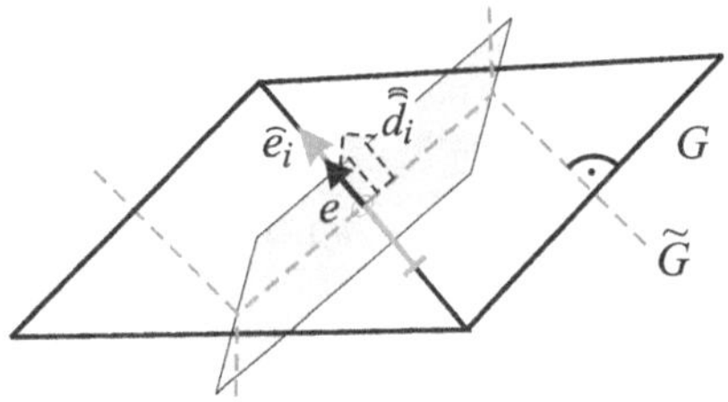

Fig. 2. Material relation for an orthogonal grid doublet: Two components $\widehat{\mathbf{e}}_i$ and fd_i that must be related for construction of a discrete permittivity matrix, and virtual electric field component e at the intersection point of primary edge and dual facet.

edges or facets of each grid need not be orthogonal to each other (cf. Fig. 2). The desired permittivity matrix $\mathbf{M}_\varepsilon$ is expected to supply the relation between the electric voltages $\widehat{\mathbf{e}}_i$ along the edges of G, and the corresponding electric fluxes $\widehat{\widehat{\mathbf{d}}}_i$ through the dual facets of $\widetilde{G}$. The normal vector of these facets is now collinear to the direction of the intersecting edge.

In order to reach this relation we introduce a virtual field component at the intersection point of the dual facet and the primary edge. As the material distribution was assumed to be homogeneous inside each cell, e is obviously allocated in such a way that it always represents a continuous quantity. This important restriction to continuous field components is not at all trivial, as in other methods, such as the Finite Element method with nodal elements, non-continuous components are computed and give consequently rise to inaccurate or even qualitatively wrong results.

To obtain the desired material relation, we approximate both the electric voltage and the electric flux based on this local field component, introducing not only the metric of the grid, but also the discretization error of the method:

$$\widehat{\mathbf{e}}_i = \int\limits_{L_i} \mathbf{E} \cdot d\mathbf{s} = e \cdot L_i + O(L^\alpha) , \tag{16}$$

$$\widehat{\widehat{\mathbf{d}}}_i = \int\limits_{\widetilde{A}_i} \varepsilon \mathbf{E} \cdot d\mathbf{A} = \varepsilon_{\mathrm{eff},i} \cdot e \cdot \widetilde{A}_i + O(\widetilde{A}_i^\beta) , \quad \varepsilon_{\mathrm{eff},i} = \frac{1}{\widetilde{A}_i} \int\limits_{\widetilde{A}_i} \varepsilon d\mathbf{A} . \tag{17}$$

Note, that the area-based averaging rule for the local effective permittivity is a direct consequence of the approximation of the dielectric flux in (17). The convergence rates α and β of the approximations in (16) and (17) can be evaluated by a Taylor expansion of the electric field (here the local continuity is needed), and depend on the type of the grid. Finally we obtain the discrete permittivity expression (one entry on the main diagonal of $\mathbf{M}_\varepsilon$):

$$\frac{\widehat{\widehat{\mathbf{d}}}_i}{\widehat{\mathbf{e}}_i} = \frac{\int\limits_{\widetilde{A}_i} \varepsilon \mathbf{E} \cdot d\mathbf{A}}{\int\limits_{L_i} \mathbf{E} \cdot d\mathbf{s}} = \frac{\varepsilon_{\mathrm{eff},i} \widetilde{A}_i}{L_i} + O(L^\gamma) \Longrightarrow \mathbf{M}_\varepsilon(i,i) = \frac{\varepsilon_{\mathrm{eff},i} \widetilde{A}_i}{L_i} . \tag{18}$$

Obviously, the orthogonality of G and $\widetilde{G}$ results in a one-to-one relation between voltages and fluxes and thus in a purely diagonal material matrix. For Cartesian grids, (18) reads (for a pair of x-components):

$$\mathbf{M}_{\varepsilon}(i,i) = \frac{\varepsilon_{\text{eff},i} \Delta\tilde{y}_i \Delta\tilde{z}_i}{\Delta\tilde{x}_i} \,. \tag{19}$$

The discrete relation between $\widehat{\widehat{\mathbf{j}}}$ and $\widehat{\mathbf{e}}$ is obtained in the same way simply by exchanging ε by the conductivity κ.

For the discrete permeability we start with approximations of the magnetic voltage along a dual edge and the magnetic flux through the related primary face:

$$\widehat{\widehat{\mathbf{b}}}_i = \int_{A_i} \mathbf{B} \cdot d\mathbf{A} = b \cdot A_i + O(A_i^{\beta}) \,, \tag{20}$$

$$\widehat{\mathbf{h}}_i = \int_{\widetilde{L}_i} \mu^{-1} \mathbf{B} \cdot d\mathbf{s} = \mu_{\text{eff},i}^{-1} \cdot b \cdot \widetilde{L}_i + O(\widetilde{L}_i^{\alpha}) \,, \quad \mu_{\text{eff},i} = \frac{1}{\widetilde{L}_i} \int_{\widetilde{L}_i} \mu^{-1} d\mathbf{s} \,. \tag{21}$$

These approximations are now based on a virtual magnetic flux density quantity b, allocated as continuous normal component at the center of the primary facet. As a consequence, the inverse permeability has to be averaged along the dual edge for a local effective material parameter. Finally, we obtain the diagonal entry of the magnetic material matrix:

$$\frac{\widehat{\mathbf{h}}_i}{\widehat{\widehat{\mathbf{b}}}_i} = \frac{\int_{\widetilde{L}_i} \mu^{-1} \mathbf{B} \cdot d\mathbf{s}}{\int_{A_i} \mathbf{B} \cdot d\mathbf{A}} = \frac{\mu_{\text{eff},i} \widetilde{L}_i}{A_i} + O(\widetilde{L}^{\gamma}) \Longrightarrow \mathbf{M}_{\mu}^{-1}(i,i) = \frac{\mu_{\text{eff},i}^{-1} \widetilde{L}_i}{A_i} \,, \tag{22}$$

or for a pair of Cartesian x-components:

$$\mathbf{M}_{\mu}^{-1}(i,i) = \frac{\mu_{\text{eff},i}^{-1} \Delta\tilde{x}_i}{\Delta\tilde{y}_i \Delta\tilde{z}_i} \,. \tag{23}$$

4 Some Properties of the Finite Integration Technique

Summarizing the numerical procedure introduced here, we have derived a complete discrete set of matrix equations that correspond one-to-one to the set of fundamental Maxwell equations. No specialization was introduced, and thus the set of these Maxwell Grid Equations is as generally applicable as Maxwell's equations are.

One of the significant characteristics of the FIT approach is the fact, that one part of the discretization step, the local integrations along edges and over facets of the grid, is a direct transformation of the corresponding operations in the continuous world and *free of any approximation errors.* The usage of a staggered grid and the corresponding allocation of FIT's state variables seems to reflect the geometric structure of Maxwell's equation in the finite world, and thus can be considered as the *natural way of disretizing electromagnetic field problems.* As a consequence, there are numerous consistency and conservation theorems concerning discrete field solutions in FIT which, in most cases can be proven in complete analogy to the proofs in classical electrodynamics. Basic elements are the topological properties (10), (13) and (14), which in the context of FIT can be interpreted as analogues to the vector-analytical identities, $\operatorname{div}\operatorname{curl} \equiv 0$ and $\operatorname{curl}\operatorname{grad} \equiv 0$.

As a very simple example we only quote here the divergence-free theorem for discrete rotational fields which, follows directly - and in an *exact* form, independently from the mesh resolution - from

$$\widehat{\widehat{\mathbf{v}}} = \mathbb{C}\widehat{\mathbf{a}} \quad (\mathbf{v} = \nabla \times \mathbf{a}) \quad \Longrightarrow \quad \mathbf{S}\widehat{\widehat{\mathbf{v}}} = \mathbf{S}\mathbb{C}\widehat{\mathbf{a}} = \mathbf{0} \;\; (\nabla \cdot \mathbf{v} = \nabla \cdot \nabla \times \mathbf{a} = 0) \quad (24)$$

For more theorems and proofs (including the conservation of energy and charges, a discrete form of Poynting's theorem, the orthogonality of cavity and waveguide modes, Courant- and long-time stability of time domain methods, etc.) we refer to Section 6 and [4, 13, 18, 19].

In practical applications, these unique features allow to qualify numerical results with a-posteriori checks on the fulfillment of such laws. They also give much more confidence in numerical results of computer codes, when e.g. it is a-priori guaranteed that no unphysical charges can occur during a calculation. On the other hand, as the consistency error of the method is concentrated in the material matrices only, it can in many cases be quantitatively analyzed and controlled.

As mentioned before, the FIT approach aims at the solution of electromagnetic field problems in all areas of industrial and scientific applications, ranging from static fields up to high frequency simulations and Particle-In-Cell (PIC) codes. Similarly to the analytical procedure of deriving second order PDE's with some specialization in it for each group of physical problems, we may now manipulate Maxwell's Grid Equations to derive specialized matrix equations for different applications. For example, introducing a scalar potential allocated on the grid nodes,

$$\widehat{\mathbf{e}} = -G\phi = \widetilde{\mathbf{S}}^{\mathrm{T}} \;\; (\mathbf{E} = -\operatorname{grad}\varphi)\,, \tag{25}$$

one can easily derive the discrete general potential equation to solve electrostatic problems:

$$\widetilde{\mathbf{S}}\mathbf{M}_{\varepsilon}\widetilde{\mathbf{S}}^{\mathrm{T}}\phi = \mathbf{q} \;\; (\operatorname{div}\varepsilon\operatorname{grad}\varphi = \varrho)\,. \tag{26}$$

For cavity analysis we can eliminate the magnetic field vector and find the eigenvalue equation:

$$\mathcal{T}\mathbf{M}_{\mu}^{-1}\mathbb{C}\widehat{\mathbf{e}} = \omega^2\mathbf{M}_{\varepsilon}\widehat{\mathbf{e}} \;\; (\operatorname{curl}\mu^{-1}\operatorname{curl}\mathbf{E} = \omega^2\varepsilon\mathbf{E})\,. \tag{27}$$

For any other possible application, whether high or low frequency, transient or static, one can easily derive dedicated discrete equations by simple matrix manipulations.

5 Explicit Time-Stepping with FIT and the Relation to FDTD

Maxwell's Grid Equations in the form (15) are still time-continuous, and no decision has been made so far how to discretize the time-dependency of transient fields. In the area of high-frequency electromagnetic fields the FDTD algorithm has become very popular, as the leapfrog time integration method incorporated in FDTD leads to a very efficient explicit recursion scheme. In the following section we will apply the leapfrog scheme to the FIT equations and compare the resulting method with FDTD. The basic idea of the leapfrog scheme is, the usage of a staggered grid also for the time-discretization, and the application of central difference approximations for the time derivatives in Maxwell's equations. For the state variables of FIT we set

$$\widehat{\mathbf{h}}^{\mathrm{m}} = \widehat{\mathbf{h}}(t_0 + m \cdot \Delta t) \ , \quad \widehat{\mathbf{e}}^{\mathrm{m}+\frac{1}{2}} = \widehat{\mathbf{e}}(t_0 + (m + \frac{1}{2}) \cdot \Delta t) \ , \tag{28}$$

and

$$\frac{d}{dt}\widehat{\mathbf{h}}^{\mathrm{m}+\frac{1}{2}} \approx \frac{\widehat{\mathbf{h}}^{\mathrm{m}+1} - \widehat{\mathbf{h}}^{\mathrm{m}}}{\Delta t} \ , \quad \frac{d}{dt}\widehat{\mathbf{e}}^{\mathrm{m}+1} \approx \frac{\widehat{\mathbf{e}}^{\mathrm{m}+\frac{3}{2}} - \widehat{\mathbf{e}}^{\mathrm{m}+\frac{1}{2}}}{\Delta t} \ . \tag{29}$$

We put these approximations in the curl-equations in (15) and obtain

$$\begin{aligned} \mathbb{C}\widehat{\mathbf{e}}^{\mathrm{m}+\frac{1}{2}} &= -\frac{d}{dt}\widehat{\mathbf{b}}^{\mathrm{m}+\frac{1}{2}} \approx \mathbf{M}_\mu \frac{\widehat{\mathbf{h}}^{\mathrm{m}+1} - \widehat{\mathbf{h}}^{\mathrm{m}}}{\Delta t} \ , \\ \mathcal{T}\widehat{\mathbf{h}}^{\mathrm{m}+1} &= -\frac{d}{dt}\widehat{\widehat{\mathbf{d}}}^{\mathrm{m}+1} + \widehat{\widehat{\mathbf{j}}}^{\mathrm{m}+1} \approx \mathbf{M}_\varepsilon \frac{\widehat{\mathbf{e}}^{\mathrm{m}+\frac{3}{2}} - \widehat{\mathbf{e}}^{\mathrm{m}+\frac{1}{2}}}{\Delta t} + \widehat{\widehat{\mathbf{j}}}^{\mathrm{m}+1} \ . \end{aligned} \tag{30}$$

From here is it only one step to the update equations of an explicit scheme:

$$\begin{aligned} \widehat{\mathbf{h}}^{\mathrm{m}+1} &= \widehat{\mathbf{h}}^{\mathrm{m}} - \Delta t \cdot \mathbf{M}_\mu^{-1}\mathbb{C}\widehat{\mathbf{e}}^{\mathrm{m}+\frac{1}{2}} \ , \\ \widehat{\mathbf{e}}^{\mathrm{m}+\frac{3}{2}} &= \widehat{\mathbf{e}}^{\mathrm{m}+\frac{1}{2}} + \Delta t \cdot \mathbf{M}_\varepsilon^{-1}(\mathbb{C}\widehat{\mathbf{h}}^{\mathrm{m}+1} - \widehat{\widehat{\mathbf{j}}}^{\mathrm{m}+1}) \ . \end{aligned} \tag{31}$$

To compare these algorithm with FDTD, we switch to Cartesian grids (the Yee-cell) and extract one line of each equation to obtain the update rule for one $\widehat{\mathbf{h}}$- and one $\widehat{\mathbf{e}}$-component. These single voltages can be scaled by the corresponding edge length to obtain some kind of averaged field components $e_i = \widehat{\mathbf{e}}_i / L_i$, $h_i = \widehat{\mathbf{h}}_i / \widetilde{L}_i$, yielding:

$$
\begin{aligned}
\widehat{\mathbf{h}}^{\mathrm{m}+1} &= h_i^{\mathrm{m}+1} \cdot \widetilde{L}_i = h_i^{\mathrm{m}} \cdot \widetilde{L}_i \\
&\quad - \Delta t \cdot \frac{\mu_{\mathrm{eff},i}^{-1} \widetilde{L}_i}{A_i} \left[\pm L_{j_1} e_{j_1}^{\mathrm{m}+\frac{1}{2}} \pm \cdots \pm L_{j_4} e_{j_4}^{\mathrm{m}+\frac{1}{2}} \right] , \\
\widehat{\mathbf{e}}^{\mathrm{m}+\frac{3}{2}} &= e_i^{\mathrm{m}+\frac{3}{2}} \cdot L_i = e_i^{\mathrm{m}+\frac{1}{2}} \cdot L_i \\
&\quad + \Delta t \cdot \frac{\varepsilon_{\mathrm{eff},i}^{-1} L_i}{\widetilde{A}_i} \left[\pm \widetilde{L}_{j_1} h_{j_1}^{\mathrm{m}+1} \pm \cdots \pm \widetilde{L}_{j_4} h_{j_4}^{\mathrm{m}+1} - \widetilde{A}_i j_i^{\mathrm{m}+1} \right] .
\end{aligned}
\tag{32}
$$

Note that each line of the curl-matrices has four non-zero entries in the Cartesian case, denoted here with indices $j_1 \ldots j_4$. If we take into consideration the metrics of a Cartesian grid with (locally)

$$
A_x = \Delta y \cdot \Delta z \; , \quad A_y = \Delta x \cdot \Delta z \; , \quad A_z = \Delta x \cdot \Delta y \; , \tag{33}
$$

we obtain the well-known update equations of FDTD, given here for the x-components:

$$
\begin{aligned}
h_{x,i}^{\mathrm{m}+1} &= h_{x,i}^{\mathrm{m}} - \Delta t \mu_{\mathrm{eff}\,(x,i)}^{-1} \left[\frac{e_{y,j_1}^{\mathrm{m}+\frac{1}{2}} - e_{y,j_2}^{\mathrm{m}+\frac{1}{2}}}{\Delta z} + \frac{e_{z,j_3}^{\mathrm{m}+\frac{1}{2}} - e_{z,j_4}^{\mathrm{m}+\frac{1}{2}}}{\Delta y} \right] , \\
e_{x,i}^{\mathrm{m}+\frac{3}{2}} &= e_{x,i}^{\mathrm{m}+\frac{1}{2}} + \Delta t \varepsilon_{\mathrm{eff}\,(x,i)}^{-1} \left[\frac{h_{y,j_1}^{\mathrm{m}+1} - h_{y,j_2}^{\mathrm{m}+1}}{\Delta \tilde{z}} + \frac{h_{z,j_3}^{\mathrm{m}+1} - h_{z,j_4}^{\mathrm{m}+1}}{\Delta \tilde{y}} - j_{x,i}^{\mathrm{m}+1} \right] .
\end{aligned}
\tag{34}
$$

The explicit time-domain algorithm of FIT on Cartesian grids is therefore computationally equivalent with FDTD, which appears as a special case the FIT approach[4]. This equivalence means, that of course the FDTD method shares all properties which have been proposed for the FIT above, such as charge conservation (provided that the discrete divergence operator is properly defined), energy conservation, the local convergence rate, etc. However, the original (and still most cited) component-wise and field-based formulation of FDTD does only poorly support a thorough analysis of many of those theoretical issues. This has important consequences for the development of FDTD extensions, as we will demonstrate in the next section by means of the important problem of local subgridding algorithms for FIT/FDTD. It was only recently that the advantages of FIT's notation have partly been recognized in the literature, and some matrix-vector formulations have been published [11, 22].

[4] The equivalence of differential and integral-based approaches has also been reported in some FDTD-publications [20, 21], and is sometimes referred to as the *Contour-path* method to derive the FDTD-equations (CFDTD). However, no integral state variables and no matrix-vector notation are introduced there, and the major part of the FDTD-literature still uses Yee's original approach.

6 Numerical Stability of FIT/FDTD and Extended Schemes

An important issue for all time domain methods is the analysis of numerical stability. From the very first FDTD papers we know the so-called Courant-stability criterion, which defines an upper limit for the time-step to be used, depending on the size of the mesh steps and the material distribution:

$$\Delta t \leq \Delta t_{\max} \approx \min_i \left\{ \varepsilon_i \mu_i \cdot \left(\frac{1}{\Delta x_i^2} + \frac{1}{\Delta y_i^2} + \frac{1}{\Delta z_i^2} \right)^{-1} \right\}^{\frac{1}{2}} \tag{35}$$

Before deriving this Courant-stability also for the FIT approach, we will first focus on a second type of stability, which is often referred to as long-time stability and which is independent of the time-step. To analyze this type of stability we consider a lossless structure ($\mathbf{M}_\kappa = \mathbf{0}$ and closed boundary conditions), and summarize the discrete curl-equations in one large system

$$\frac{d}{dt}\begin{pmatrix} \widehat{\mathbf{h}} \\ \widehat{\mathbf{e}} \end{pmatrix} = \begin{pmatrix} \mathbf{0} & -\mathbf{M}_\mu^{-1}\mathbb{C} \\ \mathbf{M}_\varepsilon^{-1}\mathcal{T} & \mathbf{0} \end{pmatrix} \begin{pmatrix} \widehat{\mathbf{h}} \\ \widehat{\mathbf{e}} \end{pmatrix} + \begin{pmatrix} \mathbf{0} \\ -\mathbf{M}_\varepsilon^{-1}\widehat{\widehat{\mathbf{j}}} \end{pmatrix} . \tag{36}$$

To symmetrize this system we introduce the scaled vectors:

$$\boldsymbol{x} = \begin{pmatrix} \widehat{\mathbf{h}}' \\ \widehat{\mathbf{e}}' \end{pmatrix} = \begin{pmatrix} \mathbf{M}_\mu^{\frac{1}{2}}\widehat{\mathbf{h}} \\ \mathbf{M}_\varepsilon^{\frac{1}{2}}\widehat{\mathbf{e}} \end{pmatrix} , \quad \boldsymbol{r} = \begin{pmatrix} \mathbf{0} \\ -\mathbf{M}_\varepsilon^{-\frac{1}{2}}\widehat{\widehat{\mathbf{j}}} \end{pmatrix} , \tag{37}$$

where for the definition of the root-matrices and with and symmetric positive definite material operators are required (here: diagonal matrices with positive entries). Using further the duality condition (10), we obtain,

$$\frac{d}{dt}\boldsymbol{x} = \boldsymbol{A}\boldsymbol{x} + \boldsymbol{r} , \tag{38}$$

with the skew-symmetric system matrix

$$\boldsymbol{A} = \begin{pmatrix} \mathbf{0} & -\mathbf{M}_\mu^{-\frac{1}{2}}\mathbb{C}\mathbf{M}_\varepsilon^{-\frac{1}{2}} \\ \mathbf{M}_\varepsilon^{-\frac{1}{2}}\mathbb{C}^{\mathrm{T}}\mathbf{M}_\mu^{-\frac{1}{2}} & \mathbf{0} \end{pmatrix} = \begin{pmatrix} \mathbf{0} & \boldsymbol{A}_{21} \\ -\boldsymbol{A}_{21}^{\mathrm{T}} & \mathbf{0} \end{pmatrix} . \tag{39}$$

From this system-view of the spatial discretization scheme (which is still time-continuous, as no time stepping has been applied so far) a very important result for the stability of the method can be derived: All eigenvalues of a skew symmetric matrix are purely imaginary numbers $\lambda_{\mathbf{A},i} = \pm j\omega_i$ (with real- valued eigenfrequencies ω_i), corresponding to undamped oscillating electromagnetic fields (eigenvectors). Any complex eigenfrequencies of this system (if the above conditions were not met) would lead to exponentially growing fields: in the time-continuous world as shown here as well as in any discrete time stepping scheme - as long as no artificial dissipation is introduced to stabilize these unstable eigensolutions. As proven by this derivation, such instable modes can not occur in the standard FIT/FDTD algorithm. However, situation may change, if extensions to the standard operators are necessary (e.g. for the implementation of open boundary conditions, non-isotropic material models, generalized grids etc.) As long as the imaginary part of such complex eigenfrequencies is very small, they lead to an exponential growth only after a long simulation time, and this type of (in)stability is therefore often referred to as *late time stability*. However, we prefer to name it *spatial stability*[5], as the key points in this derivation are the symmetry and positive definiteness of the material operators and the duality condition (10) of the curl-matrices[6].

To proceed, we now switch to the time-discrete world and the leapfrog scheme (31), which can be expressed by

$$\begin{pmatrix} \widehat{\mathbf{h}}'^{m+1} \\ \widehat{\mathbf{e}}'^{m+\frac{3}{2}} \end{pmatrix} = \underbrace{\begin{pmatrix} \boldsymbol{I} & \Delta t \boldsymbol{A}_{21} \\ -\Delta t \boldsymbol{A}_{21}^{\mathrm{T}} & \boldsymbol{I} + \Delta t^2 \boldsymbol{A}_{21}^{\mathrm{T}} \boldsymbol{A}_{21} \end{pmatrix}}_{\boldsymbol{G}(\Delta t)} \begin{pmatrix} \widehat{\mathbf{h}}'^{m} \\ \widehat{\mathbf{e}}'^{m+\frac{1}{2}} \end{pmatrix} + \begin{pmatrix} \mathbf{0} \\ -\Delta t \mathbf{M}_{\varepsilon}^{\frac{1}{2}} \widehat{\widehat{\mathbf{j}}}^{m+1} \end{pmatrix}, \tag{40}$$

featuring the iteration matrix $\boldsymbol{G}(\Delta t)$. For a stable update scheme, the eigenvalues of $\boldsymbol{G}(\Delta t)$ must lie within the unit circle of the complex plane, for an energy conserving algorithm even $|\lambda_{\mathbf{G},i}(\Delta t)| = 1$ is required for the chosen time step Δt. Provided that the system is spatially stable, the eigenvalues of $\boldsymbol{G}(\Delta t)$ and $\boldsymbol{A}$ (with $\lambda_{\mathbf{A},i} = \pm j\omega_i$) are related by

$$\lambda_{\mathbf{G},i}(\Delta t) = \frac{2 - (\Delta t \omega_i)^2}{2} \pm \sqrt{\left(\frac{2 - (\Delta t \omega_i)^2}{2}\right)^2 - 1}\,. \tag{41}$$

This finally leads to the generalized form of the Courant stability-criterion:

$$\Delta t \leq \frac{2}{\max_i |\omega_i|} \quad \Longleftrightarrow \quad |\lambda_{\mathbf{G},i}(\Delta t)| = 1 \quad \forall i\,. \tag{42}$$

For uniform grids and homogeneous material distribution, it can be easily shown by standard eigenvalue approximations that (42) is equivalent to the

[5] Note that for some non-resonant applications (e.g. the calculation of radar cross sections) even algorithms with long time instabilities sometimes can be used (and are often used), if the simulation can be terminated before the instability occurs. However, there are severe doubts about the reliability and accuracy of such simulations, as also all other eigenmodes may be affected by the violation of symmetry in the spatial discretization.

[6] Of course, these generalized symmetry-conditions of the spatial operators are only sufficient and not necessary for spatial stability. However, from a practical point of view almost any violation leads to unstable schemes; one of the rare exceptions has been presented in [23].

locally derived form (35) of the Courant criterion. However, the more general expression given here describes the *exact* stability limit also for non-uniform grids and arbitrary material fillings, where (35) is only a worst-case approximation (but of course easier to evaluate). As an additional result, (42) also guarantees that (if the stability-criterion is met) discrete energy is exactly conserved by the leapfrog scheme [19].

Of course, this analysis so far is valid not only for FIT, but also for the FDTD method, which has been shown above to be computationally equivalent to FIT for the standard Cartesian case. However, if we consider now a more complex extension of FIT/FDTD - the implementation of so-called subgridding- schemes, we will vitally need the generalized FIT-approach to understand, how these important stability properties can be preserved.

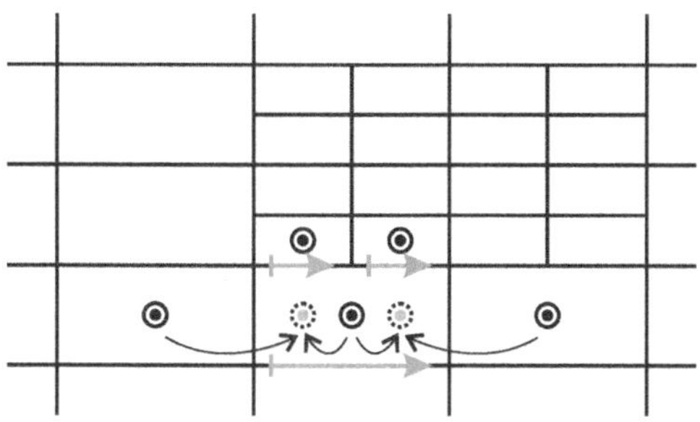

Fig. 3. A two-dimensional Cartesian mesh with a 1:2 subgrid and some state variables (electric voltages and magnetic fluxes). At the interface the topology of the grid is distorted and modified operator matrices including interpolations have to be supplied.

As shown in Fig. 3, the interface planes between such a subgrid and the underlying base grid represent a distortion of the regular topology of a Cartesian grid. As a consequence, the curl-matrices $\mathbb{C}$ and $\mathcal{T}$ have to be modified to include interpolation rules for the components in the vicinity of these interfaces. From a pure FDTD point of view, this interpolation may be nearly arbitrary, only governed by the local approximation error, and some advanced techniques to minimize that error have been presented in the literature [24, 25, 26]. However, all these approaches do not take into account the requirements of spatial stability and therefore suffer from late-time instabilities.

From this analysis it is clear, that the modified curl-matrices still have to fulfill $\mathcal{T} = \mathbb{C}^{\mathrm{T}}$ in order to ensure stability[7]. A consistent algorithm featur-

[7] Additionally, we have to define modified source matrices which still fulfill the topological properties (13) so that, charge is conserved also for subgridding schemes, whereas the symmetry and positive definiteness of the material matrices are not affected here.

ing special interpolation rules to preserve this symmetry condition has been presented in [27] and will not repeated here in detail. To our knowledge, this scheme is still the only 3D-subgridding-approach for FDTD/FIT on the market with provable long- time stability. In a similar way, this eigenvalue-based stability analysis has been used to develop provably stable FDTD extensions for nonorthogonal grids [9], more complex material models including dispersion and anisotropy [16], and higher-order material operators for FIT [28].

7 Conclusions

The Finite Integration Technique (FIT) has been reviewed as a general discretization method for all kinds of electromagnetic field problems with far-reaching consistency and conservation properties. The consequent application of a staggered grid approach also for non-Cartesian grids can be considered as a natural transformation of the geometrical structure of Maxwell's equations. An additional important feature of the FI-Technique is the separation of *exact* matrix operators, corresponding to vectoranalytical differential operators, and the unavoidable approximations of the method, which are concentrated in the so-called material matrices. This elegant formulation enables the transformation of many theorems of classical electromagnetics to corresponding properties of the discrete field solutions, which are exactly fulfilled, and not only in the limit of vanishing mesh step sizes.

It has been shown, that the FIT-approach, if applied to Cartesian grids and supplemented by the leapfrog time-stepping scheme, is computationally equivalent to the FDTD-method, which thus shares many of the before-mentioned properties. However, we consider the underlying, more general theory of FIT as indispensable for a thorough understanding of the method. Practical consequences arise for the development of all kind of FDTD extensions, where especially the important issue of long-time stability can only be fully understood using the powerful analysis tools provided by FIT. Such extensions include the implementation of generalized grids, complex material models, subgridding schemes, implicit schemes for low-frequency applications and many more.

References

1. K.S. Yee, Numerical Solution of Initial Boundary Value Problems Involving Maxwell's Equations in Isotropic Media, IEEE Transactions on Antennas and Propagation, Vol.14 (1966), pp. 302-307.
2. T. Weiland, A Discretization Method for the Solution of Maxwell's Equations for Six-Component Fields, Electronics and Communication(AEU), Vol.31 (1977), pp. 116.
3. T. Weiland: On the Numerical Solution of Maxwell's Equations and Applications in Accelerator Physics, Particle Accelerators, Vol. 15 (1984), pp. 245-291.

4. T. Weiland: Time Domain Electromagnetic Field Computation with Finite Difference Methods, International Journal of Numerical Modelling, Vol. 9 (1996), pp. 295-319.
5. U. v. Rienen, T. Weiland, Triangular Discretization Method for the Evaluation of RF-Fields in Cylindrically Symmetric Cavities, IEEE Transactions on Magnetics, Vol. 21 (1985), pp. 2317- 2320.
6. R. Schuhmann, P. Schmidt, T. Weiland: A New Whitney-based Material Operator for the Finite Integration Technique on Triangular Grids, accepted for publication in IEEE Transactions on Magnetics (2002)
7. I. Munteanu: The Finite Volume Method for Electromagnetic Field Analysis in the Time Domain, Revue Roumaine des Sciences Techniques - lectrotechnique et nergtique, Vol. 42, No. 3 (1997), pp. 321-336.
8. M. Hano, T. Itoh: Three-Dimensional Time Domain Method for Solving Maxwell's Equations based on Circumcenters of Elements, IEEE Transactions on Magnetics, Vol. 32, No. 3 (1996), pp. 946-949.
9. R. Schuhmann, T. Weiland: Stability of the FDTD Algorithm on Nonorthogonal Grids Related to the Spatial Interpolation Scheme, IEEE Transactions on Magnetics, Vol. 34, No. 5 (1998), pp. 2751-2754.
10. B. Krietenstein, R. Schuhmann, P. Thoma, T. Weiland: The Perfect Boundary Approximation Technique Facing the Big Challenge of High Precision Field Computation, Proceedings of the XIX International Linear Accelerator Conference (LINAC), Chicago, USA (1998), pp. 860-862.
11. A. Bossavit, L. Kettunen: Yee-Like Schemes on Staggered Cellular Grids: A Synthesis Between FIT and FEM Approaches, IEEE Transactions on Magnetics, Vol. 36, No. 4 (2000), pp. 861-867.
12. J.A. Kong, F.L. Teixeira (ed.): Geometric Methods for Computational Electromagnetics, Progress In Electromagnetics Research (PIER) Monograph Series, Vol. 32, EMW Publishing, Cambridge, USA, 2001.
13. M. Clemens, R. Schuhmann, T. Weiland: Algebraic Properties and Conservation Laws in the Discrete Electromagnetism, Frequenz, Vol. 53, No. 11-12 (1999),pp. 219-225.
14. S. Gutschling, H. Krger, T. Weiland: Time Domain Simulation of Dispersive Media with the Finite Integration Technique, International Journal of Numerical Modelling, Vol. 13, No. 4 (2000), pp. 329-348.
15. H. Spachmann, S. Gutschling, H. Krger, T. Weiland: FIT-Formulation for Nonlinear Dispersive Media, International Journal of Numerical Modelling, Special Issue, Vol. 12, No. 1/2 (1999), pp. 81-92.
16. T. Weiland, H. Krger, H. Spachmann: FIT-Formulation for Gyrotropic Media, International Conference on Electromagnetics in Advanced Applications (ICEAA), Torino, Italy (1999), pp. 737-740.
17. S. Drobny, T. Weiland: Iterative Algorithms For Nonlinear Transient Electromagnetic Field Calculation. Studies in Applied Electromagnetics and Mechanics, Vol. 18 (2000), IOS Press, pp. 385-388.
18. M. Clemens, T. Weiland: Discrete Electromagnetism with the Finite Integration Technique, Progress In Electromagnetics Research (PIER) Monograph Series, Vol. 32 (2001), pp. 65-87.
19. R. Schuhmann, T. Weiland: Conservation of Discrete Energy and Related Laws in the Finite Integration Technique, Progress In Electromagnetics Research (PIER) Monograph Series, Vol. 32 (2001), pp. 301-316.

20. A. Taflove, K.R. Umashankar, B. Beker, F.A. Harfoush, K.S. Yee: Detailed FD-TD Analysis of Electromagnetic Fields Penetrating Narrow Slots and Lapped Joints in Thick Conducting Screens, IEEE Transactions on Antennas and Propagation, Vol. 36, No. 2 (1988), pp. 247-257.
21. T.G. Jurgens, A. Taflove, K. Umashankar, T.G. Moore: Finite-Difference Time-Domain Modeling of Curved Surfaces, IEEE Transactions on Antennas and Propagation, Vol. 40, No. 4 (1992), pp. 357-366.
22. S. Gedney, J.A. Roden: Well posed non-orthogonal FDTD methods, Digest of the IEEE Antennas and Propagation Society International Symposium (AP-S), Atlanta, Vol. 1 (1998), pp. 596-599.
23. R. Schuhmann, M. Hilgner, T. Weiland: Convergence Properties of the Nonorthogonal FDTD Algorithm. Proceedings of the 2000 USNC/URSI National Radio Science Meeting, Salt Lake City, USA (2000), p. 22.
24. I.S. Kim, W.J.R. Hoefer: A Local Mesh Refinement Algorithm for the Time Domain-Finite Difference Method Using Maxwell's Curl Equations, IEEE Transactions on Microwave Theory and Techniques, Vol. 38, No. 6 (1990), pp. 812-815
25. M. Okoniewski, E. Okoniewska, M.A. Stuchly: Three-dimensional Subgridding Algorithm for FDTD, IEEE Transactions on Antennas and Propagation, Vol. 45, No. 3 (1997), pp. 1512-1517.
26. N. Chavannes, N. Kuster: Numerical Optimization of EM Transmitters in Complex Environments by Application of a Novel FDTD Subgrid Scheme, Proceedings of the 2000 USNC/URSI National Radio Science Meeting, Salt Lake City, USA (2000), p. 86.
27. P. Thoma, T. Weiland: A Subgridding Method in Combination with the Finite Integration Technique, Proceedings of the 25th European Microwave Conference, Vol. 2 (1995), p. 770-774.
28. H. Spachmann, R. Schuhmann, T. Weiland: Higher Order Spatial Operators for the Finite Integration Theory, submitted to: ACES Journal, Special Issue on 'Approaches to Better Accuracy / Resolution in Computational Electromagnetics' (2001).
29. P. Thoma, T. Weiland: Numerical Stability of Finite Difference Time Domain Methods, IEEE Transactions on Magnetics, Vol. 34, No. 5 (1998), pp. 2740-2743.

Appendix

Color Plates

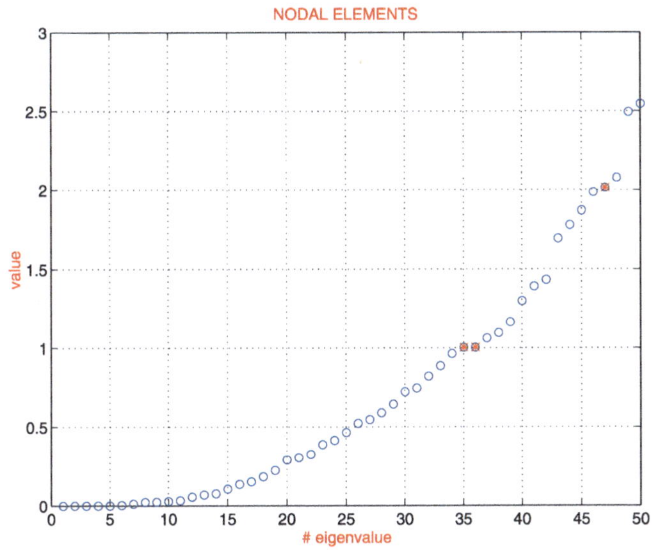

Plate 1. Eigenvalues computed with piecewise linears (Fig. 1, p. 16)

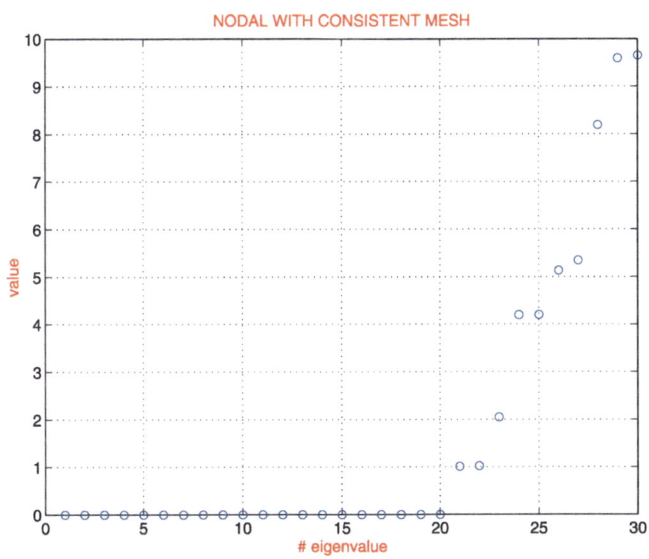

Plate 2. Eigenvalues computed with the first consistent mesh (Fig. 3, p. 17)

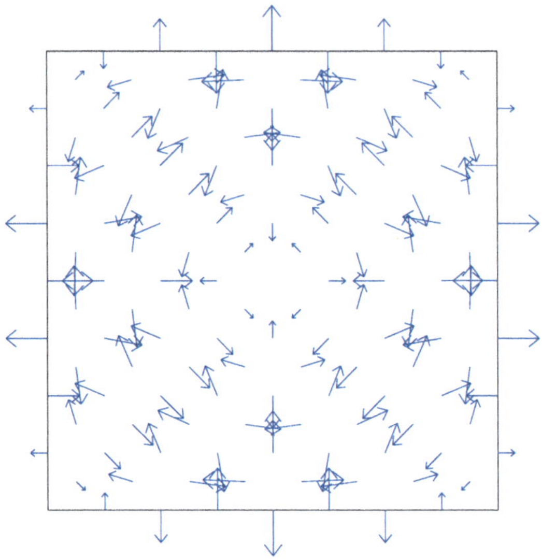

Plate 3. The spurious eigenfunction (Fig. 5, p. 18)

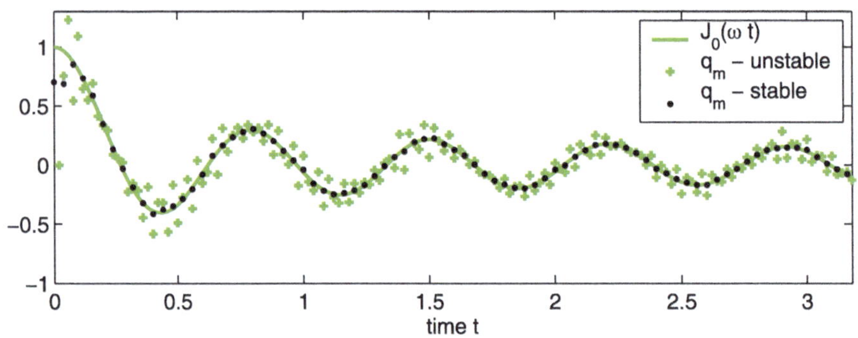

Plate 4. Comparison of normalised coefficients $2q_m(h\omega)/\Delta t$ (dots) with $J_0(\omega t^m)$ (grey solid) at frequency $\omega = (0.1\pi, 0.1\pi)$. The grey dots correspond to an unstable scheme and the black dots to a stable scheme (Fig. 3, p. 58)

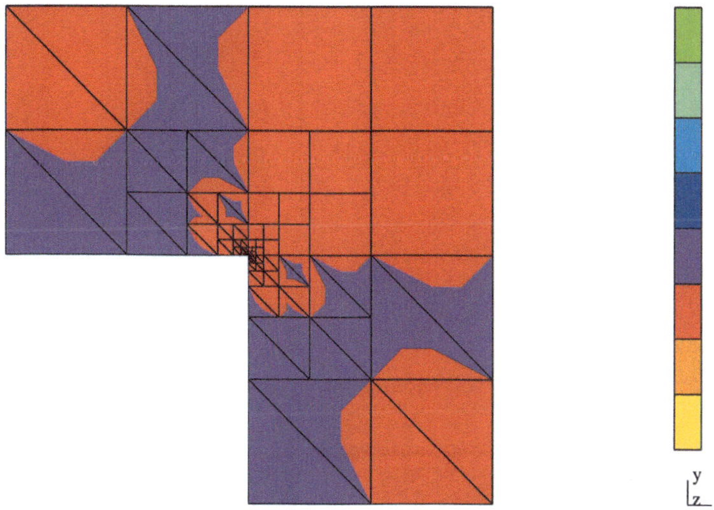

Plate 5. L-shape domain example: final, optimal *hp* mesh (Fig. 1, p. 78)

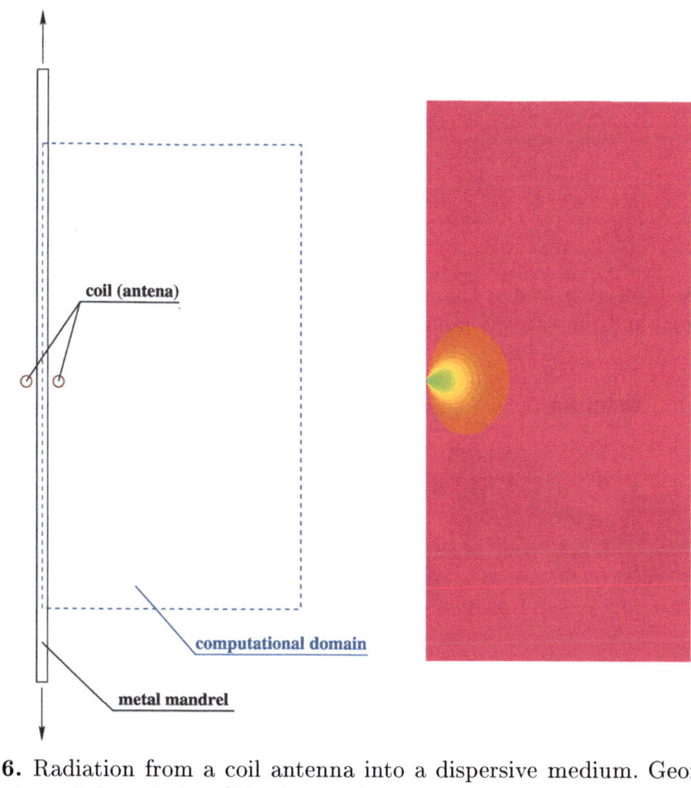

Plate 6. Radiation from a coil antenna into a dispersive medium. Geometry of the domain and the solution (Fig. 3, p. 80)

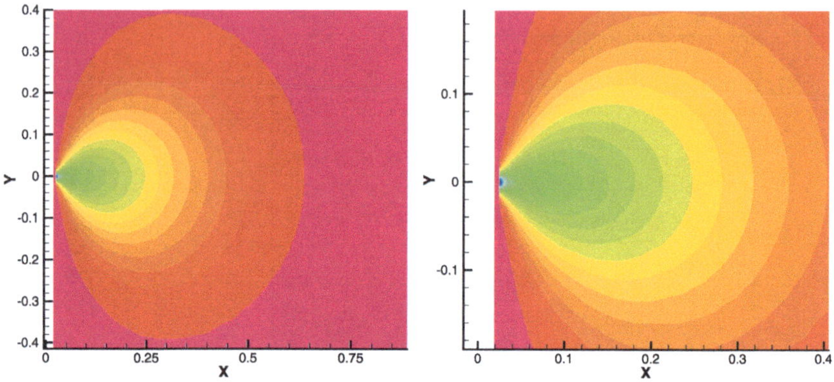

Plate 7. Radiation from a coil antenna into a dispersive medium. Zooming on the antenna (Fig. 4, p. 81)

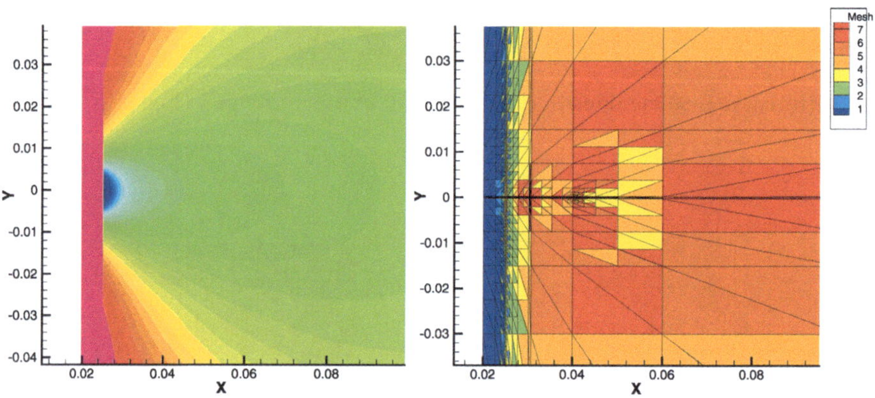

Plate 8. Radiation from a coil antenna into a dispersive medium. Final zoom on the antenna and the corresponding optimal *hp* mesh (Fig. 5, p. 81)

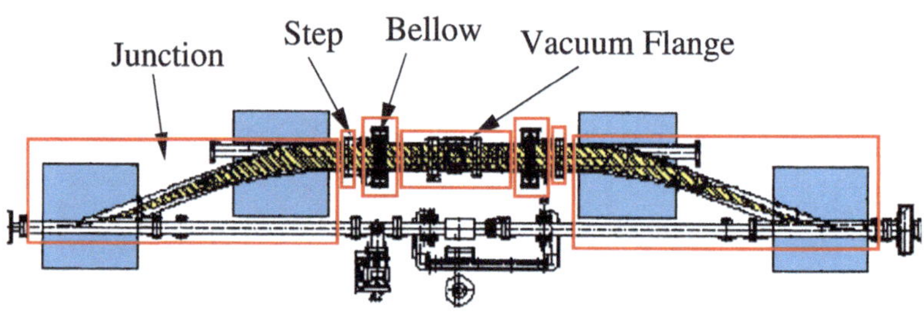

Plate 9. CAD drawing of bunch compressor II of the TESLA Test Facility (TTF). The shaded areas are inside a dipole magnet, each. The lines shall indicate the different sections chosen for the CSC calculations. Drawing: courtesy to DESY (Fig. 2, p. 87)

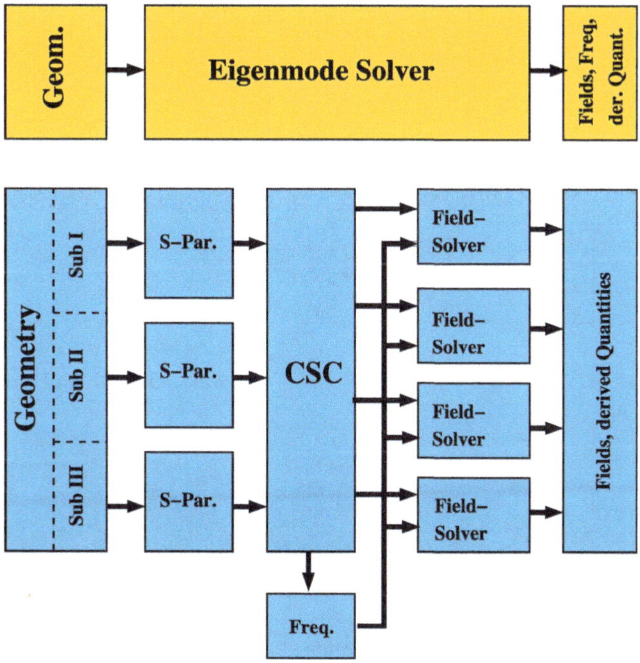

Plate 10. Calculation of eigenmodes using CSC: The geometry is divided into several sections. Then their S-parameters are determined. CSC combines them and provides the eigenfrequencies and wave amplitudes at the section's ports to compute the field distributions of the eigenmodes in the sections (Fig. 3, p. 88)

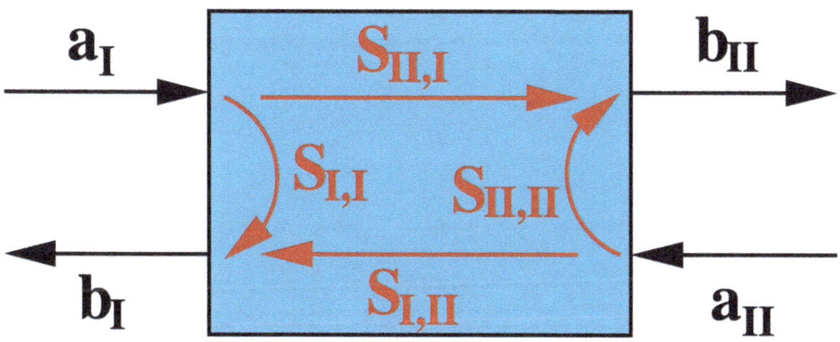

Plate 11. Description of signal reflection and transmission between two ports: Amplitudes $\underline{a}$ and $\underline{b}$ of incident and scattered waves are related by the scattering parameters, collected in the so-called S-matrix (Fig. 4, p. 89)

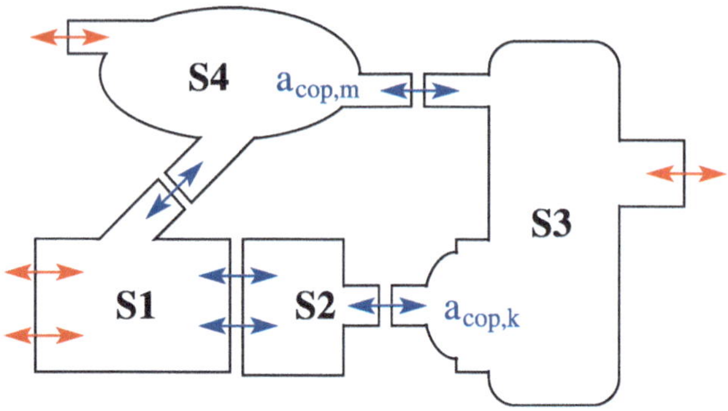

Plate 12. RF-system with four sections, several external and internal ports (Fig. 5, p. 90)

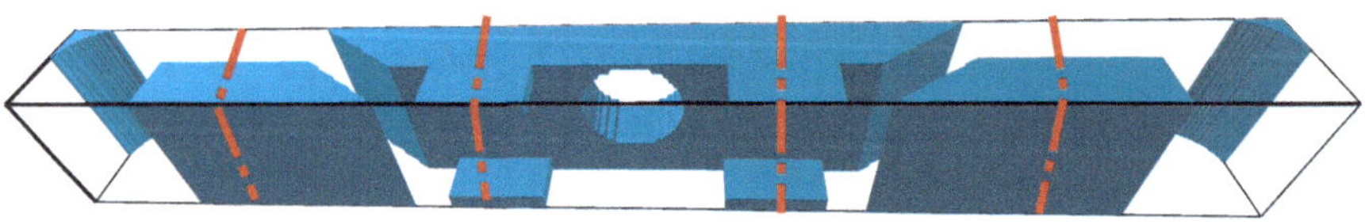

Plate 13. Geometry of the test resonator used to verify the CSC-formulation. Vertical lines indicate separation planes (Fig. 6, p. 92)

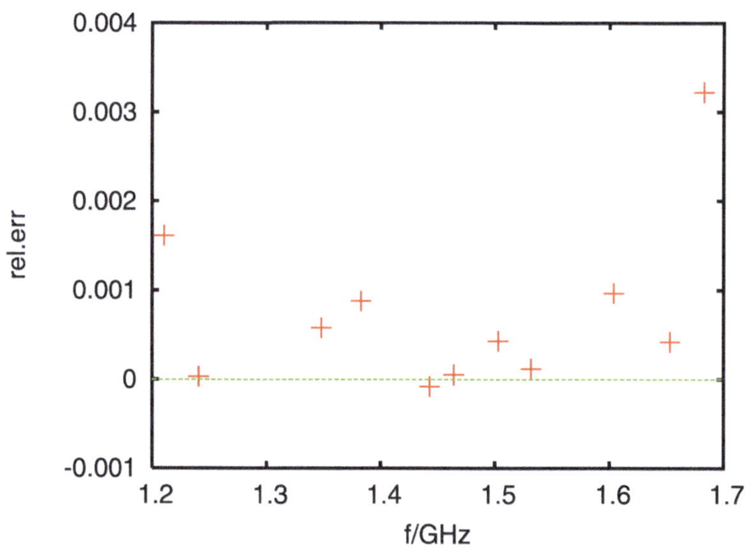

Plate 14. Relative error of eigenfrequencies of the test geometry found by the CSC-technique and by MAFIA's E-module (double precision) in the frequency range of 1.2 GHz to 1.75 GHz (Fig. 7, p. 93)

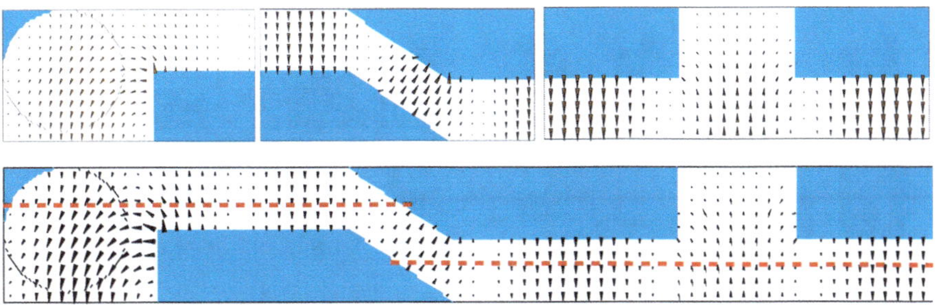

Plate 15. The electric field of the $f = 1.44258$ GHz eigenmode computed by the CSC (upper plots) and directly with MAFIA's E-module (lower plot) (Fig. 8, p. 94)

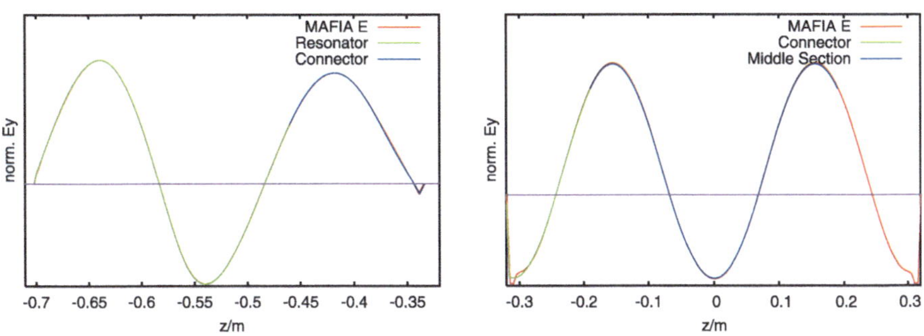

Plate 16. E_y along some path (shown dashed in Fig. 8) in z-direction shown for the eigenmode with frequency $f = 1.44258$ GHz (Fig. 9, p. 94)

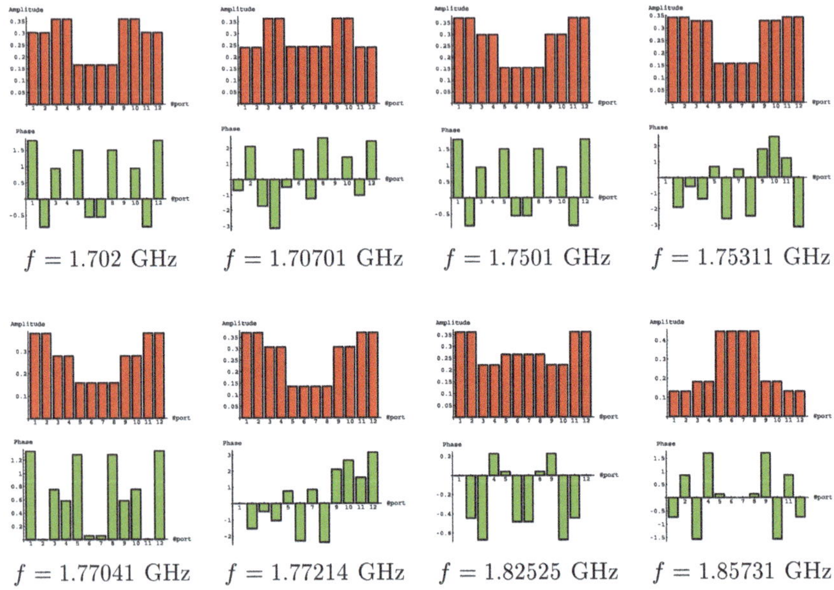

Plate 17. Schematic illustration of the first eight resonant eigenmodes of the FEL bunch compressor beam line shown in Fig. 2. Representation uses amplitudes (dark) and phases (light grey bars) of the waves incident at the 12 internal ports of the entire resonator. Modes can be clearly divided in even shapes, symmetric to the mid plane, and odd ones with complementary phases adding up to π (Fig. 10, p. 96)

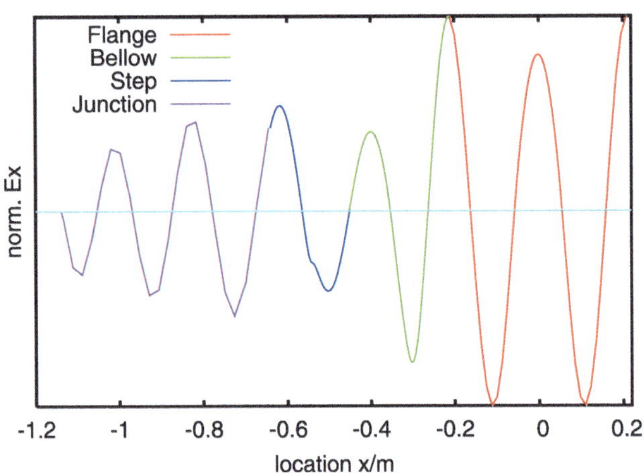

Plate 18. E_y in the FEL bunch compressor beam line along a path in z-direction along the centre line of the middle sections, shown for the eigenmode number 8 with $f = 1.85731$ GHz (Fig. 11, p. 96)

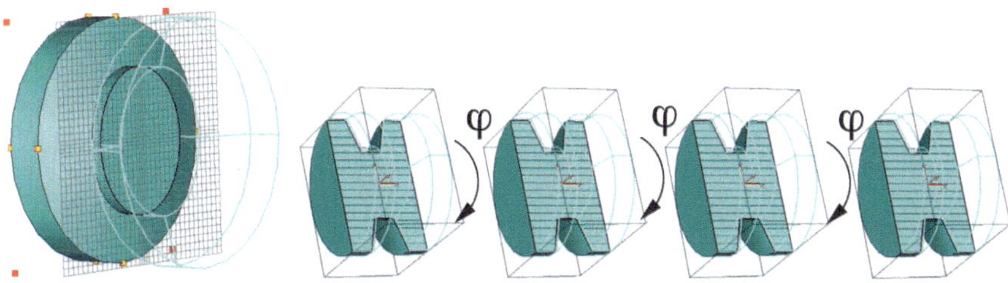

Plate 19. Cavity with rotational cross section at the outer radius and weakly elliptically deformed waist (left figure); a chain of those cells, each twisted by the angle φ with respect to the previous one (right; no intermediate space in calculation) (Fig. 12, p. 97)

Editorial Policy

§1. Volumes in the following three categories will be published in LNCSE:

i) Research monographs
ii) Lecture and seminar notes
iii) Conference proceedings

Those considering a book which might be suitable for the series are strongly advised to contact the publisher or the series editors at an early stage.

§2. Categories i) and ii). These categories will be emphasized by Lecture Notes in Computational Science and Engineering. **Submissions by interdisciplinary teams of authors are encouraged.** The goal is to report new developments – quickly, informally, and in a way that will make them accessible to non-specialists. In the evaluation of submissions timeliness of the work is an important criterion. Texts should be well-rounded, well-written and reasonably self-contained. In most cases the work will contain results of others as well as those of the author(s). In each case the author(s) should provide sufficient motivation, examples, and applications. In this respect, Ph.D. theses will usually be deemed unsuitable for the Lecture Notes series. Proposals for volumes in these categories should be submitted either to one of the series editors or to Springer-Verlag, Heidelberg, and will be refereed. A provisional judgment on the acceptability of a project can be based on partial information about the work: a detailed outline describing the contents of each chapter, the estimated length, a bibliography, and one or two sample chapters – or a first draft. A final decision whether to accept will rest on an evaluation of the completed work which should include

- at least 100 pages of text;
- a table of contents;
- an informative introduction perhaps with some historical remarks which should be accessible to readers unfamiliar with the topic treated;
- a subject index.

§3. Category iii). Conference proceedings will be considered for publication provided that they are both of exceptional interest and devoted to a single topic. One (or more) expert participants will act as the scientific editor(s) of the volume. They select the papers which are suitable for inclusion and have them individually refereed as for a journal. Papers not closely related to the central topic are to be excluded. Organizers should contact Lecture Notes in Computational Science and Engineering at the planning stage.

In exceptional cases some other multi-author-volumes may be considered in this category.

§4. Format. Only works in English are considered. They should be submitted in camera-ready form according to Springer-Verlag's specifications. Electronic material can be included if appropriate. Please contact the publisher. Technical instructions and/or TEX macros are available via http://www.springer.de/math/authors/help-momu.html. The macros can also be sent on request.

General Remarks

Lecture Notes are printed by photo-offset from the master-copy delivered in camera-ready form by the authors. For this purpose Springer-Verlag provides technical instructions for the preparation of manuscripts. See also *Editorial Policy*.

Careful preparation of manuscripts will help keep production time short and ensure a satisfactory appearance of the finished book.

The following terms and conditions hold:

Categories i), ii), and iii):
Authors receive 50 free copies of their book. No royalty is paid. Commitment to publish is made by letter of intent rather than by signing a formal contract. Springer-Verlag secures the copyright for each volume.

For conference proceedings, editors receive a total of 50 free copies of their volume for distribution to the contributing authors.

All categories:
Authors are entitled to purchase further copies of their book and other Springer mathematics books for their personal use, at a discount of 33,3 % directly from Springer-Verlag.

Addresses:

Timothy J. Barth
NASA Ames Research Center
NAS Division
Moffett Field, CA 94035, USA
e-mail: barth@nas.nasa.gov

Michael Griebel
Institut für Angewandte Mathematik
der Universität Bonn
Wegelerstr. 6
D-53115 Bonn, Germany
e-mail: griebel@iam.uni-bonn.de

David E. Keyes
Computer Science Department
Old Dominion University
Norfolk, VA 23529–0162, USA
e-mail: keyes@cs.odu.edu

Risto M. Nieminen
Laboratory of Physics
Helsinki University of Technology
02150 Espoo, Finland
e-mail: rni@fyslab.hut.fi

Dirk Roose
Department of Computer Science
Katholieke Universiteit Leuven
Celestijnenlaan 200A
3001 Leuven-Heverlee, Belgium
e-mail: dirk.roose@cs.kuleuven.ac.be

Tamar Schlick
Department of Chemistry
Courant Institute of Mathematical Sciences
New York University
and Howard Hughes Medical Institute
251 Mercer Street
New York, NY 10012, USA
e-mail: schlick@nyu.edu

Springer-Verlag, Mathematics Editorial IV
Tiergartenstrasse 17
D-69121 Heidelberg, Germany
Tel.: *49 (6221) 487-8185
e-mail: peters@springer.de
http://www.springer.de/math/peters.html

Lecture Notes in Computational Science and Engineering

Vol. 1 D. Funaro, *Spectral Elements for Transport-Dominated Equations.* 1997. X, 211 pp. Softcover. ISBN 3-540-62649-2

Vol. 2 H. P. Langtangen, *Computational Partial Differential Equations.* Numerical Methods and Diffpack Programming. 1999. XXIII, 682 pp. Hardcover. ISBN 3-540-65274-4

Vol. 3 W. Hackbusch, G. Wittum (eds.), *Multigrid Methods V.* Proceedings of the Fifth European Multigrid Conference held in Stuttgart, Germany, October 1-4, 1996. 1998. VIII, 334 pp. Softcover. ISBN 3-540-63133-X

Vol. 4 P. Deuflhard, J. Hermans, B. Leimkuhler, A. E. Mark, S. Reich, R. D. Skeel (eds.), *Computational Molecular Dynamics: Challenges, Methods, Ideas.* Proceedings of the 2nd International Symposium on Algorithms for Macromolecular Modelling, Berlin, May 21-24, 1997. 1998. XI, 489 pp. Softcover. ISBN 3-540-63242-5

Vol. 5 D. Kröner, M. Ohlberger, C. Rohde (eds.), *An Introduction to Recent Developments in Theory and Numerics for Conservation Laws.* Proceedings of the International School on Theory and Numerics for Conservation Laws, Freiburg / Littenweiler, October 20-24, 1997. 1998. VII, 285 pp. Softcover. ISBN 3-540-65081-4

Vol. 6 S. Turek, *Efficient Solvers for Incompressible Flow Problems.* An Algorithmic and Computational Approach. 1999. XVII, 352 pp, with CD-ROM. Hardcover. ISBN 3-540-65433-X

Vol. 7 R. von Schwerin, ***M**ulti **B**ody System **SIM**ulation.* Numerical Methods, Algorithms, and Software. 1999. XX, 338 pp. Softcover. ISBN 3-540-65662-6

Vol. 8 H.-J. Bungartz, F. Durst, C. Zenger (eds.), *High Performance Scientific and Engineering Computing.* Proceedings of the International FORTWIHR Conference on HPSEC, Munich, March 16-18, 1998. 1999. X, 471 pp. Softcover. 3-540-65730-4

Vol. 9 T. J. Barth, H. Deconinck (eds.), *High-Order Methods for Computational Physics.* 1999. VII, 582 pp. Hardcover. 3-540-65893-9

Vol. 10 H. P. Langtangen, A. M. Bruaset, E. Quak (eds.), *Advances in Software Tools for Scientific Computing.* 2000. X, 357 pp. Softcover. 3-540-66557-9

Vol. 11 B. Cockburn, G. E. Karniadakis, C.-W. Shu (eds.), *Discontinuous Galerkin Methods.* Theory, Computation and Applications. 2000. XI, 470 pp. Hardcover. 3-540-66787-3

Vol. 12 U. van Rienen, *Numerical Methods in Computational Electrodynamics.* Linear Systems in Practical Applications. 2000. XIII, 375 pp. Softcover. 3-540-67629-5

Vol. 13 B. Engquist, L. Johnsson, M. Hammill, F. Short (eds.), *Simulation and Visualization on the Grid.* Parallelldatorcentrum Seventh Annual Conference, Stockholm, December 1999, Proceedings. 2000. XIII, 301 pp. Softcover. 3-540-67264-8

Vol. 14 E. Dick, K. Riemslagh, J. Vierendeels (eds.), *Multigrid Methods VI.* Proceedings of the Sixth European Multigrid Conference Held in Gent, Belgium, September 27-30, 1999. 2000. IX, 293 pp. Softcover. 3-540-67157-9

Vol. 15 A. Frommer, T. Lippert, B. Medeke, K. Schilling (eds.), *Numerical Challenges in Lattice Quantum Chromodynamics.* Joint Interdisciplinary Workshop of John von Neumann Institute for Computing, Jülich and Institute of Applied Computer Science, Wuppertal University, August 1999. 2000. VIII, 184 pp. Softcover. 3-540-67732-1

Vol. 16 J. Lang, *Adaptive Multilevel Solution of Nonlinear Parabolic PDE Systems.* Theory, Algorithm, and Applications. 2001. XII, 157 pp. Softcover. 3-540-67900-6

Vol. 17 B. I. Wohlmuth, *Discretization Methods and Iterative Solvers Based on Domain Decomposition.* 2001. X, 197 pp. Softcover. 3-540-41083-X

Vol. 18 U. van Rienen, M. Günther, D. Hecht (eds.), *Scientific Computing in Electrical Engineering.* Proceedings of the 3rd International Workshop, August 20-23, 2000, Warnemünde, Germany. 2001. XII, 428 pp. Softcover. 3-540-42173-4

Vol. 19 I. Babuška, P. G. Ciarlet, T. Miyoshi (eds.), *Mathematical Modeling and Numerical Simulation in Continuum Mechanics.* Proceedings of the International Symposium on Mathematical Modeling and Numerical Simulation in Continuum Mechanics, September 29 - October 3, 2000, Yamaguchi, Japan. 2002. VIII, 301 pp. Softcover. 3-540-42399-0

Vol. 20 T. J. Barth, T. Chan, R. Haimes (eds.), *Multiscale and Multiresolution Methods.* Theory and Applications. 2002. X, 389 pp. Softcover. 3-540-42420-2

Vol. 21 M. Breuer, F. Durst, C. Zenger (eds.), *High Performance Scientific and Engineering Computing.* Proceedings of the 3rd International FORTWIHR Conference on HPSEC, Erlangen, March 12-14, 2001. 2002. XIII, 408 pp. Softcover. 3-540-42946-8

Vol. 22 K. Urban, *Wavelets in Numerical Simulation.* Problem Adapted Construction and Applications. 2002. XV, 181 pp. Softcover. 3-540-43055-5

Vol. 23 L. F. Pavarino, A. Toselli (eds.), *Recent Developments in Domain Decomposition Methods.* 2002. XII, 243 pp. Softcover. 3-540-43413-5

Vol. 24 T. Schlick, H. H. Gan (eds.), *Computational Methods for Macromolecules: Challenges and Applications.* Proceedings of the 3rd International Workshop on Algorithms for Macromolecular Modeling, New York, October 12-14, 2000. 2002. IX, 504 pp. Softcover. 3-540-43756-8

Vol. 25 T. J. Barth, H. Deconinck (eds.), *Error Estimation and Adaptive Discretization Methods in Computational Fluid Dynamics.* 2003. VII, 344 pp. Hardcover. 3-540-43758-4

Vol. 26 M. Griebel, M. A. Schweitzer (eds.), *Meshfree Methods for Partial Differential Equations.* 2003. IX, 466 pp. Softcover. 3-540-43891-2

Vol. 27 S. Müller, *Adaptive Multiscale Schemes for Conservation Laws.* 2003. XIV, 181 pp. Softcover. 3-540-44325-8

Vol. 28 C. Carstensen, S. Funken, W. Hackbusch, R. H. W. Hoppe, P. Monk (eds.), *Computational Electromagnetics.* Proceedings of the GAMM Workshop on "Computational Electromagnetics", Kiel, Germany, January 26-28, 2001. 2003. X, 209 pp. Softcover. 3-540-44392-4

Vol. 29 M. A. Schweitzer, *A Parallel Multilevel Partition of Unity Method for Elliptic Partial Differential Equations.* 2003. V, 194 pp. Softcover. 3-540-00351-7

For further information on these books please have a look at our mathematics catalogue at the following URL: `http://www.springer.de/math/index.html`

GPSR Compliance

The European Union's (EU) General Product Safety Regulation (GPSR) is a set of rules that requires consumer products to be safe and our obligations to ensure this.

If you have any concerns about our products, you can contact us on ProductSafety@springernature.com

In case Publisher is established outside the EU, the EU authorized representative is:

Springer Nature Customer Service Center GmbH
Europaplatz 3
69115 Heidelberg, Germany

Batch number: 10704713

Printed by Printforce, the Netherlands